Rat Genomics

Edited by

G. Thomas Hayman

Department of Biomedical Engineering, Rat Genome Database, Medical College of Wisconsin, Milwaukee, WI, USA

Jennifer R. Smith

Department of Biomedical Engineering, Rat Genome Database, Medical College of Wisconsin, Milwaukee, WI, USA

Melinda R. Dwinell

Genomic Sciences and Precision Medicine Center, Medical College of Wisconsin, Milwaukee, WI, USA

Department of Physiology, Rat Genome Database, Medical College of Wisconsin, Milwaukee, WI, USA

Mary Shimoyama

Department of Biomedical Engineering, Rat Genome Database, Medical College of Wisconsin, Milwaukee, WI, USA

Humana Press

Editors
G. Thomas Hayman
Department of Biomedical Engineering
Rat Genome Database
Medical College of Wisconsin
Milwaukee, WI, USA

Jennifer R. Smith
Department of Biomedical Engineering
Rat Genome Database
Medical College of Wisconsin
Milwaukee, WI, USA

Melinda R. Dwinell
Genomic Sciences and Precision
Medicine Center
Medical College of Wisconsin
Milwaukee, WI, USA

Department of Physiology
Rat Genome Database
Medical College of Wisconsin
Milwaukee, WI, USA

Mary Shimoyama
Department of Biomedical Engineering
Rat Genome Database
Medical College of Wisconsin
Milwaukee, WI, USA

ISSN 1064-3745 ISSN 1940-6029 (electronic)
Methods in Molecular Biology
ISBN 978-1-4939-9580-6 ISBN 978-1-4939-9581-3 (eBook)
https://doi.org/10.1007/978-1-4939-9581-3

This Humana imprint is published by the registered company Springer Science+Business Media, LLC, part of Springer Nature.
The registered company address is: 233 Spring Street, New York, NY 10013, U.S.A.

Dedication

This book is gratefully dedicated to the rat research community and to colleagues and friends who are gone too soon: Rat Genome Database members, Dr. Timothy F. Lowry and Dr. Victoria Petri, and founder of the Rat Resource and Research Center, Dr. John K. Critser. They are sincerely missed.

Preface

It is an exciting time to be involved in rat research. The rich history of physiological and behavioral data that is available for the rat spans over 150 years of research. This extensive body of data provides a solid foundation for the use of genomic technologies such as whole-genome and whole-exome sequencing, single-nucleotide variant discovery, and transcriptomics to explore similarities and differences between established rat models for human diseases such as kidney disease, cancer, and metabolic syndrome, as well as new models like the hybrid rat diversity panel. Recent advances in the use of genome-editing reagents and embryonic stem cells now allow researchers to produce new, more targeted models and to discover the molecular mechanisms underlying both normal and disease-related physiological processes. Thanks to the improvements in cryopreservation and rederivation, new models can be produced, studied for a period of time, and then preserved and stored to await new questions or the advent of new technologies to uncover the answers to questions we can't answer now and, in some cases, don't even know we should be asking. The emerging areas of interest, such as the microbiome, have opened up new vistas for researchers interested in the interactions between genetics and the environment.

This book provides both a historical perspective on rat research through the years and practical information to support researchers either currently involved in genomic research or planning to begin such a project. In some cases, a detailed protocol is provided for researchers looking to move into a new area of investigation or to leverage a new technology. In other cases, a detailed review of the existing models or a description of available resources can help the researcher find, understand, and/or utilize the information, the data, and the tools that they need to support their research efforts. Whatever the application, it is becoming increasingly obvious that in this so-called post-genomic era, no single type of research is sufficient to answer the increasingly complex questions of human disease and translational research. The rat as a biomedical model is uniquely poised to provide the ideal combination of established experimental models, extensive physiological data, and genomic manipulability to facilitate exploration of the underlying biology.

Milwaukee, WI, USA *Jennifer R. Smith*

Acknowledgments

We are thankful for nearly 20 years of funding from the National Heart, Lung, and Blood Institute on behalf of the National Institutes of Health. We appreciate the contributions of the authors and the assistance of Prof. John Walker and Ms. Anna Rakovsky, which helped make this book a reality.

Contents

Contributors

JAMES M. AMOS-LANDGRAF • *Department of Veterinary Pathology, College of Veterinary Medicine, University of Missouri, Columbia, MO, USA*

CARMEN BERGOM • *Department of Radiation Oncology, Medical College of Wisconsin, Milwaukee, WI, USA*

ELIZABETH R. BOLTON • *Department of Biomedical Engineering, Rat Genome Database, Medical College of Wisconsin, Milwaukee, WI, USA*

STEVEN L. BRITTON • *Department of Anesthesiology, University of Michigan, Ann Arbor, MI, USA*

SUSHEEL B. BUSI • *Department of Veterinary Pathology, College of Veterinary Medicine, University of Missouri, Columbia, MO, USA*

JEFF DE PONS • *Department of Biomedical Engineering, Rat Genome Database, Medical College of Wisconsin, Milwaukee, WI, USA*

MELINDA R. DWINELL • *Genomic Sciences and Precision Medicine Center, Medical College of Wisconsin, Milwaukee, WI, USA; Department of Physiology, Rat Genome Database, Medical College of Wisconsin, Milwaukee, WI, USA*

AARON C. ERICSSON • *Department of Veterinary Pathology, College of Veterinary Medicine, University of Missouri, Columbia, MO, USA*

TOMOO ETO • *Central Institute for Experimental Animals, Kawasaki, Kanagawa, Japan*

WANDA E. FILIPIAK • *Transgenic Animal Model Core, University of Michigan Medical School, Ann Arbor, MI, USA*

MICHAEL J. FLISTER • *Genomic Sciences and Precision Medicine Center, Medical College of Wisconsin, Milwaukee, WI, USA; Department of Physiology, Medical College of Wisconsin, Milwaukee, WI, USA*

GALINA B. GAVRILINA • *Transgenic Animal Model Core, University of Michigan Medical School, Ann Arbor, MI, USA*

G. THOMAS HAYMAN • *Department of Biomedical Engineering, Rat Genome Database, Medical College of Wisconsin, Milwaukee, WI, USA*

MASUMI HIRABAYASHI • *Center for Genetic Analysis of Behavior, National Institute for Physiological Sciences, Okazaki, Aichi, Japan*

SHINICHI HOCHI • *Faculty of Textile Science and Technology, Shinshu University, Ueda, Nagano, Japan*

MATTHEW J. HOFFMAN • *Department of Biomedical Engineering, Rat Genome Database, Medical College of Wisconsin, Milwaukee, WI, USA; Genomic Sciences and Precision Medicine Center, Medical College of Wisconsin, Milwaukee, WI, USA; Department of Physiology, Rat Genome Database, Medical College of Wisconsin, Milwaukee, WI, USA*

PAULA L. HOFFMAN • *Department of Pharmaceutical Sciences, Skaggs School of Pharmacy and Pharmaceutical Sciences, University of Colorado, Aurora, CO, USA; Department of Pharmacology, School of Medicine, University of Colorado Anschutz Medical Campus, Aurora, CO, USA*

ELIZABETH D. HUGHES • *Transgenic Animal Model Core, University of Michigan Medical School, Ann Arbor, MI, USA*

AMIT JOSHI • *Department of Radiology, Medical College of Wisconsin, Milwaukee, WI, USA*

Lauren Gerard Koch • *Department of Physiology and Pharmacology, University of Toledo, Toledo, OH, USA*

Alison J. Kriegel • *Department of Physiology, Center of Systems Molecular Medicine, Medical College of Wisconsin, Milwaukee, WI, USA*

Anne E. Kwitek • *Department of Physiology, Medical College of Wisconsin, Milwaukee, WI, USA*

Anna K. LaForest • *Transgenic Animal Model Core, University of Michigan Medical School, Ann Arbor, MI, USA*

Stanley J. F. Laulederkind • *Department of Biomedical Engineering, Rat Genome Database, Medical College of Wisconsin, Milwaukee, WI, USA*

Ronen Levi-Varadi • *Laboratory for Molecular Medicine, Israeli Rat Genome Center, Barzilai University Medical Center, Ashkelon, Israel; Department of Nephrology and Hypertension, Barzilai University Medical Center, Ashkelon, Israel; Faculty of Health Sciences, Ben-Gurion University of the Negev, Beer-Sheba, Israel*

Mingyu Liang • *Department of Physiology, Center of Systems Molecular Medicine, Medical College of Wisconsin, Milwaukee, WI, USA*

Yong Liu • *Department of Physiology, Center of Systems Molecular Medicine, Medical College of Wisconsin, Milwaukee, WI, USA*

Bruce A. Molitoris • *Division of Nephrology, Indiana University School of Medicine, Indianapolis, IN, USA; Indiana Center for Biological Microscopy, Indianapolis, IN, USA*

Abraham A. Palmer • *Department of Psychiatry, University of California San Diego, La Jolla, CA, USA; Institute for Genomic Medicine, University of California San Diego, La Jolla, CA, USA*

Oleg Palygin • *Department of Physiology, Medical College of Wisconsin, Milwaukee, WI, USA*

Yangsu Ren • *Department of Psychiatry, University of California San Diego, La Jolla, CA, USA*

Hallgeir Rui • *Department of Pathology, Medical College of Wisconsin, Milwaukee, WI, USA*

Laura M. Saba • *Department of Pharmaceutical Sciences, Skaggs School of Pharmacy and Pharmaceutical Sciences, University of Colorado, Aurora, CO, USA*

Ruben M. Sandoval • *Division of Nephrology, Indiana University School of Medicine, Indianapolis, IN, USA; Indiana Center for Biological Microscopy, Indianapolis, IN, USA*

Thomas L. Saunders • *Transgenic Animal Model Core, University of Michigan Medical School, Ann Arbor, MI, USA; Division of Genetic Medicine Genetics, Department of Internal Medicine, University of Michigan Medical School, Ann Arbor, MI, USA*

Mary Shimoyama • *Department of Biomedical Engineering, Rat Genome Database, Medical College of Wisconsin, Milwaukee, WI, USA*

Harry Smith • *Department of Biostatistics and Informatics, Colorado School of Public Health, University of Colorado, Aurora, CO, USA*

Jennifer R. Smith • *Department of Biomedical Engineering, Rat Genome Database, Medical College of Wisconsin, Milwaukee, WI, USA*

Leah C. Solberg Woods • *Department of Internal Medicine, Section on Molecular Medicine, Wake Forest University School of Medicine, Winston Salem, NC, USA*

Boris Tabakoff • *Department of Pharmaceutical Sciences, Skaggs School of Pharmacy and Pharmaceutical Sciences, University of Colorado, Aurora, CO, USA*

Akiko Takizawa • *Department of Physiology, Genomic Sciences and Precision Medicine Center, Medical College of Wisconsin, Milwaukee, WI, USA*

JYOTHI THOTA • *Department of Biomedical Engineering, Rat Genome Database, Medical College of Wisconsin, Milwaukee, WI, USA*
MAREK A. TUTAJ • *Department of Biomedical Engineering, Rat Genome Database, Medical College of Wisconsin, Milwaukee, WI, USA*
MONIKA TUTAJ • *Department of Biomedical Engineering, Rat Genome Database, Medical College of Wisconsin, Milwaukee, WI, USA*
LAUREN A. VANDERLINDEN • *Department of Biostatistics and Informatics, Colorado School of Public Health, University of Colorado, Aurora, CO, USA*
SHUR-JEN WANG • *Department of Biomedical Engineering, Rat Genome Database, Medical College of Wisconsin, Milwaukee, WI, USA*
CHANA YAGIL • *Laboratory for Molecular Medicine, Israeli Rat Genome Center, Barzilai University Medical Center, Ashkelon, Israel; Department of Nephrology and Hypertension, Barzilai University Medical Center, Ashkelon, Israel; Faculty of Health Sciences, Ben-Gurion University of the Negev, Beer-Sheba, Israel*
YORAM YAGIL • *Laboratory for Molecular Medicine, Israeli Rat Genome Center, Barzilai University Medical Center, Ashkelon, Israel; Department of Nephrology and Hypertension, Barzilai University Medical Center, Ashkelon, Israel; Faculty of Health Sciences, Ben-Gurion University of the Negev, Beer-Sheba, Israel*

Chapter 1

The Rat: A Model Used in Biomedical Research

Jennifer R. Smith, Elizabeth R. Bolton, and Melinda R. Dwinell

Abstract

The laboratory rat, *Rattus norvegicus*, has been used in biomedical research for more than 150 years, and in many cases remains the model of choice for studies of physiology, behavior, and complex human disease. This book provides detailed information on a number of methodologies that can be used in rat. This chapter gives an introduction to rat as a species and as a biomedical model, providing historical information, a brief introduction to the current state of rat research, and a perspective on the future of rat as a model for human disease.

Key words Rat, History, Models, Biomedical research, Resources, Data

1 The Evolutionary History of *Rattus norvegicus*

Rodents in general, and rats in particular, have been the subject of both affection and hatred for hundreds of years. From the rats mentioned in the *Yoso-tama-no-kakehashi* Japanese guidebook on raising rats, published in 1775, to those owned by today's "rat fanciers" (members of organizations such as the American Fancy Rat and Mouse Association (AFRMA), the National Fancy Rat Society (NFRS) in the UK, and rat clubs and rat fancier societies around the world), rats have been bred, raised, exhibited, and loved as pets. One source (http://www.petrats.org/home_.aspx) estimates that nearly half a million households in the USA own at least one pet rat or mouse.

On the other hand, across the world, rats are considered to be a major pest, destroying farm crops and spreading disease to domestic animals and humans. It is estimated that rats damage 1–5% of cereal crops worldwide (http://www.knowledgebank.irri.org/step-by-step-production/postharvest/storage/storage-pests/rodents-as-storage-pest). The ability of the rat to successfully invade, compete with existing species, and colonize new territory has led to changes in ecosystems, especially those that are relatively isolated such as islands. According to Lack et al. 2013 [1], "Of the

G. Thomas Hayman et al. (eds.), *Rat Genomics*, Methods in Molecular Biology, vol. 2018,
https://doi.org/10.1007/978-1-4939-9581-3_1,

approximately 123 island groups worldwide, about 82% have been invaded by *R. norvegicus*, *R. rattus*, or the Polynesian rat (*R. exulans*; [2]), and recent reports estimated that introduced rats have been responsible for 40–60% of all bird and reptile extinctions since 1600 (Island Conservation 2006)." In addition, rats are well documented to transmit a substantial number of zoonotic diseases including Hantavirus-related diseases such as hemorrhagic fever with renal syndrome, salmonellosis, typhus, and, of course, plague [3–5].

With few exceptions (notably some areas of the Arctic and New Zealand, the continent of Antarctica, and the Canadian province of Alberta), the species *Rattus norvegicus* is essentially ubiquitous in its global distribution [6], and is considered a true commensal with respect to humans (https://www.aciar.gov.au/node/8241), with the two species uneasily coexisting in a relationship where the rats reap the benefits of warmth, shelter, and abundant food supplies while the humans are, more or less, not directly harmed by the association.

Although the general spread of the rat coinciding with the spread of humans is well documented, the evolutionary origin of the species *R. norvegicus* remains unclear. There is general agreement that the species first developed in Asia. Traditional understanding placed central Asia, for example, Kazakhstan and/or south central Russia [7], or northern China and southern Mongolia [8], as the place of origin for the species, whereas more recent estimates based on archeological evidence, fossils, and bone remains suggested an origin in southwestern China. Recent studies based on mitochondrial DNA sequencing [9] and whole-genome sequencing [10] gave strong evidence for southeastern Asia as the area where the species arose and used these analyses to predict the subsequent migration of *R. norvegicus* and other rat species. Both groups postulated early migrations from southern East Asia to northern China, followed by later migrations from southern East Asia to the Middle East, Europe, and Africa. As an example, Zeng et al. [10] used multiple analysis methods including the construction of phylogenetic trees using multiple algorithms, principal component analysis, Bayesian clustering, demographic modeling, and haplotype sharing analyses to demonstrate that rat populations from non-Asian locales are more closely related to rats from southern East Asia than they are to rats from northern China. Interestingly, their analysis suggested that the migration from southern Asia to northern Asia was far earlier (~202,000 years ago) than the westward migration. They dated the spread of the brown rat from southern Asia to the Middle East at approximately 3600 years ago, to Africa at about 2600 years ago, and to Europe at ~1800 years ago, far earlier than previous estimates for the arrival of *R. norvegicus* in Europe (estimated to be between 1500 and 1750 CE).

Mitochondrial DNA analysis has also been used to elucidate the colonization and spread of two rat species, *R. rattus* and *R. norvegicus*, in the United States [1]. Haplotype comparisons between samples from within and outside of the USA suggested that, although *R. rattus* appeared to have expanded from relatively few (perhaps only one or two) colonization events, *R. norvegicus* populations appeared to be substantially more diverse, suggesting multiple colonization and migration events. Although the authors admitted that their conclusions were somewhat speculative, they suggested that there could have been two invasions of rats from East Asia into the US west coast, in addition to one from southern locations such as the Lesser Antilles, South America, and South Africa, and another from Europe into the US East Coast. In each case, the rat populations spread from their coastal locations across the neighboring inland areas, and in some cases, across the country. Relatively high genetic diversity in coastal locations suggests a continuing influx of individuals being integrated into the established coastal populations.

2 The Early History of the Rat in Research

The use of rats in biomedical research began more than 160 years ago. The first recorded use of rats for scientific investigation was a study by J. M. Philipeaux in 1856 [11] on the results of adrenalectomy in albino rats (the 1836 article by Samuel Moss, "Notes on the habits of a domesticated White Rat and a Terrier Dog that lived in harmony together" notwithstanding). Since that time, rats have been used as models to study a wide variety of biological, physiological, and medical subjects.

2.1 Nutrition

In 1993, Dr. Janet R. Hunt [12] stated, "Rats were the principal animal used to discover most of the vitamins, the essential trace elements, and the essential amino acids. As a result, more is known about the nutritional requirements of the rat than about any other species."

The first study of the nutritional quality of proteins in a mammal was an article published in both The Lancet [13] and the Proceedings of the Royal Society of London [14] in 1863 entitled "Experiments on food; its destination and uses." In it, William S. Savory detailed how he fed rats "nitrogenous" (high protein, very low fat), "non-nitrogenous" (high carbohydrate and fat, very low protein), and mixed diets to ascertain whether nitrogenous materials were utilized for "heat production," tissue formation, or both. Savory explained his use of rats in a footnote: "Rats were chosen as subjects for these experiments because they are omnivorous and will readily feed on almost any kind of diet. Moreover from their size they are very convenient to manage."

In 1912, F. G. Hopkins [15] published a study in which he fed young rats, those at a stage when rapid growth is expected, carefully controlled diets of protein, starch, sugar, lard, and "salts" consisting of the ash of incinerated "oats and dog-biscuits" with or without the addition of a small ration of milk each day. He found that the rats without the milk grew slowly for the first 13 days, then began to lose weight. Within 4 weeks, 5 of the 6 rats that did not receive milk had died. By contrast, all of the rats given milk grew normally for the entire experimental period. In additional experiments, he found that addition of milk to the diets of the malnourished rats "rescued" their growth rate and that addition of a larger amount of milk in the diet did not significantly improve the growth rate. From these results, Hopkins concluded that milk contained some "accessory factors," possibly with a "catalytic or stimulative function," since the amount needed was so small. These "factors" we now know to be vitamins, and on the basis of his early work in this field, Hopkins was awarded a Nobel Prize in 1929 "for his discovery of the growth-stimulating vitamins."

E. V. McCollum was an organic chemist who, through a number of circumstances, became interested in biochemistry and animal nutrition. Early in his career he noted that in all published cases, animals fed simplified diets restricted to isolated proteins, carbohydrates, fats, and mineral salt mixtures failed to thrive. He decided then that "the most important discovery to be made in nutrition would be the elucidation of the cause or causes of these failures" [16].

In 1907 McCollum began a colony of albino rats originally purchased from a "pet-stock dealer," the first rat colony maintained for nutrition studies. He selected rats as his research subjects because their size made them manageable and their shorter life span made it easier to follow the effects of changes in diet throughout the life of the animal. In 1913, McCollum and Davis published a paper on an ether-soluble extract of egg yolk or butter that is necessary for healthy growth and reproduction in rats [17]. This "factor," separately discovered by T.B. Osborne and L. B. Mendel in the same year (also using rats) and published in the same issue of the Journal of Biological Chemistry [18], was later termed "Vitamin A" [19].

McCollum continued his studies with various simplified diets to determine what components could be combined to provide the nutrition necessary for health in rats. Two years later McCollum, again with Marguerite Davis, discovered a water-soluble factor which promoted growth and was found to correct polyneuritis in pigeons fed an inadequate diet of purified food stuffs [20, 21]. In 1916, he and Cornelia Kennedy lobbied to eliminate the increasingly popular term "vitamine" which they maintained implied an indispensability (*vita*) and suggested a specific chemical structure (*amine*), neither of which had been shown [22]. Rather, they suggested the neutral terms "fat-soluble A" and "water-soluble B." By 1917 McCollum had begun to generalize his findings to the American diet "as being of poor quality because it was derived too largely from white flour or cornmeal, muscle meats, potatoes, and sugar" because the "foods listed. . .were not constituted to supplement each other by making good their deficiencies." [16].

Like McCollum, T. B. Osborne and L. B. Mendel were early pioneers in the study of animal nutrition using rats. In addition to the essentially simultaneous (with McCollum and Davis) discovery of a factor in butter that increased growth in rats, Osborne, Mendel et al. meticulously documented the effects on rats fed diets composed of minimal purified components supplemented with various amino acids to determine which should be considered essential and which could be manufactured by the body without direct supplementation [23–25], with cod liver oil as a source of fat-soluble factors, especially "fat-soluble A," and trace minerals such as phosphorus to demonstrate that certain mineral are essential for health. Osborne, Mendel et al. demonstrated in 1914 that cod liver oil was able to reverse xerophthalmia in rats [26]. In addition to numerous advancements in the field of nutrition, Osborne and Mendel, with their colleague E. L. Ferry, were the first to design and utilize "metabolic cages" for their rats that allowed them to closely monitor the animals' intake and excretion, and to prevent coprophagy which interfered with their attempts to do so [27].

Other discoveries in the field of nutrition for which the use of rats was instrumental include the discovery by H. Steenbock that ultraviolet irradiation increased the content of vitamin D in foods and other organic materials [28, 29]. By irradiating the rodent food he was able to reverse the symptoms of rickets in his rats. H.C. Sherman made extensive use of rats in his studies, not only of the types of supplementation experiments that McCollum and Davis, Osborne and Mendel, and others performed (e.g., Sherman and Pappenheimer [30]), but of the effects of diet on life span and longevity [31, 32]. He is also cited as an early pioneer in the use of statistics for the analysis of biological data [7]. As an illustration of the importance of rat research to advances in nutrition research, in the Journal of Nutrition's 2008 Symposium on Animal Models in Nutrition Research [33], rats were named as being instrumental in the discovery of cures for six of the nine vitamin deficiency diseases listed.

2.2 Breeding, Genetics, and Characterization

The practice of breeding rats for variations in coat color substantially predates their use in laboratory science. Rather, it was common in Japan at least as far back as the 1700s as evidenced by the first guidebook for breeding fancy rats entitled *Yoso-tama-no-kake-hashi*, published in 1775 [34]. Text and illustrations in the book describe and show a variety of coat colors and patterns, many of which are still seen in modern laboratory rat strains and fancy rat lines. In addition, the author gave advice for breeding these rats so as to not lose their "special characteristics."

In his 1947 paper on the domestication of the rat, W. E. Castle described a number of early studies of coat color inheritance in rat [35]. Between 1877 and 1885, H. Crampe published a series of articles detailing extensive breeding experiments beginning with a tame albino female rat and a wild gray male and continuing with successive rounds of interbreeding of the offspring. This pre-Mendelian study showed inheritance patterns which we now know to be controlled by three coat color mutations occurring at the c (albino), a (non-agouti), and h (hooded) loci, although this explanation for the various patterns of coat color and of inheritance was far from clear at the time. Crampe's prodigious dataset was reviewed and reanalyzed using Mendelian principles by Bateson in 1903. Doncaster used Crampe's data to categorize the offspring color patterns from his breeding experiments of brown/gray, black, and albino rats, published in 1906. A publication by MacCurdy and Castle in 1907 [36], which also detailed breeding experiments to determine coat color in rats and guinea pigs, was beginning to move closer to the idea that the inheritance patterns of color and markings are controlled by more than one factor, although obviously without an understanding of the specific genes or mutations.

W. E. Castle began studying rat coat color mutations in 1907, reporting the development of a "pink-eyed yellow" and a "red-eyed yellow" mutant in England at that time [35]. Among other pursuits, he and colleagues continued their exploration of the genetics of coat color, mutations, and linkage in the rat through 1951, resulting in the publication of more than 20 articles on the subject (e.g., [37–54]). Several of these were published with Dr. Helen Dean King of the Wistar Institute, looking at the linkage groups for the mutations—both those determining coat color and physiological mutations such as "waltzing"—which were found or confirmed by Dr. King during her domestication studies of wild gray rats and breeding studies of the Wistar albinos. In the aforementioned article on the domestication of the rat [35], Castle lists 23 known mutations, 14 of which he was able to place into four linkage groups, representing four of the 20 rat autosomes.

Dr. Henry H. Donaldson was a professor of neurology at the University of Chicago while John B. Watson was working on his degree. Donaldson had previously published papers on the nervous systems of human and frog but during his tenure at the University of Chicago he was introduced to the rat as a model for human neurology, probably through the influence of Swiss neuropathologist Adolf Meyer. After careful consideration, Donaldson chose to begin working with the albino rat. E. G. Conklin quoted him in his 1938 biography of Donaldson [55] as saying "It was found that the nervous system of the rat grows in the same manner as that of man—only some thirty times as fast. Further, the rat of three years may be regarded as equivalent in age to a man of ninety years, and this equivalence holds through all portions of the span of life, from birth to maturity. By the use of the equivalent ages, observations on the nervous system of the rat can be transferred to man and tested. The results so obtained show a satisfactory agreement and indicate that the rat may be used for further studies in this field."

In 1905, Dr. Donaldson, his assistant Dr. Shinkishi Hatai and his rat colony moved to the Wistar Institute. Believing that it was necessary to establish a standardized stock of rats to ensure accurate quantitative results by minimizing individual differences, Donaldson began to systematically breed albino rats under controlled conditions, giving rise to the Wistar and Wistar-derived rat strains. During this time, detailed records of morphological characteristics such as body and organ sizes, and "life processes" such as reproduction and growth were maintained for both the albino stocks and wild "Norways." In 1915, Donaldson published the first edition of these extensive records as "The Rat: Reference Tables and Data for the Albino Rat and the Norway Rat" [56]. A second and even more extensive edition was published in 1924 [57] and consisted of 496 pages containing 212 tables, 72 charts, and more than 2000 references. The intent of these publications was to create a detailed reference record that could be used by researchers to comparatively assess the characteristics of their own rats.

Dr. Helen Dean King joined the Wistar Institute in 1908, starting as a volunteer before being officially hired [58]. She began working with Dr. Donaldson in 1909, helping him to characterize the albino rats and beginning a project to further standardize the rats by inbreeding them. The rats inbred by Dr. King and others at the Wistar Institute have given rise to a number of strains and substrains of the "Wistar rat," including the PA, WKA, WF, WKY, and LEW, to name a few [7].

King and Donaldson were particularly interested in the process of domestication and the physiological changes that resulted from the process. In this regard they first attempted to plant "Albinos where they might lead a wild life, in order to see how far, under these conditions, they would return toward the ancestral type" [59]. Unfortunately, in all five cases the colonies failed. Rather than try again, the decision was made to recapitulate the domestication process under controlled conditions and follow the resulting changes in the physiology of the rats over time. The project began in 1909 with 16 male and 20 female wild gray rats captured in the Philadelphia area and continued for years, with detailed reports on the physiology and morphology of the captive rats released at the tenth generation in 1929 [59] and the 25th generation in 1939 [60]. It was noted in the first of these articles that "in the eleventh generation the strain 'broke' and several mutant varieties appeared, so that these ten generations are marked off by this event from those that follow."

Changes reported in the domesticated rats vs. their wild progenitors include "(1) accelerated growth rate resulting in increased body size; (2) decreased 'nervous tension' resulting in tamableness when the animals were handled frequently in early life; (3) mutations in color or structure of the hair." [35]. These changes are among those listed in a recent publication about the "domestication syndrome" [61] which also cited the project by Donaldson and King as the "first set of experimental domestication studies."

2.3 Behavior

Reports about studies on rat behavior began in 1898 with a paper in the inaugural issue of the American Journal of Physiology, "Variations in daily activity produced by alcohol and by changes in barometric pressure and diet, with a description of recording methods," by Colin C. Stewart [62]. The paper described a fairly sophisticated system for recording the total daily activity of the rats. A drum-shaped cage that the rat rotated by running in it was connected to a clock modified to show the count of the number of rotations of the cage and display it on the dial, possibly the first semi-automated activity monitor. This paper was probably also one of the first papers to look at the behavioral impact of nutrition and of addictive substances such as alcohol.

The paper, "An experimental study of the mental processes of the rat. II." by Willard S. Small [63], published in the American Journal of Psychology, appears to be the first description of the use of a maze to test rat behavior. Small mentioned that the "Hampton Court Maze served as model for the apparatus. The diagram given in the Encyclopedia Britannica was corrected to a rectangular form, as being easier of construction." Since the goal of the study was to examine "the method of animal intelligence" rather than being a quantitative study of the time needed for a rat to traverse the maze, the author gave extensive descriptions of the movements of the rats in the maze during two series of trials, one consisting of five experiments and the other consisting of nine. Not surprisingly, he noted that overall, the time to achieve the reward at the center of the maze and the number of errors along the way both diminished over time, although there were some variations.

Two years later, in 1903, the psychologist John B. Watson, who is best known for establishing the "behaviorism" theory of psychology, earned his PhD from the University of Chicago on the basis of his work on the relationship between brain myelination and learning in rats of various ages [64]. Watson also employed mazes as a test of learning ability. In the book "Behavior: an introduction to comparative psychology" [65], published in 1914, Watson began his description of the modified Hampton Court maze by saying that it was "too well known to require description," implying that they were commonly in use.

Although best known for his work on human behaviors and behaviorism as a branch of psychology, Watson was a strong advocate for studying animal behavior: "Psychology, as the behaviorist views it, is a purely objective, experimental branch of natural science. . .Heretofore the viewpoint has been that such data have value only in so far as they can be interpreted by analogy in terms of consciousness. The position is taken here that the behavior of man and the behavior of animals must be considered on the same plane; as being equally essential to a general understanding of behavior" ([65], p. 27).

Dr. Curt Richter began his career in psychobiology in 1919 by studying the innate behaviors of rats, particularly variations in and periodicity of their running activity [66]. He is well known for his early descriptions of circadian rhythms, of gender differences in activity, and of the influence of the estrous cycle on the running behavior of female rats [67]. He did considerable work on rats' ability to maintain their homeostatic balances when allowed to select freely from a "cafeteria" of proteins, carbohydrates, minerals, and vitamins, and demonstrated that animals deprived of an essential component, such as sodium, would compensate by eating more of that component. He also showed that an adrenalectomized rat, which had been shown to consistently lose sodium as a result of the loss of the sodium-retaining hormones secreted by the adrenals, had a substantially increased "salt appetite" and that the increase in salt consumption led to an increase in the survival rate following adrenalectomy [68]. Interestingly, he later discovered that this sodium-balancing ability, though present in domesticated rats, was not found in their wild progenitors. Such changes in both the physiology and behavior of rats during domestication was of interest to Richter and was a matter of study, for example, in connection with his efforts during World War II to control the rat population in Baltimore [69]. He noted that rats displayed "bait shyness" and "learned-poison-avoidance," now referred to, and frequently studied, as neophobia and conditioned taste aversion. Richter also developed several pieces of equipment for measuring rat behavior, including the "Richter tube" for measuring fluid consumption and the running wheel for monitoring activity.

In an often-quoted passage in his 1968 article "Experiences of a Reluctant Rat-Catcher: The Common Norway Rat – Friend or Enemy?" [70], Dr. Richter stated "During my almost half-century in behavioral and neurological research I have chiefly used rats, but also many different animals such as cats, dogs, monkeys, sloths, rabbits, beavers, porcupines, honey bears, alligators, and others. If someone were to give me the power to create an animal most useful for all types of studies on problems concerned directly or indirectly with human welfare, I could not possibly improve on the Norway rat."

2.4 Endocrinology

As mentioned previously, the first recorded use of rats for the study of endocrinology was in an 1856 study by J. M. Philipeaux on the results of adrenalectomy in albino rats. Like Philipeaux's experiments, many of the earlier studies of endocrinology in rats were studies of the effects of wholly removing endocrine glands, including J. M. Stotsenburg's studies at the Wistar Institute of the effects of castration [71] and spaying [72] on growth in rats, and F. S. Hammett's 1924 examination of the effects of the removal of the thyroid and/or parathyroid glands on the growth of the brain and spinal cord [73].

J. A. Long and H. M. Evans made observations using both intact animals and those with various endocrine glands removed or ablated, including removal of the ovaries, total hysterectomy and removal of the mammary glands, as well as ovarian transplantation studies, published in the sixth volume of the Memoirs of the University of California in 1922 [74]. In the methods section of this publication they included a brief description of the rats that they used—a cross between "several white females and a wild gray male caught in Berkeley, black, gray and hooded varieties resulting." The white females were said to have come from the Wistar Institute albino stock [7]. This is the first mention of the "Long Evans" rat which was later inbred to give the "LE" strains and substrains.

In 1921, Drs. Long and Evans published two short articles on the administration of anterior pituitary—one in which they fed the rats fresh or dried whole glands [75] and another in which they injected an extract from the glands intraperitoneally [76]. The feeding study showed no results, but following injection of the extract, the authors reported both the lengthening or cessation of the rats' normally regular estrous cycle and a marked increase in growth compared to litter mate controls. H. M. Evans and his collaborators subsequently continued this work, characterizing the hormones of the anterior pituitary (for examples, *see* Refs. [77–94].

3 "The Rat Toolbox"

3.1 Rat as a Precision Model for Disease

From Crampe's first adrenalectomy studies in the 1850s, rats have been used as models for human physiology and disease. In "Animal Models of Human Nutrition" Dr. Janet R. Hunt references E. V. McCollum's decision to use rats in his nutrition studies "because of their convenient size, omnivorous feeding habits and lack of economic value" and goes on to state that the "omnivorous feeding patterns of the rat usually make it a better model for human nutrition questions than a strict herbivore such as the rabbit." [12]. Likewise, H. C. Sherman, who was the first to develop quantitative measures of nutrients based on their ability to correct diseases stemming from nutritional deficiencies, said of his research with rats, "These animals are my burettes and balances. They give quantitative answers in chemical terms to many of man's greatest problems!" [95].

The same is true today. Rats are used as pinch hitters for humans in the study of debilitating and life-threatening diseases. The ability to quickly breed rats and select for specific traits and, in recent years, to genomically manipulate rats to produce traits, correct deficiencies or explore the influence of specific genes or genomic regions in the development of diseases, along with their extensive history of physiological measurements, makes them an ideal model for targeting specific diseases in human. In part because the ability to do genomic manipulation in rats was delayed, researchers traditionally concentrated on development of a wide variety of rat strains to model human disease. We will mention some of these here and explore the resources and techniques for obtaining or developing more models throughout the rest of this book.

3.2 The Development of Rat Strains to Study Disease Mechanisms

3.2.1 Cardiovascular Diseases

In 1958, Smirk and Hall reported results of ongoing breeding of genetically hypertensive (GH) rats from the Wistar-derived rat colony at the University of Otago Medical School [96]. They developed several lines of rats by both cross-breeding and brother-sister mated inbreeding. In the first report, they showed that the line produced by cross-breeding had a higher average blood pressure than the lines produced by inbreeding (141.95 ± 12.53 mmHg for cross-bred males vs. 135.81 ± 8.2 mmHg for one strain of inbred males and 124.14 ± 10.64 mmHg for control males). The differences reported are not as great as for the SHR rats; however they are statistically significant. In a subsequent publication [97], in which they reported on the development of cardiac hypertrophy in the B strain of their genetically hypertensive rats, they stated that more than 50% of the male rats in that strain "have blood pressures exceeding 150 mmHg."

In 1962, L. K. Dahl et al. published the first report of two strains of rats, selected from an outbred (unselected) colony of Sprague-Dawley rats [98]. At the time of publication, these rats had been selectively bred by brother-sister matings for three generations and already displayed a divergence in the effects of salt on their blood pressure. The rats were selected based on blood pressure measurements after being fed a diet containing 11.6% sea salt and administration of triiodothyronine which had been shown to accelerate the development of hypertension in these rats. Later studies demonstrated that the "S," i.e., salt sensitive (SS), rats developed increased blood pressure whether on T3 + high salt or on high salt alone whereas the blood pressure of the "R," or salt resistant, rats was less than or equal to the BP of the parental Sprague Dawley rats regardless of the conditions. By contrast, when fed a low salt diet, both lines remained normotensive. In addition to increased blood pressure, SS rats displayed cardiovascular abnormalities including left ventricular hypertrophy [99] and susceptibility to myocardial ischemia [100]. These abnormal traits could be a direct result of the hypertension or could be separate but linked genetic traits.

In 1963, Okamoto and Aoki reported on their work to develop a rat strain with spontaneous hypertension [101]. Starting with rats from an inbred Wistar strain, they selected one male "that had shown persistent high blood pressures (150–175 mmHg) since 7 weeks after birth and a female rat with a blood pressure slightly above the average (from 130 to 140 mmHg)." The two were mated and progeny with blood pressure > 150 mmHg for more than a month were selected and further brother-sister mated through the sixth generation. Testing of rats in the sixth generation of inbreeding showed that the control rats remained relatively normotensive throughout the test period of 60 weeks, whereas the blood pressure of the hypertensive rats rose steadily with age. In males, the blood pressure was 136 ± 13.4 mmHg at 5 weeks of age, rising to 206 ± 16.2 mmHg at 55 weeks of age. At this point in the breeding, the incidence of spontaneous hypertension was 100% in this strain and the authors began to refer to them as "spontaneously hypertensive rats (SHR)."

In the 1970s Okamoto and his colleagues began breeding sublines of the SHR rat [102, 103]. In one case, they selected for rats with a tendency to develop cerebral hemorrhage and/or cerebral infarction (stroke) and in the other selecting for those that were hypertensive but did not develop cerebrovascular lesions. They referred to these lines as stroke-prone (SHRSP) and stroke-resistant (SHRSR) spontaneously hypertensive rats. A 1975 article [104] demonstrated that, although the rats did not have as substantial a response to salt loading as the Dahl SS rats, both blood pressure and incidence of cerebrovascular lesions increased with salt.

In addition to strains that were developed specifically as controls for cardiovascular disease model strains, such as the Dahl SR and the stroke-resistant SHRSR strains, several additional strains are widely accepted as control strains for cardiovascular diseases. Among these are the Brown Norway (BN) strain which has been shown to be resistant to salt-sensitive hypertension [105] and to myocardial ischemia [100], Lewis rats (LEW) which are highly resistant to salt-induced hypertension [106], and the Wistar Kyoto (WKY) strain which is considered a normotensive control for SHR [107]. It should be noted, however, that an inbred strain that is considered a control in one set of studies could be "affected" for a study which focuses on a different trait.

3.2.2 Metabolic Syndrome

The Lyon hypertensive (LH), normotensive (LN), and hypotensive (LL) rat strains were first developed in 1973 in Lyon, France, from a colony of Sprague Dawley rats [108]. The rats for breeding were selected on the basis of the mean systolic blood pressure at 6–12 weeks of age and the slope of systolic blood pressure vs. age. Although only selected for blood pressure traits, the inbred LH strain was found to also display an increased body weight and increased plasma lipids. Plasma phospholipids, total cholesterol, HDL-cholesterol, and VLDL+ LDL-cholesterol were all elevated in the LH strain relative to LN and LL. Interestingly, at 5 weeks of age, the LL strain had the highest plasma triglyceride level. However, as the rats aged, the triglyceride level in the LH rats increased so that at 32 weeks of age LH was significantly higher than either LN or LL, but LL was still significantly higher than LN. Additional studies showed that the LH strain displayed additional metabolic disorders, namely an increase in both the insulin level and the insulin:glucose ratio [109].

Like the LH rat strain, the SHR strain was selected for increased blood pressure, but was later found to display symptoms of metabolic disfunction. Fasting glucose was greater in SHR rats than WKY [110] and the insulin response to an oral glucose challenge was higher in SHR than WKY, suggesting possible insulin resistance [111].

The OLETF strain was developed at the Tokushima Research Institute in Japan from a spontaneously diabetic rat discovered in 1984 in an outbred colony of Long-Evans rats [112, 113]. The rats were mildly obese and developed spontaneous hyperglycemia. Sex differences were noted in the course of the disease. Males developed hyperglycemia much earlier than females (25 weeks of age for males vs. 65 weeks for females). Over time, male rats also became hypoinsulinemic and required insulin therapy to survive which was not seen in females. Histopathological changes in the pancreatic islets and in the kidney were also seen in males but not females.

The Zucker "fatty" rat was first discovered as a mutant in the 13M rat stock, an outbred line derived from black offspring of albinos from the colony of Dr. H. C. Sherman (Columbia University), crossed with wild males (the "M" line) and additional rats from the Sherman colony [114] at the Harriet G. Bird Memorial Laboratory in Stow, Massachusetts. The mutation was named "fatty" because when present in a homozygous form the rat became extremely obese as a juvenile. Since heterozygous litter mates were lean and phenotypically indistinguishable from the non-mutant homozygotes, the mutation was understood to be a recessive allele in a single gene and has since been shown to be a p.Gln269Pro mutation in the extracellular domain of the leptin receptor. The original description of the Zucker rat includes severe hyperlipidemia and kidney lesions, but not hyperglycemia [115].

Although the Zucker rats were not routinely hyperglycemic, it was later found that they did show some signs of glucose intolerance and occasionally animals with high blood glucose were observed. Peterson et al. reported in 1990 the development of an inbred diabetic strain from the Zucker rats, the ZDF strain [116]. Male ZDF rats displayed substantial increases in blood glucose with age, leveling off in the 400–600 mg% range at 10–15 weeks of age. They also showed increased levels of glycosylated hemoglobin, free fatty acids, triglycerides, and cholesterol compared to lean controls. Insulin was increased in younger animals but was reduced in older animals due to fatty acid-induced apoptosis of the pancreatic beta cells [117].

3.2.3 Behavior and Addiction

As noted earlier in this chapter, the use of rats for studies of behavior began over 100 years ago with studies by C. C. Stewart on the effect of alcohol consumption, diet, and barometric pressure on activity in captive rats. Since that time, rats have been the model of choice for the study of behavior and addiction, and for testing treatments for psychiatric disorders. The body of literature covering these topics is enormous: a search in PubMed for "rat behavior" in 2018 returned over 150,000 articles, including almost 6500 review articles.

Often rat strains developed for the study of a non-behavioral phenotype are found to also show behavioral abnormalities. For instance, the Spontaneously Hypertensive Rat (SHR), established as a model of age-related hypertension, was found to develop vascular brain disorder with associated behavioral changes as a result of its increased blood pressure, and has also been used as a model for Attention Deficit Hyperactivity Disorder (ADHD) as a result of observed changes to the catecholaminergic transmission system [118]. Similarly, the WAG/Rij rat, a model for absence epilepsy, also showed depression-like symptoms [119]. The Wistar Kyoto (WKY) strain was originally bred as a normotensive control for the SHR but has been shown to display "depressive-like symptoms," including increased immobility in the forced swim test [120] and anhedonia characterized by lower consumption of a sweet-tasting solution in response to acute or chronic mild stress [121]. Physiologically, the WKY exhibited abnormalities in dopaminergic and noradrenergic responses and the HPA axis and TSH systems [122]. The WKY rat, in addition to being considered a model for depression, also displayed traits considered to be indicative of anxiety such as reduced activity in the open field as well as development of stress-induced ulcers. Another model of depression, the Flinders Sensitive Line (FSL) rat, on the other hand, displayed similar immobility in the forced swim test but did not appear to have an increased tendency toward anxiety [120]. It was shown, however, that young FSL rats engaged in more "intrusive" social play than Sprague-Dawley controls whereas as adults displayed less non-play social interaction/investigation behaviors than controls, both of which were interpreted as depressive-like behaviors.

The Fawn Hooded Hypertensive (FHH) inbred strain was also developed as a model of hypertension and was later found to display both depressive-like behaviors and abnormally high voluntary alcohol consumption, leading to the suggestion that the FHH rat could be used as a model for comorbid depression and alcoholism [122]. Interestingly, it was found that immobility in the forced swim test and alcohol consumption were not correlated, and in fact, administration of antidepressant drugs reduced the immobility in the swim test without affecting the alcohol consumed.

A number of rat strains have been developed specifically for high and low consumption of alcohol. These include the UChA and UChB lines from the University of Chile [123], the "Alko-accepting" (AA) and "Alko-nonaccepting" (ANA) lines, developed at the Research Laboratories of the State Alcohol Monopoly (Alko), in Helsinki, Finland [124], the alcohol Preferring and Non-preferring (P/NP) [125, 126] and the High Alcohol Drinking and Low Alcohol Drinking (HAD1/LAD1 and HAD2/LAD2) strains [127] produced at the Indiana University School of Medicine, and the Sardinian Preferring and Non-preferring (sP/sNP) lines [128, 129].

The AA/ANA, P/NP, and HAD/LAD strain pairs were all selected by giving only alcohol for a period of time, then measuring ethanol consumption in a two-choice paradigm. Both AA/ANA and P/NP started with outbred Wistar rats, then high and low drinkers were inbred. Looking to develop lines with more genetic diversity than the Wistar-derived P/NP rats, Li et al. developed two replicate sets of model strains, HAD1/LAD1 and HAD2/LAD2, using the same selection process with the National Institutes of Health's Heterogeneous Stock (N:HS, also referred to as N:NIH) outbred rats as the initial breeding stock. HS rats are descended from eight genetically and phenotypically diverse founder strains (ACI/N, BN/SsN, BUF/N, F344/N, M520/N, MR/N, WKY/N, and WN/N) [130]. The genetic diversity of the rats is maintained through a rotational breeding scheme to avoid inbreeding, drift, and fixation. Since the resulting rats are a random mosaic of the genomes of the founder strains, they provided far more diversity for the development of models for drinking behavior than any one strain could provide. The P strain and both HAD lines met the criteria for animal models of alcoholism [131]. Given a choice, their voluntary consumption of ethanol resulted in pharmacologically meaningful blood ethanol concentrations. They displayed alcohol-seeking behavior for the pharmacological effects, rather than for the taste or other properties, as evidenced by their willingness to dispense an ethanol solution intragastrically (although a recent article reported that P rats did not show a preference for intravenous ethanol [132]). In addition, they have been shown to develop a tolerance to the effects of ethanol, to display signs of physical dependence when the ethanol was withdrawn after long-term exposure, and to display relapse behavior following two or more weeks of abstinence.

3.2.4 Cancer

In 1919, Dr. F. D. Bullock and Dr. M. R. Curtis at the Crocker Institute of Cancer Research (Columbia University, New York, NY, USA) began a project to produce a number of inbred strains of rat for use in their studies of cancer. Previous work to induce neoplasms using tapeworm infestation as a chronic irritant had demonstrated that rats from some sources were more susceptible to tumor induction than those from other sources [133]. The group began inbreeding rats from four commercial breeders—Fischer, Zimmerman, Marshall, and August—in 1919 and expanded their efforts to include rats originally sourced from Copenhagen, Denmark, in 1920 [7]. From these efforts, at least ten inbred strains, including ACI, COP, Marshall 520 (M520), Fischer 344 (F344), and the now extinct Fischer 230, were developed, a number of which are still in active use for cancer research. Between 1920 and 1970, Dr. Curtis with her colleague Dr. W. F. Dunning and coworkers produced a substantial number of articles (for example, [134–142]) on tapeworm-induced sarcomas, strain differences in susceptibility to a variety of spontaneous tumors, chemical carcinogenesis, and chemotherapy using these inbred strains.

Because of its history of use for cancer research, the F344/N rat strain was the model of choice for the National Cancer Institute, and subsequently the National Toxicology Program (NTP), for standardized bioassays of carcinogenicity for chemical compounds. During the more than three decades of use for these studies, the NCI and NTP amassed an immense, publicly available dataset derived from the testing of thousands of possibly carcinogenic compounds [143, 144]. However, in 2006, an NTP workshop, "Animal Models for the NTP Rodent Cancer Bioassay: Strains and Stocks—Should We Switch?" [145] reviewed the use of the F344/N strain for future studies. Workshop participants concluded that, due to problems with infertility, seizures and chylothorax in the F344/N colony in particular, and a variable but relatively high incidence of spontaneous tumors, particularly testicular interstitial cell tumors and mononuclear cell leukemia, inherent to the F344 strain in general [146], the F344/N strain should no longer be used for NTP bioassays. After considering several alternatives, including using a different substrain of F344 or using F1 rats from a cross between F344 and BN, the recommendation was to replace the inbred strain with outbred rats. The initial recommendation was to use the outbred Wistar Han rat as the "default" for most toxicological bioassays. However, after some experimentation it was found that reproductive performance of the Wistar Han rats was less than ideal and the outbred Hsd:SD Sprague-Dawley strain became the default.

The Wistar Furth (WF) strain was inbred in the hopes of producing a rat strain that developed leukemia at high frequency [147]. Dr. J. Furth began with a partially inbred line of Wistar rats which had been shown to develop malignant lymphomas. However, despite the large number of lymphomas found in the progenitor stock, the resulting fully inbred strain did not show particularly high susceptibility to either lymphomas or leukemia. The incidences of leukemias and malignant lymphomas in the WF strain were reported as 9% and 7%, respectively, as of 1960. Instead of leukemias/lymphomas, the rats were documented to spontaneously develop pituitary and mammary tumors at high frequency (27% and 21%, respectively). In addition, adrenal and uterine neoplasms and lipomas occurred at much lower frequency (2–3%), and a general category of "unclassified tumors" were seen at a rate of approximately 4%. Despite the lower-than-expected frequency of leukemias in the WF strain, the authors still noted that the intractability of the WF leukemias, that is, their relative resistance to the chemotherapeutic agents of the time, made WF a better model for human disease than the Fischer strain, where the leukemias proved more sensitive to chemotherapeutics.

4 Resources for Rat Researchers

4.1 Where to Find Rat Strain Models

There are a number of both commercial and non-commercial sources for rat model strains. Some of these will be covered more thoroughly later in this book (*see* Chapter 3). By far, the widest variety of strains, including inbred, outbred, mutant, and transgenic strains, can be found at the Rat Resource and Research Center (RRRC) in the United States and the National BioResource Project for the Rat (NBRP-Rat) in Japan.

4.1.1 RRRC

The Rat Resource and Research Center (RRRC, http://www.rrrc.us/, [148]) was established at the University of Missouri in 2001 by Dr. John Critser. The Center was developed to supply the rat research community with "high quality, well-characterized inbred, hybrid and genetically engineered rat[s]." In 2011, the leadership of the RRRC was transferred to Dr. Elizabeth Bryda. RRRC continues to maintain a limited number of live strains and a much larger collection of cryopreserved germplasm for submitted strains. Researchers submitting strains for distribution and/or cryopreservation by the RRRC supply information about any diseases for which that strain is a model, and in what research area(s) the strain has been used. In addition to this, strain pages show information about the history of the strain, the researcher who submitted it, associated references, product availability (e.g., available as live animals, cryopreserved embryos or sperm), and descriptions of the genetics, breeding and husbandry where this is available. RRRC works with the Rat Genome Database (RGD) to ensure correct nomenclature of the strains (*see* below) and links to additional information at RGD.

4.1.2 NBRP-Rat

NBRP-Rat (http://www.anim.med.kyoto-u.ac.jp/nbr/Default.aspx, [148]) is one branch of the larger National BioResource Projects (NBRP) in Japan. The NBRP was established in 2002 by the Ministry of Education, Culture, Sports, Science and Technology (MEXT) to "collect, preserve, and provide bioresources (such as experimental animals and plants) that are essential experimental materials for life sciences research" (http://www.nbrp.jp/about/about.jsp). As such, NBRP-Rat's stated mission is "collection of rat strains and genetic sub strains, phenotypic and genotypic characterization, cryopreservation of embryos/sperm, supply of the collected rat strains and a publicly accessible database of all assembled data."(http://www.anim.med.kyoto-u.ac.jp/nbr/about.aspx).
The strains available at NBRP-Rat include inbred, congenic, and recombinant strains, as well as spontaneous mutants and transgenic and mutagenized rats. Additionally, NBRP-Rat has undertaken an extensive standardized phenotyping project, the goal of which is to phenotype all of the standard strains and many of the mutant strains submitted. Six male and/or female rats of each strain undergo a battery of tests between 5 and 10 weeks of age. Tests include morphological measurements such as body and organ weights, and physiological tests including blood pressure, blood chemistry, hematology and urine chemistry, as well as locomotor and neurobehavioral tests. The data provide an invaluable survey of phenotypes across a wide variety of strains measured under control, pathogen-free conditions, and can be accessed both at NBRP and in RGD's PhenoMiner tool.

4.1.3 GERRC

With the advent of genome editing technologies for the rat, the demand for genetically modified rats for use as disease models has skyrocketed. However, for many researchers, production of a gene-edited rat to confirm the involvement of a gene or genomic region in their disease or phenotype of interest was out of reach due to the lack of expertise and/or funding to produce such models. In 2013, Dr. Howard Jacob and his colleagues at the Medical College of Wisconsin were awarded a grant to begin the MCW Gene Editing Rat Resource Center (GERRC, https://rgd.mcw.edu/wg/gerrc/). The GERRC was designed to leverage existing infrastructure and expertise in gene editing to support the needs of the rat research community by designing, producing, and distributing rats with modifications in specific genes generated on specific genetic backgrounds (i.e., existing, well-characterized inbred strains) at low cost to researchers. The project was funded by the National Heart, Lung and Blood Institute (NHLBI) to produce 250 gene-targeted rat models over a period of five years to support and accelerate research of complex diseases, particularly those of interest to the NHLBI. Researchers nominated a gene to be edited on one or more background strains and an external advisory board (EAB) reviewed the nominations and approved, rejected, or deferred the nomination for editing. As of this writing, the GERRC, now under the leadership of Dr. Melinda Dwinell and Dr. Aron Geurts, has received a list of 109 genes for which nominations have been accepted by the EAB, and has produced 134 publicly available strains with genomic modifications in 53 unique genes. Strains are made available, first to the nominating group, then to any researcher via links on the GERRC webpage.

4.1.4 Commercial Vendors

In addition to the large non-commercial repositories already mentioned, there are several commercial sources for rat strains. Sprague Dawley, Inc. was started in 1925 by Robert Dawley near Madison, Wisconsin. The original breeding stock was purported to have come from mating a hooded male of unclear origin with albino females from Wistar stock, and subsequently with the albino offspring of that mating. The original male was described as "a hybrid hooded male rat of exceptional size and vigor which genetically was half-white" [149]. The line was partially inbred, then changed to random breeding and the parental strain was considered outbred. Sprague Dawley, Inc. was obtained by Harlan in 1980 to form Harlan Sprague Dawley, which was in turn acquired by Envigo, Inc. in 2015 (http://www.envigo.com/). Envigo sells 14 strains of rats, two of which are offered specifically as aged animals—Sprague Dawley® outbred rats (SD) and Fischer 344 inbred rats. The rats offered by Envigo include direct descendants of a number of the original laboratory rat stocks, including Holtzman rats (an offshoot of SD), Lewis rats, and Wistar outbred rats, in addition to the aforementioned SD rats.

Charles River Laboratories (https://www.criver.com/) was started in 1947 by veterinarian Dr. Henry Foster to breed rats and supply them to laboratories in the Boston, Massachusetts area. Their catalog currently lists 40 rat strains that are available, including 22 inbred strains and 16 outbred.

Originally Sage Laboratories, Horizon Discovery's stock of "off the shelf" knockout rats (https://www.horizondiscovery.com/in-vivo-models) includes models for several research areas. These include knockouts of xenobiotic sensors and drug transporters for toxicology/ADMET (absorption, distribution, metabolism, and excretion-toxicity) studies. For researchers interested in translational studies, Horizon offers knockout models for Alzheimer's disease, Autism, and Parkinson's disease, as well as models for oncology and cardiovascular research. For optogenetics studies, Horizon offers ready-made Cre-driver, fluorescence reporter, and opsin-expressing rats.

For researchers in need of specific genomic alterations, there are a number of companies that offer modifications as a service. A list of these can be found on the RGD Laboratory Resources webpage at https://rgd.mcw.edu/wg/resource-links/laboratory-resources/#strains. The companies listed include Applied StemCell, Cyagen Biosciences, Transposagen, and PolyGene Transgenetics. Horizon Discovery also offers generation of genome-modified rats in addition to their "off the shelf" lines. Several of these companies also supply reagents for researchers to generate their own models.

4.2 Where to Find Data for Rat Model Strains

The first two of the FAIR principles for data management [150, 151] require that data be both Findable and Accessible. This requires the development and maintenance of data stores and knowledgebases to consolidate and integrate data from various sources, and in many cases, to expand, interpret, and/or analyze the data through processes such as manual curation of the literature. Such resources also create environments in which researchers can access and utilize the data for their own analyses and download both the original data and their analysis results for their own records. The resources available for rat data will be covered more completely in Chapter 3, but we will touch on some of these sources here.

4.2.1 RGD

Arguably, the most diverse and inclusive source for rat data is the Rat Genome Database (RGD, https://rgd.mcw.edu, [152]). RGD was started in 1999 "to collect, consolidate and integrate data generated from ongoing rat genetic and genomic research efforts and make these data widely available to the scientific community." From the beginning, RGD was intended as a multiple datatype and cross-species resource, including data for rat genes, markers, quantitative trait loci (QTLs), and strains, as well as homologous mouse and human genes for comparative purposes. This cross-disciplinary focus has continued. RGD now houses data which associate disease, phenotype, molecular function, biological process, subcellular localization, molecular pathway, gene-chemical interactions, and protein-protein interactions with the genomes of rat, human, mouse, dog, squirrel, chinchilla, pig, and bonobo. Many of these associations are via the genes for these species. In addition, RGD imports data for disease and phenotype associations for human variants, as well as the extensive phenotype data for mouse genes and QTLs to assist with comparative genomic analyses.

In addition to genome-associated data, RGD houses a comprehensive listing of rat strains. Strain data include information about the origin of the strain and its characteristics such as any documented disease or phenotype associations, information about breeding/husbandry, and drug or chemical responses. More recently, RGD has begun a major project to curate the results of quantitative phenotype measurements. These data can be accessed using RGD's PhenoMiner tool (https://rgd.mcw.edu/rgdweb/phenominer/home.jsp, [153, 154]). For each measurement, information about the strain, number of animals, their age and sex, the conditions under which the measurement was made, and the method used to make the measurement are captured in a standardized format so that results can be directly compared across not only strains and conditions, but also across studies. Researchers interested in accessing the data can search based on strain, clinical measurement, measurement method, and/or experimental condition [155–157] to retrieve their data of interest. When a result set of interest is found, the data can be downloaded for the researcher's records or for further analysis in other tools. RGD's PhenoMiner includes quantitative phenotype data from published research papers as well as from high-throughput phenotyping projects such as the PhysGen Program for Genomic Applications (PGA) and NBRP-Rat.

Researchers interested in disease can access information about disease models either through the Phenotypes and Models portal or through the Disease Portals [158]. The Strains and Models section of the former portal contains links to established models for six disease categories—cardiovascular, neurological, respiratory and immune/inflammatory diseases, as well as mammary cancer and diabetes. The assignment of strains as models for these diseases is done manually and is based on an extensive review of the literature. RGD's Disease Portals also list strains that have been associated with the applicable disease categories. As of this writing, there are 12 portals covering the following disease categories: Aging and Age-Related Disease, Cancer, Cardiovascular Disease, Developmental Disease, Diabetes, Hematologic Disease, Immune and Inflammatory Disease, Neurological Disease, Obesity and Metabolic Syndrome, Renal Disease, Respiratory Disease, and Sensory Organ Disease. Each portal is a consolidated view of the genes, QTLs, and strains which are associated with any disease in the category across rat, mouse, and human. Researchers can view all of the associated data objects or drill down to more specific groups of diseases.

4.2.2 NBRP-Rat

As mentioned previously, NBRP-Rat stores and disseminates both the rats themselves and data about those rats. The NBRP-Rat strain pages give information about genetic status of the strain, what research categories that strain has been used for, strain characteristics and breeding performance, and references where applicable. In many cases, there is a picture of the rat and an image of representative organs for that strain. As also mentioned previously, NBRP-Rat performs extensive phenotyping on strains submitted to their repository. These data are available on their website under the "Phenome" tab (http://www.anim.med.kyoto-u.ac.jp/NBR/phenome.aspx) and are available in both graphical and tabular format.

4.2.3 NCBI, Ensembl, and UCSC Genome Browser for Genes and Genomics

The National Center for Biotechnology Information (NCBI, https://www.ncbi.nlm.nih.gov/), the European Bioinformatics Institute's Ensembl (https://www.ensembl.org/index.html), and the University of California, Santa Cruz's (UCSC, https://genome.ucsc.edu/) Genome resources are multispecies resources with diverse datasets that include genome sequences, gene, transcript and protein records, protein domain information, functional data, and more. Much of the data provided are consolidated from other resources, including the Rat Genome Database in terms of functional annotations for rat genes as well as curated records for QTLs. NCBI and Ensembl both do gene predictions for whole genome assemblies. Because the algorithms they use are disparate, the predicted gene sets are not the same although there is substantial overlap. For more information about the NCBI and Ensembl genome annotation pipelines, *see* Chapter 2 in this book. These resources also supply tools for analysis of the data they provide, including genome browsers for viewing genes and other genomic elements in the wider genomic context.

4.2.4 UniProtKB

UniProtKB (https://www.uniprot.org/, [159, 160]) is a cross-species resource that provides data for rat proteins, including protein sequences, protein structure and domains, post-translational modifications, tissue-level expression and subcellular localization, protein-protein interaction data, and protein family assignments. Manual gene ontology annotations for rat proteins at UniProtKB are imported from the Rat Genome Database. UniProtKB also employs an automated pipeline to predict GO annotations based on domains, keywords, etc.

4.2.5 dbSNP/EBI's European Variation Archive

The dbSNP database at NCBI and EBI's European Variation Archive (EVA) have, in the past, both accepted submissions of rat genomic variant data to be included in their multispecies variant resources. As of 2017, however, dbSNP is no longer storing or presenting variant data for nonhuman species, making EVA the major source for nonhuman variants. The data presented include genomic positions, affected genes, where applicable, and predicted or validated variant consequences for corresponding transcripts.

dbSNP in the past and EVA going forward consolidate multiple records corresponding to the same variant into a single "reference" variant record, eliminating what can at times be substantial redundancy in variant data from multiple sources. The "rs ID" has been and will continue to be the gold standard for nonredundant variant designation.

4.3 Strain and Allele Nomenclature

The ability to identify a specific strain and/or its associated allele (s) is absolutely essential for disseminating and accessing the appropriate data for that strain or allele. Without proper identification, it is difficult, if not impossible to reproduce any results associated with it. As such, committees for determining guidelines for strain, gene, and allele nomenclature were assigned for both mouse (International Committee on Standardized Genetic Nomenclature for Mice) and rat (Rat Genome and Nomenclature Committee, RGNC) strains. The RGNC has tasked RGD with assigning the correct nomenclature to these objects. In terms of emerging rat strains, RGD curators review the available information, including the origin of the strain and what, if any, genomic modifications have been made and assign nomenclature based on the "Guidelines for Nomenclature of Mouse and Rat Strains" (https://rgd.mcw.edu/nomen/nomen.shtml). Authors are asked to contact RGD before publication to register their strains in order to receive proper nomenclature and RGD IDs, and to use the appropriate nomenclature and ID(s) in their publications for maximum traceability and reproducibility of their results.

4.4 Molecular Genetic Tools

The molecular genetic toolbox for rat includes genetic and genomic data as well as a variety of tools for using them. Interest in rat genetics has been a foundational research focus since the first studies on the genetics of coat color linkage and inheritance in the late nineteenth and early twentieth centuries. Since that time, the field has evolved to incorporate the use of genetic markers and single nucleotide variants as markers for QTLs, establishment of a reference genome sequence for the rat, and whole genome sequencing (WGS) of a number of inbred rat strains which are considered to be either established models of human disease or control strains for those models. Many of the chapters that follow outline research, methods and resources that came as a direct result of this molecular genetic toolbox and/or associated investments into infrastructure for the rat.

4.4.1 Genetic Markers

Until the advent of whole genome sequencing, assignment of genes and genetic markers to relative chromosomal positions on chromosomes was accomplished by somatic cell hybrids, genetic or radiation hybrid mapping or in situ hybridization. In 1990 [161] and again in 1991 [162], Levan et al. published the then-current rat gene map, consisting of 214 genes and 11 linkage groups with assignments to 20 of the 22 rat chromosomes. In 1991, Jacob et al. [163] published the first QTL analysis in rat, calculating linkage between blood pressure and the genome. They developed a set of 112 polymorphic simple sequence repeat (SSR) markers and estimated that approximately 90% of the genome should lie within 30 cM of a marker. Using these markers, they were able to find statistical associations between two genomic regions and blood pressure, Bp1 on chromosome 10 for salt-loaded hypertension and Bp2 on chromosome 18 for baseline diastolic blood pressure.

SSRs, later termed simple sequence length polymorphisms (SSLPs)—stretches of tandem repeats of two, three, or four nucleotides which are often polymorphic between individuals, and in the case of rat, between strains—can be followed through genetic crosses to assign genomic regions in the progeny to one or the other progenitor strain. SSLPs have been used extensively as markers to delineate regions of linkage, as peak and flanking markers, for QTLs. Between 1991 and 2004 when the first draft of the rat genome sequence was published, the Jacob laboratory and other groups released high density genetic and RH maps for the rat [164–167], increasing the number of markers available for analyses and decreasing the distance between markers to facilitate the paring down of QTL sizes in order to target the actual gene or genes responsible for the observed effect on a phenotype. Data about these markers were deposited into the Rat Genome Database, including the centimorgan chromosomal position, the relative order of markers on each chromosome, the PCR primer sequences for each, and, in many cases, the sequence of the amplified PCR product. Steen et al. [167] also characterized 4328 SSLPs in 48 commonly used inbred rat strains and supplied the expected sizes of the PCR products for each strain. As noted in that work, "These maps provide the basic tools for rat genomics. They will facilitate studies of multifactorial disease and functional genomics, allow construction of physical maps, and provide a scaffold for both directed and large-scale sequencing efforts and comparative genomics in this important experimental organism."

4.4.2 Genome Sequence

The project to sequence the rat genome was initiated in 2000 with a Request for Application (RFA) to form a "Network for Large-Scale Sequencing of the Rat Genome" (https://grants.nih.gov/grants/guide/rfa-files/RFA-HG-00-002.html). The initiative was funded jointly by the National Human Genome Research Institute (NHGRI) and the National Heart, Lung and Blood Institute (NHLBI). The stated goal was to produce "a working draft version (3-4 fold sequence coverage) of the rat genome sequence in two years or less." The specification that the result would be a working draft version indicated that there was no intention to "finish" the sequence, meaning that, although the draft was expected to be of high quality, it was intended that a certain proportion of the errors would never be corrected. As of this writing, this is still the case. Chapter 2 of this book contains a detailed analysis of the available genome assemblies for rat. We will introduce the subject here.

Following the release of several early versions of the rat genome between 2001 and 2003, the release of the RGSC 3.1 assembly in June of 2003 made rat the third mammalian genome to be available as a high-quality draft assembly. In April of 2004, Gibbs et al. published an article announcing the completion of the first high quality draft of the rat assembly and reported on their extensive analysis of the rat genome sequence and their comparisons of the rat, human, and mouse reference sequences [168]. The 3.x assembly was generated using a new dual-method approach, utilizing both whole genome shotgun (WGS) sequencing and low-coverage BAC sequencing. Additional methods used to support assembly of the sequence included the development of "fingerprint contig (FPC)" maps of clones from the Children's Hospital Oakland Research Institute (CHORI-230) rat BAC library, BAC end sequencing, and construction of a yeast artificial chromosome (YAC)-based physical map. The methods were developed and utilized in parallel, reducing the amount of time needed to produce the assembly. In addition, existing genetic and radiation hybrid maps were used to support assembly and to assess the quality of the resulting reference, allowing for robust draft assembly. The sequence was further enhanced over time with the incorporation of finished BAC sequences, e.g., of the Encyclopedia of DNA Elements (ENCODE) regions [169], constituting minor upgrades of the assembly to RGSC 3.4 in November 2004.

In 2008, Baylor College of Medicine released a new, independent assembly of the rat genome sequence, the Rnor 4.0 assembly, and an upgrade of that assembly approximately a year later. In addition to incorporating additional Solexa, SOLiD, and 454 sequencing reads, the sequence was assembled using a new version of the Atlas software which employed a different method for merging overlapping eBACs and WGS scaffolds (ftp://ftp.hgsc.bcm.edu/Rnorvegicus/Rnor4.0/README_Rnor4.txt, ftp://ftp.hgsc.bcm.edu/Rnorvegicus/Rnor4.1/READMErat4.1.txt). However, although NCBI began the work of annotating the latter assembly and fully integrating it into their data resources, the 4.x versions were not generally accepted by the rat research community and the assembly was not promoted to RefSeq assembly status at NCBI (https://www.ncbi.nlm.nih.gov/assembly/GCA_000001895.2, note that there is no GCF reference accession).

Many of the underlying problems with the assembly were addressed in the Rnor_5.0 version of the rat reference. The Atlas assembler used to build the initial rat reference assembly was improved over time to better handle the BAC sequences that had been used to build Rnor_3.1 [169]. Rnor_5.0 also featured more cDNAs for annotating untranslated regions of protein-coding genes and an updated rat-specific repeat library (http://mar2015.archive.ensembl.org/info/genome/genebuild/2012_04_rat_5_genebuild.pdf). Additional improvements included the addition of annotations of genes for small structured RNAs from RFAM and miRBase, and a substantial increase in the number of genes at NCBI with more than one transcript (from 303 genes for the v3.4 assembly to almost 7400 genes for the v5.0 assembly).

The most recent assembly, Rnor_6.0, was released in 2014, and includes additional BAC sequences, manual corrections, and PacBio data to fill scaffold gaps. It also includes the Y chromosome from the SHR rat. More information on rat genome assemblies and annotation pipelines can be found in Chapter 2 of this book.

4.5 Genome-Related Data

The availability of whole genome sequence for the rat has resulted in increased availability of genome-dependent data such as variants and RNA sequencing data.

4.5.1 Strain-Specific Variants

Along with upgrading and maintaining the BN reference, there have been many attempts to identify and catalog regions of genetic variation in strains and substrains. As detailed earlier in this chapter, the rat has served as a physiological model for over 100 years, and as multiple strains have been developed to study pathophysiological processes, comparative analyses between phenotypically similar and phenotypically disparate rat models may yield insights into the underlying mechanisms. As such, a number of whole genome sequencing projects were undertaken in order to find and catalog these genetic differences between rat strains.

Atanur et al. [170] sequenced the genomes of 27 rat strains, including both cardiovascular and metabolic models of disease as well as control strains, and determined variations using the Rnor_3.4 assembly. Two years later, Hermsen et al. [171] not only reanalyzed the sequence data for these 27 strains, but also expanded the analysis by including an additional 13 strains, aligning all of the sequences against the Rnor_5.0 assembly, and calling the variants using updated software.

Findings from these and other similar studies suggested that in pathophysiological models of diseases like hypertension, relevant phenotypes may arise due to different combinations of genetic factors, and this, in turn, reflects the complex nature of these diseases in humans [170, 172].

When the Rnor 6.0 version of the rat reference genome was released, an interim remapping of variants from some of the strains was done, utilizing UCSC's Batch Coordinate Conversion, or "LiftOver" tool to convert the coordinates of SNPs from the v5.0 assembly to the corresponding coordinates on the v6.0 assembly. This, however, does not always give uniformly reliable results when used for single nucleotide variants and small indels. Because of this, a more recent project was undertaken at the Rat Genome Database to again reanalyze the sequences for the various strains, aligning the raw sequence reads to the Rnor 6.0 assembly and using the latest software to call the variants.

The full set of strain-specific single nucleotide variants (SNVs), small insertions and deletions and copy number variants where these were called for all assemblies is available as variant call format (.vcf) files from the RGD ftp site (ftp://ftp.rgd.mcw.edu/pub/strain_specific_variants/). In addition, researchers can query the RGD database for strain-specific SNVs and small insertions and deletions using the Variant Visualizer tool (https://rgd.mcw.edu/rgdweb/front/config.html). A more complete description of this tool can be found in Chapter 3 of this book.

4.5.2 Variant Archives

In May of 2017, NCBI announced that they would no longer support the submission, storage, and presentation of nonhuman variants, including single nucleotide polymorphisms in dbSNP and larger structural variations in dbVar. Responsibility for all nonhuman variants was transitioned to the European Variation Archive (EVA) at EMBL-EBI as of November of 2017 (https://www.ebi.ac.uk/eva/?Help#key-steps-transitional-process). During and immediately following the initial transition period, all nonhuman variants were transferred from NCBI's dbSNP database to the EVA. The EVA now accepts variant submissions, assigns unique IDs to each variant and periodically consolidates redundant variant records into single "reference variant" records with the same rs-formatted IDs previously assigned by dbSNP. In addition, EVA normalizes the data to ensure that variant positions are standardized, annotates variant effects, and calculates allele frequencies. Variants, whether from small studies with only a few variants or from high-throughput studies producing millions of variants, are submitted in VCF files. A VCF validation software suite is provided so submitters can ensure their files are ready for loading to expedite the process. Rat researchers unfamiliar with the validation and submission process can submit their variants to the Rat Genome Database. RGD can incorporate the variants into the Variant Visualizer tool and concurrently help with the submission process to the EVA.

4.5.3 RNA-Seq

The advent of high-throughput sequencing technologies was closely followed by the application of those technologies to mapping and quantification of the transcriptome via RNA sequencing (RNA-Seq) [173]. Total or fractionated RNA from a tissue or a population of cells, or more recently from a single cell, is sequenced and the resulting "reads" are either aligned to a reference genome or to a list of reference transcripts, or assembled de novo to form a genome-wide transcription map consisting of the structure of the transcripts (including alternatively spliced transcripts) and/or the level of expression at either the gene or transcript level. RNA-Seq's wide dynamic range for detecting differences in expression, and its potential for revealing differences in transcript structure make it a powerful tool for comparison of transcriptomes between organs, between species, between individuals or strains of the same species, between developmental stages, or between conditions.

4.5.4 PhenoGen

PhenoGen Informatics (https://phenogen.org) comprises a database and website that provide tools and data sets to explore DNA variants, RNA expression, and QTLs. PhenoGen provides a substantial body of data for download, as well as a genome/transcriptome browser to explore specific regions of the genome and tools for gene list analysis. Functionality includes tools for weighted gene co-expression network analysis, pathway analysis, exon expression correlations, and promoter analysis. PhenoGen also has data and tools for eQTL analysis.

4.5.5 GEO/ArrayExpress

The Gene Expression Omnibus at NCBI (https://www.ncbi.nlm.nih.gov/geo/, [174]) and ArrayExpress at EMBL-EBI (https://www.ebi.ac.uk/arrayexpress/, [175]) are the major public repositories for high-throughput functional genomics datasets, including expression data derived from microarray and RNA-Seq analyses, and ChIP-Seq and methylation profiling data based on DNA sequencing, among others. Both repositories accept data submissions that comply with the "Minimum Information About A Microarray Experiment" (MIAME) and "Minimum Information About a Sequencing Experiment" (MINSEQE) guidelines. For all studies, these groups store the experimental metadata (that is, information about the experimental design, the samples and the submitters, with links to applicable publications where available) as well as the processed data, e.g., normalized expression values for array experiments or FPKM/RPKM values for RNA-Seq experiments. The raw sequence data for RNA-Seq, ChIP-Seq, and other high-throughput sequencing assays, on the other hand, are stored in NCBI's Sequence Read Archive (SRA, https://trace.ncbi.nlm.nih.gov/Traces/sra/sra.cgi) or EBI's European Nucleotide Archive (ENA, https://www.ebi.ac.uk/ena).

GEO and ArrayExpress both offer analysis tools and pre-analyzed results based on the data they are storing. GEO's DataSet records pages include functionality to compare samples within a dataset to produce several types of cluster heatmaps for gene expression, and the "Experiment design and value distribution" tool which shows box plots for the distribution of expression values for each sample in a dataset, grouped into subsets determined by specific experimental variables, for example, samples grouped by age or disease status. Along the same lines, GEO2R is an online tool that uses R to compare groups of samples within a dataset to identify differentially expressed genes. In addition to the study-based presentation of data, GEO provides gene-level expression profiles. Each profile displays a chart showing the expression level of a gene for each sample in a dataset. GEO Profiles make it easy to see whether a particular gene was differentially expressed across conditions. ArrayExpress provides an R Bioconductor package to access records and build Bioconductor data structures. In addition, ArrayExpress data provide the basis for EBI's Expression Atlas tool (https://www.ebi.ac.uk/gxa/home). Users can search for genes and/or conditions within or across organisms, or browse experiments to view and download expression results. As of this writing, the Expression Atlas included 141 experiments for rat.

4.6 Gene Manipulation in the Rat

Transgenic rat models have been generated for more than 30 years by DNA microinjection of donor DNA into embryos [176, 177] and used to study the function of a gene of interest. Early methods to create mutations within specific genes include ENU (N-ethyl-N-nitrosourea) mutagenesis or through introduction of a Sleeping Beauty transposon [178]. Both of these methods successfully identified mutations in targeted genes but are limited in efficiency and specificity. Many pups need to be screened using a variety of strategies to identify positive mutant founders [179]. Although these random mutagenesis strategies require large-scale screening efforts to identify specific mutations, ENU-induced mutant models have been archived at the Rat Resource and Research Center (http://www.rrrc.us/) and the PhysGen Knockout Program for subsequent follow-up phenotyping [178]. Additionally, an archive of ENU-induced mutant sperm was created by the Kyoto University Mutant Rat Archive [180] to be used for large-scale screening or rederivation of models using intracytoplasmic sperm injection. The Sleeping Beauty transposon system was implemented in rats after successful use in mice [181, 182]. This strategy had several advantages over ENU mutagenesis, including the ability to modify both the transposon and transposases and also that only a few transposon insertions were made in each founder [178]. Similar to the ENU models, the Rat Resource and Research Center repository has nearly 100 transposon-derived models available to investigators.

The advent of site-directed nucleases to target specific genes in rat embryos has rapidly changed the gene engineering landscape for investigators using rat models [183]. These sequence-specific nucleases, including zinc finger nucleases (ZFN), TALENs, and CRISPR/Cas approaches, have rapidly allowed the rat genome to be manipulated in ways previously only available in the mouse. These strategies were initially used to create a double-stranded break in the genomic DNA at specific locations in the genome. These breaks are typically repaired by nonhomologous end joining

(NHEJ) which can lead to a loss or gain of DNA. Such insertions or deletions can result in a knockout of the gene through removal of protein coding information or introduction of a frameshift mutation [183, 184]. In addition, a knockin can be created through homology-directed repair (HDR) and the inclusion of a homologous DNA template. Genetically modified rats have been developed using ZFNs, TALENs, and the CRISPR/Cas9 approaches using both NHEJ and HRD events. CRISPR/Cas is currently the most widely used approach due the ease of use, efficiency, and cost. In addition, the CRISPR/Cas system can be multiplexed, allowing multiple founders to be generated from a single round of embryo injections [184]. However, current reports suggested that the CRISPR/Cas system results in more frequent off-target effects than ZFNs and TALENs [183].

For laboratories without access to or expertise in embryo microinjection, a new technique for in vivo genome editing, *g*enome editing via *o*viductal *n*ucleic *a*cids *d*elivery (GONAD), has shown promising results in rats. This new strategy (referred to as i-GONAD [185] and rGONAD [186]) eliminates the need to handle the embryos to manipulate the genome. "Improved GONAD" (i-GONAD) was developed and tested in mice to bypass the need for collection of embryos and microinjection (e.g., pronuclear injection). Rather, early preimplantation embryos are modified by intraoviductal injection of Cas9 protein and synthetic gRNAs, followed by in vivo electroporation. This technology has been extended to rats with demonstration of highly efficient knockout and knockin modifications in two rat strains (DA and WKY).

Unlike in mouse, rat embryonic stem cells (ESCs) have been used less frequently for gene engineering, primarily due to technical challenges [187]. Rat ESCs have been established for several different rat strains [188–191] and have been successfully used for genome editing [191]. Recent work has demonstrated the successful use of the CRISPR/Cas9 system in rat ESCs to create both in vitro and in vivo models [187]. These techniques have the potential to create many new models to study early embryonic development, focus on single specific cell types in vitro, and potentially to humanize regions of the rat genome as has been done in the mouse [184, 192–194].

4.7 Rat in the Larger Context

Rat is an excellent model for human disease, and in many cases of complex disease it is the model of choice. But for cases where another model is preferred, or where additional information is needed, a researcher needs to be able to leverage the research done in other organisms and apply that information to their system. In these cases, it is helpful, possibly even necessary, to be able to access data for multiple species on a single site, and even in a single view. To meet this need, groups like the Rat Genome Database (RGD) and the Alliance of Genome Resources offer sites that incorporate data across multiple species. For comparative purposes, RGD has always provided data for mouse and human, in addition to rat. RGD curators manually assign disease and pathway terms for all three species and import phenotype and disease annotations for mouse and human to complement their annotation of rat genes, QTLs and strains. In recent years, however, RGD has expanded their repertoire of species to also include chinchilla, 13-lined ground squirrel, dog, pig, and bonobo. In this way, a researcher interested in otitis media, for which the chinchilla is the model of choice, or retinal diseases, for which squirrel is the better model, can easily compare data for that species to a rich set of functional, disease, pathway, phenotype, and gene-chemical interaction data for rat, mouse, and human.

The Alliance of Genome Resources, on the other hand, is a collaboration between six model organism databases and the Gene Ontology Consortium. The founding databases for this effort include Saccharomyces Genome Database (SGD, [195]), WormBase [196], FlyBase [197], Mouse Genome Database (MGD, [198, 199]), Rat Genome Database (RGD), and the Zebrafish Information Network (ZFIN, [200]). The Alliance is developing an infrastructure and website to standardize data storage and presentation across seven species: human, mouse, rat, zebrafish, fly, nematode, and yeast. The goal is a federated view which highlights the diverse types of research performed using the various organisms and pulls together these diverse data into a unified presentation of the most complete information possible for a given gene, genomic region, disease, phenotype or network to drive research into human diseases forward.

5 Future Directions

Several large initiatives are building and extending rat resources to accelerate the understanding of disease mechanisms, mapping biology to the genome, and creating new tools to build better animal models, datasets, and analytic tools. Two rat-centric centers have been funded by the National Institute on Drug Abuse (NIDA) to develop Centers of Excellence with a focus on drug abuse and addiction. The NIDA Center for GWAS in Outbred Rats (https://projectreporter.nih.gov/project_info_description.cfm?aid=9464528&icde=42187407) uses heterogenous stock rats (HS) [130] and sophisticated genomic approaches to identify associations, eQTLs, and genes influencing behavior. The new NIDA Core Center of Excellence in Omics, Systems Genetics, and the Addictome (https://projectreporter.nih.gov/project_info_description.cfm?aid=9531327&icde=42187459) is focused on providing resources to the community for analysis of omics data sets to link sequence variation, epigenetic factors, and environmental factors to phenotypes in order to further understanding of how the phenotype is altered by the genomic variation [201]. Although this new center has a focus on rodent models of addiction, the new methods, tools, and assembled datasets will be useful for many complex diseases. A third rat-centric program, the Hybrid Rat Diversity Program (https://projectreporter.nih.gov/project_info_description.cfm?aid=9488560&icde=42187641), was funded by the Office of the Director at the National Institutes of Health to build a panel similar to the Hybrid Mouse Diversity Panel [202] involving 96 rat strains to be used broadly to study the genetic and phenotypic diversity in complex diseases. This 96-strain panel, consisting of 33 classic inbred strains and two panels of recombinant inbred strains, will include rederived rat models, basic phenotypic characterization, complete genomic sequencing for all strains, and data analysis, integration, and dissemination through the HRDP Portal at the Rat Genome Database. Saba and Tabakoff have demonstrated how the first stages in the HRDP are a renewable resource that can be used by investigators across many disciplines to generate connectomes to further investigate pathways and gene interactions associated with complex traits [203].

On the horizon are genome editors and delivery mechanisms to enable repair in somatic cells. The new NIH Somatic Cell Genome Editing (SCGE) Program, launched in January 2018, has funded projects to develop cell-specific and tissue-specific delivery vehicles to target genome editing to specific cells [204]. Although rat is currently not a validation model, the breakthroughs made within this program should be translatable to other model organisms with the goal to develop treatments for human disorders.

References

1. Lack JB, Hamilton MJ, Braun JK, Mares MA, Van Den Bussche RA (2013) Comparative phylogeography of invasive Rattus rattus and *Rattus norvegicus* in the U.S. reveals distinct colonization histories and dispersal. Biol Invasions 15(5):1067–1087. https://doi.org/10.1007/s10530-012-0351-5
2. Courchamp F, Chapuis JL, Pascal M (2003) Mammal invaders on islands: impact, control and control impact. Biol Rev Camb Philos Soc 78(3):347–383
3. Centers for Disease Control and Prevention. https://www.cdc.gov/rodents/diseases/index.html. Accessed 15 Aug 2018
4. Lin XD, Guo WP, Wang W, Zou Y, Hao ZY, Zhou DJ et al (2012) Migration of Norway rats resulted in the worldwide distribution of Seoul hantavirus today. J Virol 86(2):972–981. https://doi.org/10.1128/jvi.00725-11
5. Kosoy M, Khlyap L, Cosson J-F, Morand S (2015) Aboriginal and invasive rats of genus Rattus as hosts of infectious agents. Vector Borne Zoonotic Dis 15(1):3–12. https://doi.org/10.1089/vbz.2014.1629
6. Harper GA, Bunbury N (2015) Invasive rats on tropical islands: their population biology and impacts on native species. Glob Ecol Conserv 3:607–627. https://doi.org/10.1016/j.gecco.2015.02.010
7. Lindsey JR (1979) Historical foundations. In: Baker HJ, Lindsey JR, Weisbroth SH (eds) The laboratory rat: biology and diseases, American College of Laboratory Animal Medicine Series, vol 1. Academic Press, New York, pp 1–36
8. Puckett EE, Park J, Combs M, Blum MJ, Bryant JE, Caccone A et al (2016) Global population divergence and admixture of the brown rat (*Rattus norvegicus*). Proc Biol Sci

283(1841):20161762. https://doi.org/10.1098/rspb.2016.1762

9. Song Y, Lan Z, Kohn MH (2014) Mitochondrial DNA phylogeography of the Norway rat. PLoS One 9(2):e88425. https://doi.org/10.1371/journal.pone.0088425
10. Zeng L, Ming C, Li Y, Su L-Y, Su Y-H, Otecko NO et al (2016) Evolutionary history of the brown rat: out of southern East Asia and selection. bioRxiv. https://doi.org/10.1101/096800
11. Philipeaux JM (1856) Note sur l extirpation des capsules surrénales chez les rats albinos (Mus ratus), vol XLIII. Comptes Rendus, Paris, pp 904–906
12. Hunt JR (1993) Animal models of human nutrition. In: Macrae R, Robinson RK, Sadler MJ (eds) Encyclopaedia of food science, food technology, and nutrition, vol 1. Academic Press, Ltd., San Diego, CA, pp 188–194
13. Savory W (1863) Experiments on food; its destination and uses. Lancet 81 (2066):381–383. https://doi.org/10.1016/S0140-6736(02)65694-6
14. Savory WS (1863) III. Experiments on food; its destination and uses. Proc R Soc Lond 12:121–123. https://doi.org/10.1098/rspl.1862.0019
15. Hopkins FG (1912) Feeding experiments illustrating the importance of accessory factors in normal dietaries. J Physiol 44 (5–6):425–460
16. McCollum EV (1953) My early experiences in the study of foods and nutrition. Annu Rev Biochem 22:1–16. https://doi.org/10.1146/annurev.bi.22.070153.000245
17. McCollum EV, Davis M (1913) The necessity of certain lipins in the diet during growth. J Biol Chem 15(1):167–175
18. Osborne TB, Mendel LB, Ferry EL, Wakeman AJ (1913) The relation of growth to the chemical constituents of the diet. J Biol Chem 15(2):311–326
19. Semba RD (2012) On the 'discovery' of vitamin A. Ann Nutr Metab 61(3):192–198. https://doi.org/10.1159/000343124
20. McCollum EV, Davis M (1915) The nature of the dietary deficiencies of rice. J Biol Chem 23 (1):181–230
21. McCollum EV, Davis M (1915) The essential factors in the diet during growth. J Biol Chem 23(1):231–246
22. McCollum EV, Kennedy C (1916) The dietary factors operating in the production of polyneuritis. J Biol Chem 24(4):491–502
23. Osborne TB, Mendel LB (1916) The growth of rats upon diets of isolated food substances. Biochem J 10(4):534–538
24. Osborne TB, Mendel LB, Ferry EL, Wakeman AJ (1914) Amino-acids in nutrition and growth. J Biol Chem 17(3):325–349
25. Osborne TB, Mendel LB, Ferry EL, Wakeman AJ (1917) The relative value of certain proteins and protein concentrates as supplements to corn gluten. J Biol Chem 29 (1):69–92
26. Osborne TB, Mendel LB, Ferry EL, Wakeman AJ (1914) The influence of cod liver oil and some other fats on growth. J Biol Chem 17(3):401–408
27. Ferry EL (1919) Nutrition experiments with rats. A description of methods and technic. J Lab Clin Med 5(11):735–745. https://doi.org/10.5555/uri:pii:S0022214320900161
28. Nelson EM, Steenbock H (1925) Fat-soluble vitamins: XXI. Observations bearing on the alleged induction of growth-promoting properties in air by irradiation with ultra-violet light. J Biol Chem 62(3):575–593
29. Steenbock H (1924) The induction of growth promoting and calcifying properties in a ration by exposure to light. Science 60 (1549):224–225. https://doi.org/10.1126/science.60.1549.224
30. Sherman HC, Pappenheimer AM (1921) Experimental rickets in rats: I. A diet producing rickets in white rats, and its prevention by the addition of an inorganic salt. J Exp Med 34(2):189–198
31. Sherman HC, Campbell HL (1928) The influence of food upon longevity. Proc Natl Acad Sci U S A 14(11):852–855
32. Sherman HC, Trupp HY (1949) Further experiments with vitamin A in relation to aging and to length of life. Proc Natl Acad Sci U S A 35(2):90–92
33. Baker DH (2008) Animal models in nutrition research. J Nutr 138(2):391–396
34. Kuramoto T (2011) Yoso-tama-no-kakehashi; the first Japanese guidebook on raising rats. Exp Anim 60(1):1–6
35. Castle WE (1947) The domestication of the rat. Proc Natl Acad Sci U S A 33(5):109–117
36. MacCurdy H, Castle WE (1907) Selection and cross-breeding in relation to the inheritance of coat-pigments and coat-patterns in rats and guinea-pigs. The Carnegie Institution of Washington, Washington, DC
37. Castle WE, Phillips JC (1914) Piebald rats and selection; an experimental test of the

effectiveness of selection and of the theory of gametic purity in Mendelian crosses. The Carnegie Institution of Washington, Washington, DC

38. Castle WE, Wright S (1915) Two color mutations of rats which show partial coupling. Science 42(1075):193–195. https://doi.org/10.1126/science.42.1075.193
39. Castle WE (1919) Piebald rats and the theory of genes. Proc Natl Acad Sci U S A 5 (4):126–130
40. Castle WE, Wachter WL (1924) Variations of linkage in rats and mice. Genetics 9(1):1–12
41. Castle WE (1925) A sex difference in linkage in rats and mice. Genetics 10(6):580–582
42. King HD, Castle WE (1935) Linkage studies of the rat (*Rattus norvegicus*). Proc Natl Acad Sci U S A 21(6):390–399
43. King HD, Castle WE (1937) Linkage studies of the rat (*Rattus norvegicus*): II. Proc Natl Acad Sci U S A 23(2):56–60
44. Castle WE, King HD (1940) Linkage studies of the rat (*Rattus norvegicus*): III. Proc Natl Acad Sci U S A 26(9):578–580
45. Castle WE, King HD (1941) Linkage studies of the rat (*Rattus norvegicus*): V. Proc Natl Acad Sci U S A 27(8):394–398
46. Castle WE, King HD, Daniels AL (1941) Linkage studies of the rat (*Rattus norvegicus*): IV. Proc Natl Acad Sci U S A 27(6):250–254
47. Castle WE (1941) Influence of certain color mutations on body size in mice, rats, and rabbits. Genetics 26(2):177–191
48. Castle WE (1944) Linkage of Waltzing in the rat. Proc Natl Acad Sci U S A 30(9):226–230
49. Castle WE, King HD (1944) Linkage studies of the rat (*Rattus norvegicus*): VI. Proc Natl Acad Sci U S A 30(4):79–82
50. Castle WE (1946) Linkage in the albino chromosome of the rat. Proc Natl Acad Sci U S A 32(2):33–36
51. Castle WE, King HD (1947) Linkage studies of the rat. J Hered 38(11):341–344
52. Castle WE, King HD (1948) Linkage studies of the rat: IX. Cataract. Proc Natl Acad Sci U S A 34(4):135–136
53. Castle WE, King HD (1949) Linkage studies of the rat: X. Proc Natl Acad Sci U S A 35 (9):545–546
54. Castle WE (1951) Variation in the hooded pattern of rats, and a new allele of hooded. Genetics 36(3):254–266
55. Conklin EG (1938) Biographical memoir of Henry Herbert Donaldson 1857–1938. Biogr Mem Natl Acad Sci 20:227–243
56. Donaldson HH (1915) The rat: reference tables and data for the albino rat and the Norway rat. The Wistar Institute of Anatomy and Biology, Philadelphia
57. Donaldson HH (1924) The rat: data and reference tables for the albino rat (Mus norvegius albinus) and the Norway rat (Mus norvegius). The Wistar Institute of Anatomy and Biology, Philadelphia
58. Ogilvie MB (2007) Inbreeding, eugenics, and Helen Dean King (1869–1955). J Hist Biol 40(3):467–507. https://doi.org/10.1007/s10739-006-9117-1
59. King HD, Donaldson HH (1929) Life processes and size of the body and organs of the gray Norway rat during ten generations in captivity. In: Stockard CR, Evans HM (eds) The American Anatomical Memoirs, vol 14. The Wistar Institute of Anatomy and Biology, Philadelphia
60. King HD (1939) Life processes in gray Norway rats during fourteen years in captivity. In: Stockard CR, Evans HM (eds) The American Anatomical Memoirs, vol 17. The Wistar Institute of Anatomy and Biology, Philadelphia
61. Wilkins AS, Wrangham RW, Fitch WT (2014) The "domestication syndrome" in mammals: a unified explanation based on neural crest cell behavior and genetics. Genetics 197 (3):795–808. https://doi.org/10.1534/genetics.114.165423
62. Stewart CC (1898) Variations in daily activity produced by alcohol and by changes in barometric pressure and diet, with a description of recording methods. Am J Phys 1(1):40–56. https://doi.org/10.1152/ajplegacy.1898.1.1.40
63. Willard SS (1901) Experimental Study of the Mental Processes of the Rat. II. Am J Psychol 12(2):206–239. https://doi.org/10.2307/1412534
64. Watson JB (1903) Animal education; an experimental study on the psychical development of the white rat, correlated with the growth of its nervous system. The University of Chicago Press, Chicago, p 122
65. Watson JB (1914) Behavior: an introduction to comparative psychology. Holt, New York
66. Schulkin J, Rozin P, Stellar E (1994) Curt P. Richter – February 20, 1894–December 21, 1988. Biogr Mem Natl Acad Sci 65:311–320
67. Richter CP (1922) A behavioristic study of the activity of the rat. Williams & Wilkins Company, Baltimore

68. Richter CP (1936) Increased salt appetite in adrenalectomized rats. Am J Phys 115 (1):155–161. https://doi.org/10.1152/ajplegacy.1936.115.1.155
69. Richter CP (1953) Experimentally produced behavior reactions to food poisoning in wild and domesticated rats. Ann N Y Acad Sci 56 (2):225–239
70. Richter CP (1968) Experiences of a reluctant rat-catcher: the common Norway rat-friend or enemy? Proc Am Philos Soc 112 (6):403–415
71. Stotsenberg J (1909) On the growth of the albino rat (Mus norvegicus var. albus) after castration. Anat Rec 3(4):233–244. https://doi.org/10.1002/ar.1090030410
72. Stotsenburg JM (1913) The effect of spaying and semi-spaying young albino rats (Mus norvegicus albinus) on the growth in body weight and body length. Anat Rec 7(6):183–194. https://doi.org/10.1002/ar.1090070602
73. Hammett FS (1924) Studies of the thyroid apparatus. XX. The effect of thyroparathyroidectomy and parathyroidectomy at 75 days of age on the growth of the brain and spina cord of male and female albino rats. J Comp Neurol 37(1):15–30. https://doi.org/10.1002/cne.900370103
74. Long JA, Evans HM (1922) The oestrous cycle in the rat and its associated phenomena, vol 6. Memoirs of the University of California. University of California Press, Berkeley, CA
75. Evans HM, Long JA (1921) The effect of feeding the anterior lobe of the hypophysis on the oestrous cycle of the rat. Anat Rec 21 (1):62. https://doi.org/10.1002/ar.1090210105
76. Evans HM, Long JA (1921) The effect of the anterior lobe administered intraperitoneally upon growth, maturity, and oestrous cycles of the rat. Anat Rec 21(1):62–63. https://doi.org/10.1002/ar.1090210105
77. Simpson ME, Evans HM (1946) Comparison of the spermatogenic and androgenic properties of testosterone propionate with those of pituitary ICSH in hypophysectomized 40-day old male rats. Endocrinology 39(5):281–285. https://doi.org/10.1210/endo-39-5-281
78. Li CH, Kalman C, Evans HM (1947) The effect of the hypophyseal growth hormone on the alkaline phosphatase of rat plasma. J Biol Chem 169(3):625–629
79. Evans HM, Becks H, Asling CW, Li CH (1947) Gigantism produced in normal female rats by chronic treatment with pure pituitary growth hormone. Anat Rec 97(3):333
80. Evans HM, Becks H (1948) The gigantism produced in normal rats by injection of the pituitary growth hormone; skeletal changes; tibia, costochondral junction, and caudal vertebrae. Growth 12(1):43–54
81. Li CH, Simpson ME, Evans HM (1948) The gigantism produced in normal rats by injection of the pituitary growth hormone; main chemical components of the body. Growth 12 (1):39–42
82. Evans HM, Sompson ME, Li CH (1948) The gigantism produced in normal rats by injection of the pituitary growth hormone; body growth and organ changes. Growth 12 (1):15–32
83. Li CH, Simpson ME, Evans HM (1949) Influence of growth and adrenocorticotropic hormones on the body composition of hypophysectomized rats. Endocrinology 44 (1):71–75. https://doi.org/10.1210/endo-44-1-71
84. Asling CW, Walker DG, Simpson ME, Evans HM (1950) Differences in the skeletal development attained by 60-day-old female rats hypophysectomized at ages varying from 6 to 28 days. Anat Rec 106(4):555–569
85. Walker DG, Simpson ME, Asling CW, Evans HM (1950) Growth and differentiation in the rat following hypophysectomy at 6 days of age. Anat Rec 106(4):536–554
86. Koneff AA, Moon HD, Simpson ME, Li CH, Evans HM (1951) Neoplasms in rats treated with pituitary growth hormone. IV Pituitary gland. Cancer Res 11(2):113–117
87. Van Dyke DC, Garcia JF, Simpson ME, Huff RL, Contopoulos AN, Evans HM (1952) Maintenance of circulating red cell volume in rats after removal of the posterior and intermediate lobes of the pituitary. Blood 7 (10):1005–1016
88. Asling CW, Walker DG, Simpson ME, Li CH, Evans HM (1952) Deaths in rats submitted to hypophysectomy at an extremely early age and the survival effected by growth hormone. Anat Rec 114(1):49–65
89. Walker DG, Asling CW, Simpson ME, Li CH, Evans HM (1952) Structural alterations in rats hypophysectomized at six days of age and their correction with growth hormone. Anat Rec 114(1):19–47
90. Fels IG, Simpson ME, Evans HM (1953) The effect of magnesium ion upon the alkaline phosphatase activity in the thyroid of the hypophysectomized rat. J Biol Chem 204 (2):807–814
91. Nelson MM, Lyons WR, Evans HM (1953) Comparison of ovarian and pituitary

hormones for maintenance of pregnancy in pyridoxine-deficient rats. Endocrinology 52 (5):585–589. https://doi.org/10.1210/endo-52-5-585

92. Contopoulos AN, Van Dyke DC, Ellis S, Simpson ME, Lawrence JH, Evans HM (1955) Prevention of neonatal anemia in the rat by the pituitary erythropoietic factor. Blood 10(2):115–119
93. Wooten E, Nelson MM, Simpson ME, Evans HM (1955) Effect of pyridoxine deficiency on the gonadotrophic content of the anterior pituitary in the rat. Endocrinology 56 (1):59–66. https://doi.org/10.1210/endo-56-1-59
94. Wooten E, Nelson MM, Simpson ME, Evans HM (1958) Response of vitamin B6-deficient rats to hypophyseal follicle-stimulating and interstitial-cell-stimulating hormones. Endocrinology 63(6):860–866. https://doi.org/10.1210/endo-63-6-860
95. King CG (1975) Biographical memoir of Henry Clapp Sherman. Natl Acad Sci Biograph Memoirs. 395–429
96. Smirk FH, Hall WH (1958) Inherited hypertension in rats. Nature 182(4637):727–728
97. Phelan EL, Smirk FH (1960) Cardiac hypertrophy in genetically hypertensive rats. J Pathol Bacteriol 80:445–448
98. Dahl LK, Heine M, Tassinari L (1962) Effects of chronia excess salt ingestion. Evidence that genetic factors play an important role in susceptibility to experimental hypertension. J Exp Med 115:1173–1190
99. Moreno C, Dumas P, Kaldunski ML, Tonellato PJ, Greene AS, Roman RJ et al (2003) Genomic map of cardiovascular phenotypes of hypertension in female Dahl S rats. Physiol Genomics 15(3):243–257. https://doi.org/10.1152/physiolgenomics.00105.2003
100. Baker JE, Konorev EA, Gross GJ, Chilian WM, Jacob HJ (2000) Resistance to myocardial ischemia in five rat strains: is there a genetic component of cardioprotection? Am J Physiol Heart Circ Physiol 278(4): H1395–H1400. https://doi.org/10.1152/ajpheart.2000.278.4.H1395
101. Okamoto K, Aoki K (1963) Development of a strain of spontaneously hypertensive rats. Jpn Circ J 27:282–293
102. Okamoto K, Yamori Y, Oshima A, Tanaka T (1972) Development of substrains in spontaneously hypertensive rats: genealogy, isozymes and effect of hypercholesterolemic diet. Jpn Circ J 36(5):461–470
103. Nagaoka A, Iwatsuka H, Suzuoki Z, Okamoto K (1976) Genetic predisposition to stroke in spontaneously hypertensive rats. Am J Phys 230(5):1354–1359. https://doi.org/10.1152/ajplegacy.1976.230.5.1354
104. Hazama F, Ooshima A, Tanaka T, Tomimoto K, Okamoto K (1975) Vascular lesions in the various substrains of spontaneously hypertensive rats and the effects of chronic salt ingestion. Jpn Circ J 39(1):7–22
105. Kunert MP, Drenjancevic-Peric I, Dwinell MR, Lombard JH, Cowley AW Jr, Greene AS et al (2006) Consomic strategies to localize genomic regions related to vascular reactivity in the Dahl salt-sensitive rat. Physiol Genomics 26(3):218–225. https://doi.org/10.1152/physiolgenomics.00004.2006
106. Garrett MR, Dene H, Walder R, Zhang QY, Cicila GT, Assadnia S et al (1998) Genome scan and congenic strains for blood pressure QTL using Dahl salt-sensitive rats. Genome Res 8(7):711–723
107. Ye P, West MJ (2003) Cosegregation analysis of natriuretic peptide genes and blood pressure in the spontaneously hypertensive rat. Clin Exp Pharmacol Physiol 30(12):930–936
108. Dupont J, Dupont JC, Froment A, Milon H, Vincent M (1973) Selection of three strains of rats with spontaneously different levels of blood pressure. Biomedicine 19(1):36–41
109. Bilusic M, Bataillard A, Tschannen MR, Gao L, Barreto NE, Vincent M et al (2004) Mapping the genetic determinants of hypertension, metabolic diseases, and related phenotypes in the lyon hypertensive rat. Hypertension 44 (5):695–701. https://doi.org/10.1161/01.HYP.0000144542.57306.5e
110. Yamori Y, Ohta K, Ohtaka M, Nara Y, Horie R, Ooshima A (1978) Glucose metabolism in spontaneously hypertensive rats. Jpn Heart J 19(4):559–560
111. Mondon CE, Reaven GM (1988) Evidence of abnormalities of insulin metabolism in rats with spontaneous hypertension. Metabolism 37(4):303–305
112. Kawano K, Hirashima T, Mori S, Saitoh Y, Kurosumi M, Natori T (1992) Spontaneous long-term hyperglycemic rat with diabetic complications. Otsuka Long-Evans Tokushima Fatty (OLETF) strain. Diabetes 41 (11):1422–1428
113. Kawano K, Hirashima T, Mori S, Natori T (1994) OLETF (Otsuka Long-Evans Tokushima Fatty) rat: a new NIDDM rat strain. Diabetes Res Clin Pract 24(Suppl): S317–S320
114. Zucker LM (1960) Two-way selection for body size in rats, with observations on

simultaneous changes in coat color pattern and hood size. Genetics 45(4):467–483
115. Zucker LM, Zucker TF (1961) Fatty, a new mutation in the rat. J Hered 52(6):275–278. https://doi.org/10.1093/oxfordjournals.jhered.a107093
116. Peterson RG, Shaw WN, Neel M-A, Little LA, Eichberg J (1990) Zucker diabetic fatty rat as a model for non-insulin-dependent diabetes mellitus. ILAR J 32(3):16–19. https://doi.org/10.1093/ilar.32.3.16
117. Shimabukuro M, Zhou YT, Levi M, Unger RH (1998) Fatty acid-induced beta cell apoptosis: a link between obesity and diabetes. Proc Natl Acad Sci U S A 95(5):2498–2502
118. Tayebati SK, Tomassoni D, Amenta F (2012) Spontaneously hypertensive rat as a model of vascular brain disorder: microanatomy, neurochemistry and behavior. J Neurol Sci 322 (1–2):241–249. https://doi.org/10.1016/j.jns.2012.05.047
119. Sarkisova K, van Luijtelaar G (2011) The WAG/Rij strain: a genetic animal model of absence epilepsy with comorbidity of depression [corrected]. Prog Neuro-Psychopharmacol Biol Psychiatry 35 (4):854–876. https://doi.org/10.1016/j.pnpbp.2010.11.010
120. Malkesman O, Weller A (2009) Two different putative genetic animal models of childhood depression--a review. Prog Neurobiol 88 (3):153–169. https://doi.org/10.1016/j.pneurobio.2009.03.003
121. Malkesman O, Braw Y, Zagoory-Sharon O, Golan O, Lavi-Avnon Y, Schroeder M et al (2005) Reward and anxiety in genetic animal models of childhood depression. Behav Brain Res 164(1):1–10. https://doi.org/10.1016/j.bbr.2005.04.023
122. Overstreet DH (2012) Modeling depression in animal models. Methods Mol Biol 829:125–144. https://doi.org/10.1007/978-1-61779-458-2_7
123. Mardones J, Segovia-Riquelme N (1983) Thirty-two years of selection of rats by ethanol preference: UChA and UChB strains. Neurobehav Toxicol Teratol 5(2):171–178
124. Eriksson K (1968) Genetic selection for voluntary alcohol consumption in the albino rat. Science 159(3816):739–741. https://doi.org/10.1126/science.159.3816.739
125. Bell RL, Rodd ZA, Lumeng L, Murphy JM, McBride WJ (2006) The alcohol-preferring P rat and animal models of excessive alcohol drinking. Addict Biol 11(3–4):270–288. https://doi.org/10.1111/j.1369-1600.2005.00029.x
126. Li TK, Lumeng L, Doolittle DP (1993) Selective breeding for alcohol preference and associated responses. Behav Genet 23 (2):163–170
127. Li TK, Lumeng L, Doolittle DP, Carr LG (1991) Molecular associations of alcohol-seeking behavior in rat lines selectively bred for high and low voluntary ethanol drinking. Alcohol Alcohol Suppl 1:121–124
128. Colombo G, Agabio R, Lobina C, Reali R, Zocchi A, Fadda F, Gessa GL (1995) Sardinian alcohol-preferring rats: a genetic animal model of anxiety. Physiol Behav 57 (6):1181–1185
129. Fadda F, Mosca E, Colombo G, Gessa GL (1989) Effect of spontaneous ingestion of ethanol on brain dopamine metabolism. Life Sci 44(4):281–287
130. Parker CC, Chen H, Flagel SB, Geurts AM, Richards JB, Robinson TE et al (2014) Rats are the smart choice: Rationale for a renewed focus on rats in behavioral genetics. Neuropharmacology 76(Pt B):250–258. https://doi.org/10.1016/j.neuropharm.2013.05.047
131. Murphy JM, Stewart RB, Bell RL, Badia-Elder NE, Carr LG, McBride WJ et al (2002) Phenotypic and genotypic characterization of the Indiana University rat lines selectively bred for high and low alcohol preference. Behav Genet 32(5):363–388
132. Windisch KA, Kosobud AE, Czachowski CL (2014) Intravenous alcohol self-administration in the P rat. Alcohol 48 (5):419–425. https://doi.org/10.1016/j.alcohol.2013.12.007
133. Bullock FD, Curtis MR (1920) The experimental production of sarcoma of the liver of rats. Proc Meet NY Pathol Soc 20:149–175
134. Curtis MR, Dunning WF, Bullock FD (1933) Is malignancy due to a process analogous to somatic mutation? Science 77 (1989):175–176. https://doi.org/10.1126/science.77.1989.175
135. Dunning WF, Curtis MR (1946) The respective roles of longevity and genetic specificity in the occurrence of spontaneous tumors in the hybrids between two inbred lines of rats. Cancer Res 6:61–81
136. Dunning WF, Curtis MR (1954) Further studies on the relation of dietary tryptophan to the induction of neoplasms in rats. Cancer Res 14(4):299–302
137. Dunning WF, Curtis MR (1952) The incidence of diethylstilbestrol-induced cancer in reciprocal F hybrids obtained from crosses between rats of inbred lines that are susceptible and resistant to the induction of

mammary cancer by this agent. Cancer Res 12 (10):702–706

138. Dunning WF, Curtis MR, Madsen ME (1947) The induction of neoplasms in five strains of rats with acetylaminofluorene. Cancer Res 7(3):134–140

139. Dunning WF, Curtis MR, Maun ME (1949) The effect of dietary fat and carbohydrate on diethylstilbestrol-induced mammary cancer in rats. Cancer Res 9(6):354–361

140. Dunning WF, Curtis MR, Segaloff A (1953) Strain differences in response to estrone and the induction of mammary gland, adrenal, and bladder cancer in rats. Cancer Res 13 (2):147–152

141. Dunning WF, Curtis MR, Stevens M (1968) Comparative carcinogenic activity of dimethyl and trimethyl derivatives of benz(a)anthracene in Fischer line 344 rats. Proc Soc Exp Biol Med 128(3):720–722

142. Dunning WF, Curtis MR, Stevens ML, Dumenigo F (1967) Five transplantable leukemias in the Fischer rat, and their responsiveness to steroids. Cancer Res 27(6 Pt 2):696–727

143. Zeiger E (2017) Reflections on a career and on the history of genetic toxicity testing in the National Toxicology Program. Mutation Res 773:282–292. https://doi.org/10.1016/j.mrrev.2017.03.002

144. Bucher JR (2002) The National Toxicology Program rodent bioassay: designs, interpretations, and scientific contributions. Ann N Y Acad Sci 982:198–207

145. King-Herbert A, Thayer K (2006) NTP workshop: animal models for the NTP rodent cancer bioassay: stocks and strains--should we switch? Toxicol Pathol 34(6):802–805. https://doi.org/10.1080/01926230600935938

146. King-Herbert AP, Sills RC, Bucher JR (2010) Commentary: update on animal models for NTP studies. Toxicol Pathol 38 (1):180–181. https://doi.org/10.1177/0192623309356450

147. Kim U, Clifton KH, Furth J (1960) A highly inbred line of Wistar rats yielding spontaneous mammo-somatotropic pituitary and other tumors. J Natl Cancer Inst 24:1031–1055

148. Shimoyama M, Smith JR, Bryda E, Kuramoto T, Saba L, Dwinell M (2017) Rat genome and model resources. ILAR J 58 (1):42–58. https://doi.org/10.1093/ilar/ilw041

149. Poiley SM (1955) History and information concerning the rat colonies in the animal section of the National Institutes of Health, in "Rat Quality: A Consideration of Heredity, Diet and Disease." Proceedings of the Symposium Held at Columbia University, College of Physicians and Surgeons, New York, January 31, 1952. Q Rev Biol 30:4. https://doi.org/10.1086/401094

150. Wilkinson MD, Dumontier M, Aalbersberg IJ, Appleton G, Axton M, Baak A et al (2016) The FAIR Guiding Principles for scientific data management and stewardship. Sci Data 3:160018. https://doi.org/10.1038/sdata.2016.18

151. FAIR principles for data stewardship (2016) Nat Genetics 48(4):343. https://doi.org/10.1038/ng.3544

152. Shimoyama M, De Pons J, Hayman GT, Laulederkind SJ, Liu W, Nigam R et al (2015) The Rat Genome Database 2015: genomic, phenotypic and environmental variations and disease. Nucleic Acids Res 43(Database issue):D743–D750. https://doi.org/10.1093/nar/gku1026

153. Laulederkind SJ, Liu W, Smith JR, Hayman GT, Wang SJ, Nigam R et al (2013) PhenoMiner: quantitative phenotype curation at the rat genome database. Database 2013:bat015. https://doi.org/10.1093/database/bat015

154. Wang SJ, Laulederkind SJ, Hayman GT, Petri V, Liu W, Smith JR et al (2015) PhenoMiner: a quantitative phenotype database for the laboratory rat, *Rattus norvegicus*. Application in hypertension and renal disease. Database 2015:bau128. https://doi.org/10.1093/database/bau128

155. Shimoyama M, Nigam R, McIntosh LS, Nagarajan R, Rice T, Rao DC, Dwinell MR (2012) Three ontologies to define phenotype measurement data. Front Genet 3:87. https://doi.org/10.3389/fgene.2012.00087

156. Smith JR, Park CA, Nigam R, Laulederkind SJ, Hayman GT, Wang SJ et al (2013) The clinical measurement, measurement method and experimental condition ontologies: expansion, improvements and new applications. J Biomed Semantics 4(1):26. https://doi.org/10.1186/2041-1480-4-26

157. Nigam R, Munzenmaier DH, Worthey EA, Dwinell MR, Shimoyama M, Jacob HJ (2013) Rat Strain Ontology: structured controlled vocabulary designed to facilitate access to strain data at RGD. J Biomed Semantics 4 (1):36. https://doi.org/10.1186/2041-1480-4-36

158. Hayman GT, Laulederkind SJ, Smith JR, Wang SJ, Petri V, Nigam R et al (2016) The disease portals, disease-gene annotation and the RGD disease ontology at the Rat Genome

Database. Database 2016:baw034. https://doi.org/10.1093/database/baw034

159. The UniProt Consortium (2018) UniProt: the universal protein knowledgebase. Nucleic Acids Res 46(5):2699. https://doi.org/10.1093/nar/gky092
160. UniProt Consortium (2015) UniProt: a hub for protein information. Nucleic Acids Res 43 (Database issue):D204–D212. https://doi.org/10.1093/nar/gku989
161. Levan G, Klinga K, Szpirer C, Szpirier J (1990) Gene map of the rat (*Rattus norvegicus*). In: O'Brien Stephen J (ed) Genetic maps: locus maps of complex genomes, vol 4, 5th edn. Cold Spring Harbor Laboratory, Cold Spring Harbor NY, pp 4.80–84.87
162. Levan G, Szpirer J, Szpirer C, Klinga K, Hanson C, Islam MQ (1991) The gene map of the Norway rat (*Rattus norvegicus*) and comparative mapping with mouse and man. Genomics 10(3):699–718
163. Jacob HJ, Lindpaintner K, Lincoln SE, Kusumi K, Bunker RK, Mao YP et al (1991) Genetic mapping of a gene causing hypertension in the stroke-prone spontaneously hypertensive rat. Cell 67(1):213–224
164. Brown DM, Matise TC, Koike G, Simon JS, Winer ES, Zangen S et al (1998) An integrated genetic linkage map of the laboratory rat. Mamm Genome 9(7):521–530
165. Jacob HJ, Brown DM, Bunker RK, Daly MJ, Dzau VJ, Goodman A et al (1995) A genetic linkage map of the laboratory rat, *Rattus norvegicus*. Nat Genetics 9(1):63–69. https://doi.org/10.1038/ng0195-63
166. Kwitek AE, Gullings-Handley J, Yu J, Carlos DC, Orlebeke K, Nie J et al (2004) High-density rat radiation hybrid maps containing over 24,000 SSLPs, genes, and ESTs provide a direct link to the rat genome sequence. Genome Res 14(4):750–757. https://doi.org/10.1101/gr.1968704
167. Steen RG, Kwitek-Black AE, Glenn C, Gullings-Handley J, Van Etten W, Atkinson OS et al (1999) A high-density integrated genetic linkage and radiation hybrid map of the laboratory rat. Genome Res 9(6): AP1–AP8
168. Gibbs RA, Weinstock GM, Metzker ML, Muzny DM, Sodergren EJ, Scherer S et al (2004) Genome sequence of the Brown Norway rat yields insights into mammalian evolution. Nature 428(6982):493–521. https://doi.org/10.1038/nature02426
169. Worley KC, Weinstock GM, Gibbs RA (2008) Rats in the genomic era. Physiol Genomics 32 (3):273–282. https://doi.org/10.1152/physiolgenomics.00208.2007
170. Atanur SS, Diaz AG, Maratou K, Sarkis A, Rotival M, Game L et al (2013) Genome sequencing reveals loci under artificial selection that underlie disease phenotypes in the laboratory rat. Cell 154(3):691–703. https://doi.org/10.1016/j.cell.2013.06.040
171. Hermsen R, de Ligt J, Spee W, Blokzijl F, Schafer S, Adami E et al (2015) Genomic landscape of rat strain and substrain variation. BMC Genomics 16:357. https://doi.org/10.1186/s12864-015-1594-1
172. Baud A, Hermsen R, Guryev V, Stridh P, Graham D, McBride MW et al (2013) Combined sequence-based and genetic mapping analysis of complex traits in outbred rats. Nat Genetics 45(7):767–775. https://doi.org/10.1038/ng.2644
173. Wang Z, Gerstein M, Snyder M (2009) RNA-Seq: a revolutionary tool for transcriptomics. Nat Reviews Genet 10(1):57–63. https://doi.org/10.1038/nrg2484
174. Barrett T, Wilhite SE, Ledoux P, Evangelista C, Kim IF, Tomashevsky M et al (2013) NCBI GEO: archive for functional genomics data sets--update. Nucleic Acids Res 41(Database issue):D991–D995. https://doi.org/10.1093/nar/gks1193
175. Kolesnikov N, Hastings E, Keays M, Melnichuk O, Tang YA, Williams E et al (2015) ArrayExpress update--simplifying data submissions. Nucleic Acids Res 43(Database issue):D1113–D1116. https://doi.org/10.1093/nar/gku1057
176. Dwinell MR, Lazar J, Geurts AM (2011) The emerging role for rat models in gene discovery. Mamm Genome 22(7–8):466–475. https://doi.org/10.1007/s00335-011-9346-2
177. Yoshimi K, Mashimo T (2018) Application of genome editing technologies in rats for human disease models. J Hum Genet 63 (2):115–123. https://doi.org/10.1038/s10038-017-0346-2
178. Jacob HJ, Lazar J, Dwinell MR, Moreno C, Geurts AM (2010) Gene targeting in the rat: advances and opportunities. Trends Genet 26 (12):510–518. https://doi.org/10.1016/j.tig.2010.08.006
179. Smits BM, Mudde JB, van de Belt J, Verheul M, Olivier J, Homberg J et al (2006) Generation of gene knockouts and mutant models in the laboratory rat by ENU-driven target-selected mutagenesis. Pharmacogenet Genomics 16(3):159–169.

https://doi.org/10.1097/01.fpc.0000184960.82903.8f

180. Mashimo T, Yanagihara K, Tokuda S, Voigt B, Takizawa A, Nakajima R et al (2008) An ENU-induced mutant archive for gene targeting in rats. Nat Genetics 40(5):514–515. https://doi.org/10.1038/ng0508-514
181. Kitada K, Ishishita S, Tosaka K, Takahashi R, Ueda M, Keng VW et al (2007) Transposon-tagged mutagenesis in the rat. Nat Methods 4 (2):131–133. https://doi.org/10.1038/nmeth1002
182. Lu B, Geurts AM, Poirier C, Petit DC, Harrison W, Overbeek PA et al (2007) Generation of rat mutants using a coat color-tagged Sleeping Beauty transposon system. Mamm Genome 18(5):338–346. https://doi.org/10.1007/s00335-007-9025-5
183. Flister MJ, Prokop JW, Lazar J, Shimoyama M, Dwinell M, Geurts A (2015) 2015 Guidelines for establishing genetically modified rat models for cardiovascular research. J Cardiovasc Transl Res 8 (4):269–277. https://doi.org/10.1007/s12265-015-9626-4
184. Meek S, Mashimo T, Burdon T (2017) From engineering to editing the rat genome. Mamm Genome 28(7–8):302–314. https://doi.org/10.1007/s00335-017-9705-8
185. Takabayashi S, Aoshima T, Kabashima K, Aoto K, Ohtsuka M, Sato M (2018) i-GONAD (improved genome-editing via oviductal nucleic acids delivery), a convenient in vivo tool to produce genome-edited rats. Sci Rep 8(1):12059. https://doi.org/10.1038/s41598-018-30137-x
186. Kobayashi T, Namba M, Koyano T, Fukushima M, Sato M, Ohtsuka M et al (2018) Successful production of genome-edited rats by the rGONAD method. BMC Biotechnol 18(1):19. https://doi.org/10.1186/s12896-018-0430-5
187. Chen Y, Spitzer S, Agathou S, Karadottir RT, Smith A (2017) Gene editing in rat embryonic stem cells to produce in vitro models and in vivo reporters. Stem Cell Reports 9 (4):1262–1274. https://doi.org/10.1016/j.stemcr.2017.09.005
188. Buehr M, Meek S, Blair K, Yang J, Ure J, Silva J et al (2008) Capture of authentic embryonic stem cells from rat blastocysts. Cell 135 (7):1287–1298. https://doi.org/10.1016/j.cell.2008.12.007
189. Li P, Tong C, Mehrian-Shai R, Jia L, Wu N, Yan Y et al (2008) Germline competent embryonic stem cells derived from rat blastocysts. Cell 135(7):1299–1310. https://doi.org/10.1016/j.cell.2008.12.006
190. Men H, Bryda EC (2013) Derivation of a germline competent transgenic Fischer 344 embryonic stem cell line. PLoS One 8 (2):e56518. https://doi.org/10.1371/journal.pone.0056518
191. Yamamoto S, Nakata M, Sasada R, Ooshima Y, Yano T, Shinozawa T et al (2012) Derivation of rat embryonic stem cells and generation of protease-activated receptor-2 knockout rats. Transgenic Res 21 (4):743–755. https://doi.org/10.1007/s11248-011-9564-0
192. Wallace HA, Marques-Kranc F, Richardson M, Luna-Crespo F, Sharpe JA, Hughes J et al (2007) Manipulating the mouse genome to engineer precise functional syntenic replacements with human sequence. Cell 128(1):197–209. https://doi.org/10.1016/j.cell.2006.11.044
193. Lee EC, Liang Q, Ali H, Bayliss L, Beasley A, Bloomfield-Gerdes T et al (2014) Complete humanization of the mouse immunoglobulin loci enables efficient therapeutic antibody discovery. Nat Biotechnol 32(4):356–363. https://doi.org/10.1038/nbt.2825
194. Macdonald LE, Karow M, Stevens S, Auerbach W, Poueymirou WT, Yasenchak J et al (2014) Precise and in situ genetic humanization of 6 Mb of mouse immunoglobulin genes. Proc Natl Acad Sci U S A 111(14):5147–5152. https://doi.org/10.1073/pnas.1323896111
195. Cherry JM, Hong EL, Amundsen C, Balakrishnan R, Binkley G, Chan ET et al (2012) Saccharomyces Genome Database: the genomics resource of budding yeast. Nucleic Acids Res 40(Database issue):D700–D705. https://doi.org/10.1093/nar/gkr1029
196. Lee RYN, Howe KL, Harris TW, Arnaboldi V, Cain S, Chan J et al (2018) WormBase 2017: molting into a new stage. Nucleic Acids Res 46(D1):D869–D874. https://doi.org/10.1093/nar/gkx998
197. Thurmond J, Goodman JL, Strelets VB, Attrill H, Gramates LS, Marygold SJ et al (2018) FlyBase 2.0: the next generation. Nucleic Acids Res. https://doi.org/10.1093/nar/gky1003
198. Bult CJ, Blake JA, Smith CL, Kadin JA, Richardson JE (2018) Mouse Genome Database (MGD) 2019. Nucleic Acids Res 47 (D1):D801–D806. https://doi.org/10.1093/nar/gky1056

199. Smith CM, Hayamizu TF, Finger JH, Bello SM, McCright IJ, Xu J et al (2018) The mouse Gene Expression Database (GXD): 2019 update. Nucleic Acids Res 47(Database-Issue):D774–D779. https://doi.org/10.1093/nar/gky922

200. Ruzicka L, Bradford YM, Frazer K, Howe DG, Paddock H, Ramachandran S et al (2015) ZFIN, The zebrafish model organism database: updates and new directions. Genesis 53(8):498–509. https://doi.org/10.1002/dvg.22868

201. Ashbrook DG, Mulligan MK, Williams RW (2018) Post-genomic behavioral genetics: from revolution to routine. Genes Brain Behav 17(3):e12441. https://doi.org/10.1111/gbb.12441

202. Bennett BJ, Farber CR, Orozco L, Kang HM, Ghazalpour A, Siemers N, Neubauer M, Neuhaus I, Yordanova R, Guan B, Truong A, Yang WP, He A, Kayne P, Gargalovic P, Kirchgessner T, Pan C, Castellani LW, Kostem E, Furlotte N, Drake TA, Eskin E, Lusis AJ (2010) A high-resolution association mapping panel for the dissection of complex traits in mice. Genome Res 20(2):281–290. https://doi.org/10.1101/gr.099234.109

203. Saba L, Hoffman P, Tabakoff B (2017) Using baseline transcriptional connectomes in rat to identify genetic pathways associated with predisposition to complex traits. Methods Mol Biol 1488:299–317. https://doi.org/10.1007/978-1-4939-6427-7_14

204. Perry ME, Valdes KM, Wilder E, Austin CP, Brooks PJ (2018) Genome editing to 're-write' wrongs. Nat Rev Drug Discov 17(10):689–690. https://doi.org/10.1038/nrd.2018.91

Chapter 2

Rat Genome Assemblies, Annotation, and Variant Repository

Monika Tutaj, Jennifer R. Smith, and Elizabeth R. Bolton

Abstract

The first and only published version of the rat reference genome sequence was RGSC3.1, accomplished by the Rat Genome Sequencing Project Consortium. Here we present the history of the community effort in the correction of sequence errors and filling missing gaps in the process of refining and providing researchers with a high-quality rat reference sequence. The genome assembly improvements, addition of different evidence resources over time, such as RNA-Seq data, and software development methodologies had a positive impact on the gene model annotations. Over the years we observed a great increase in the numbers of genes, protein coding sequences, predicted transcripts and transcript features. Before the sequencing of the rat genome was possible, first biochemical and next genomic markers like RAPD, AFLP, RFLP, and SSLP were fundamental in research studies involving cross-breeding between different rat strains, in finding the level of polymorphism, linkage mapping, and phylogeny. Linkage maps provide information on recombination rates, give insight into intra- and interspecies gene rearrangements, and help to identify Mendelian loci and Quantitative Trait Loci (QTL). In the 1990s many reports were published on the construction of rat linkage maps that incorporated increasing numbers of markers and facilitated the localization of disease loci. Current genetic monitoring and linkage mapping relies on single nucleotide polymorphisms (SNPs). The Rat Genome Database collects information on genetic variation from the worldwide community of rat researchers and provides tools for searching and retrieving these data. As of today we show details about almost 605 million variants coming from many studies in our Variant Visualizer tool.

Key words Reference genome, Gene model annotations, Genomic markers, Variants, SNPs

1 Rat Genome Assemblies

1.1 History of the Rat Genome Sequencing Project

The laboratory rat (*Rattus norvegicus*) became the third mammalian genome to be sequenced when the Rat Genome Sequencing Project Consortium (RGSPC) published a high-quality draft sequence of the rat genome. The project was a collaborative effort involving sequencing and analyses by researchers at 40 organizations from seven countries, coordinated by the Baylor College of Medicine Human Genome Sequencing Center (BCM-HGSC). Funding was primarily supplied by the National Human Genome

G. Thomas Hayman et al. (eds.), *Rat Genomics*, Methods in Molecular Biology, vol. 2018,
https://doi.org/10.1007/978-1-4939-9581-3_2,

Research Institute (NHGRI) and the National Heart, Lung, and Blood Institute (NHLBI) supplemented by additional funding for individual researchers from a variety of sources [1].

The animals selected for sequencing were from a substrain of the Brown Norway (BN) rat strain. BN/SsNHsd rats had been obtained by the Medical College of Wisconsin from Harlan Sprague Dawley but were found to be incompletely inbred, that is, containing regions of heterozygosity when subjected to genetic testing. Additional inbreeding was carried out at MCW to create a fully inbred substrain. Two female rats taken from the MCW colony at the 13th generation were used for most of the sequencing. The BN strain was selected by the rat research community based on the fact that it was considered genetically diverse and had been used as a control strain in a wide variety of studies. In addition, BN rats were being used to develop congenic and consomic strains, as founders for the LXB recombinant inbred panel of strains, and as one of eight founder strains for the Heterogeneous Stock rats (*see* Chapter 11).

The RGSPC used a combination of whole-genome shotgun sequencing (WGS) and a bacterial artificial chromosome (BAC) sequencing approach to generate the first draft of the rat genome. BAC clones combined with WGS data formed intermediate products called "enriched BACs" (eBACs). The researchers developed the Atlas Genome Assembly system [2] to generate almost 19,000 eBACs. At the same time, a large number of clones from the Children's Hospital Oakland Research Institute BAC library (CHORI-230) were used to create a fingerprint contig (FPC) map. Clones were "fingerprinted" by restriction enzyme digestion and assembled by overlapping these segments into a genome-wide contig map. This map was then used to assemble the sequenced eBACs into larger super bactigs and even larger ultrabactigs. In addition to the FPC map, a yeast artificial chromosome (YAC)-based physical map was constructed to improve clone positioning [2, 3]. BAC-end and in silico coordinates were used together to localize the contigs to their correct chromosomal region in the assembly (NCBI RefSeq accession number prefix NC_). The contigs without assigned genomic positions were annotated as unlocalized and unplaced sequences (NCBI RefSeq accession number prefix NW_). Although some sequence for the Y chromosome was generated in the first round of sequencing, the Y chromosome was not included in the initial versions of the assembly. Attempts to purify the Y chromosome for sequencing revealed that this chromosome in the BN strain was almost twice the size of the Y in other strains [4]. Because of the large size, it was found that insufficient material had been isolated, resulting in unacceptably low coverage of the sequence (about 2×). The sequence was therefore excluded from the assembly.

The first public release of the rat genome sequence was designated RGSC/Rnor2.0. Its release in November of 2002 was followed shortly thereafter by the release of Rnor version 2.1 in January of 2003. Release notes for the v2.1 assembly noted that a "reduction of assembly artifacts has slightly reduced the number of bases assembled while increasing the size of contigs, scaffolds, and ultrabactigs" (reports available at the BCM-HGSC ftp site). The total size of the assembly mapped onto chromosomes was 2.72 Gbp for the 2.0 release and 2.66 Gbp for the 2.1 release, while the average size of ultrabactigs and scaffolds increased by 5.9% and 17.2%, respectively.

Rnor3.0 represented a complete reassembly of the genome sequence. Improvements incorporated into this version of the sequence included the addition of new sequences (over 1100 new BACs to cover gaps), better software accuracy and relevance, utilization of an improved marker set from the Medical College of Wisconsin, and a new FPC map from the BC Cancer Agency Genome Sciences Centre [5]. The new FPC map was based on automated assembly of BAC clones based on the "fingerprinting," followed by a process of manual curation and sequencing of clones, and the use of human and mouse orthology information to resolve conflicts and to correct placement of sequence units. This process of automated and manual editing expanded the contiguity of the rat fingerprint map and in turn allowed for targeted BAC clone selection and filling of contig gaps, as well as linking some of the unlocalized segments of the rat assembly to chromosomes [3]. Rnor3.0 was rapidly followed by the release of version 3.1 in June of 2003. Rnor3.1 was considered a minor update to the previous assembly with changes only affecting chromosomes 7 and X. Although not the first public version of the sequence, Rnor3.1 was the first (and only) published version of the rat reference genome sequence [1]. Analysis of the assembly revealed that it had an average overall of approximately sevenfold sequence coverage, with 60% provided by WGS and 40% by BACs. The assembly covered about 90% of the estimated 2.75 Gbp rat genome and contained a similar number of genes as described for human and mouse (20,000–25,000) [1, 6].

In 2004, a series of minor updates to the assembly brought the designation to Rnor3.4. Updates included three additions of finished BAC sequences and the correction of several alignment switch points. Despite additional work done to improve the assembly during the interim, the Rnor3.4 assembly remained the de facto reference assembly for the rat for almost eight years. After 2004 a number of improvements were proposed and funded by the NHGRI to provide a more complete genome with improved accuracy. New assemblies were released in March of 2008 (Rnor4.0) and November of 2009 (Rnor4.1), both of which utilized reads downloaded from NCBI in January of 2007. The assemblies were constructed using updated Atlas genome assembly software which had been extended to include better methods to deal with repeats and improve the alignment of BAC sequences, and included new modules for handling heterozygosity. However, although some work was done at NCBI to incorporate and begin to annotate the Rnor4.1, neither assembly was accepted by the rat research community as an improvement over version 3.4 and therefore were never completely incorporated into the major sequence databases.

The BN/NHsdMcwi strain was initially sequenced by the traditional capillary sequencing method. With advancing sequencing methodology, SOLiD (Sequencing by Oligonucleotide Ligation and Detection) sequence reads were generated and combined with the previous WGS-plus-BAC assembly data for a more complete representation of the genome. Highly repetitive reads were omitted and the SOLiD data were used for revision of the scaffolding. This resulted in the Rnor5.0 version of the rat reference assembly in 2012 that showed improvements at both the nucleotide and structural level.

The latest version of the assembly—Rnor6.0—was manually checked and corrected for potential mis-assemblies by comparison to the human and mouse genomes as well as to the previous version 3.4 assembly. Scaffold gaps were filled using long-read PacBio sequences with the PBjelly software and high-quality BAC clone sequences from phase 2 and phase 3 sequencing were spliced into the assembly to replace matched locations. In addition, a draft sequence of the Y chromosome from a male rat of the SHR/NHsdAkr strain which had been produced as part of a collaborative project to sequence Y chromosomes from several mammalian species, including rat [7, 8], was added. The Rnor6.0 genome reference is longer and more contiguous than previous versions, and many gaps have been closed (Total assembly gap length, *see* Table 1). The ungapped assembly lengths improved with each release version: v6.0 ungapped length is 2.730 Gb, versus v5.0 at 2.573 Gb and v3.4 at 2.568 Gb. The Rnor 6.0 assembly comprises 75,697 contigs in comparison with 132,131 contigs for v5.0 and 238,325 contigs for v3.4. Older versions of the rat genome assemblies can be found in the archive sites (*see* Table 2).

1.2 Other Rat Genome Assemblies

Since 2004, the genomes of a number of other rat strains have been sequenced by the rat community. In 2008, the STAR consortium used a combination of shotgun sequencing, low coverage WGS, and BAC end sequencing to discover almost three million single nucleotide variants from the SS/Jr., WKY/Bbb, GK/Ox, SHRSP/Bbb, Sprague Dawley, and F344/Stm strains [9]. A subset of approximately 20,000 of these were used to genotype 167 inbred strains and 2 recombinant inbred (RI) panels (the 31 HXB-BXH strains and the 33 FXLE-LEXF strains). Almost 10,000 of the SNVs were then used to genotype an additional 89 F2 animals

Table 1
Statistics of rat genome assemblies

NCBI assembly information	RGSC_v3.4	Rnor4.1-scaffold	Rnor_4.1	Rnor_5.0	Rnor_6.0
RefSeq Assembly Accession	GCF_000001895.3	—	—	GCF_000001895.4	GCF_000001895.5
GenBank Assembly Accession	—	GCA_000001895.1	GCA_000001895.2	GCA_000001895.3	GCA_000001895.4
Release Date	12/13/2004	2/5/2010	3/12/2012	3/16/2012	7/1/2014
Total sequence length	2,826,224,306	2,472,228,416	2,796,368,354	2,909,698,938	2,870,184,193
Ungapped length	2,567,937,207	2,471,873,957	2,480,163,185	2,573,083,111	2,729,985,404
Total assembly gap length	258,270,786	354,459	316,205,169	336,599,514	140,198,789
Spanned gaps	249,838	10,043	207,756	121,283	74,303
Gaps between scaffolds	270	0	46	8109	440
Number of scaffolds	741	187,024	3238	10,848	1395
Scaffold N50	18,621,810	45,127	126,646,908	2,178,346	14,986,627
Scaffold L50	46	14,621	10	387	65
Number of contigs	238,325	197,067	210,974	132,131	75,697
Contig N50	36,847	40,558	35,614	52,491	100,461
Contig L50	18,984	16,657	18,076	13,663	7356
Unlocalized sequences count	246	—	3106	1278	354
Unplaced sequences	203	—	65	1439	578
Total number of chromosomes and plasmids	21	0	21	22	23

Table 2
Archived references

Baylor release	UCSC version	Release date
v3.1	rn3	Jun. 2003
v2.1	rn2	Jan. 2003
v1.0	rn1	Nov. 2002

from a cross between the BN/Par and GK/Ox strains. In 2010 the SHR/OlaIpcv rat genome was sequenced at 10.7-fold coverage by paired-end sequencing on the Illumina platform. Initially 681.8 million reads were mapped to the BN reference genome (v3.4) and covered 97.7% of the reference assembly by at least three reads [10]. Subsequently, the NGS data set of the SHR/OlaIpcv strain was expanded, thereby increasing the median coverage of this genome to 23-fold in 2012. The researchers, in the same study, also generated whole genome NGS data from the same genetic material that was used to create the BN reference sequence (referred to in that study as "Eve"), as well as from the BN-*Lx*, a mutant strain closely related to BN. The data from the new BN sequence (32-fold NGS coverage) were used to search for Rnor3.4 assembly errors [11].

In 2013, two more genomes were sequenced, becoming the first non-reference de novo assemblies of rat genomes [12]. The DA and F344 rat strains were sequenced with an average depth of 32× using Illumina technology. Researchers employed a reference-aided assembly method (RAM), using the BN reference genome as well as the Short Oligonucleotide Alignment Program (SOAP), and GapCloser, an algorithm for contig-end extension and gap filling. First, a semi-finished genome was constructed by aligning sequencing reads to Rnor3.4 using SOAPaligner [13], then an independent de novo assembly of contigs and scaffolds was generated using SOAPdenovo [14]. Finally, the genome draft was assembled by anchoring scaffolds onto the semi-finished genome. The read alignment of each strain with the BN genome covered 98% of the reference (three reads or more). The DA and F344 genome drafts were 1.94% and 1.91% larger than the BN genome, respectively. The DA and F344 genome drafts contained more than 49 million novel base pairs for each genome that bridged around 400,000 gaps of the BN genome [12].

In 2013 the eight inbred strains (ACI/N, BN/SsN, BUF/N, F344/N, M520/N, MR/N, WKY/N, and WN/N) which had been used as the founder/progenitor strains for the NIH's heterogeneous stock (N:HS) rats were sequenced using SOLiD technology at an average of 22-fold sequence coverage, that represents

~88% of the reference genome [15]. The same year another large-scale sequencing project was completed. The sequencing of 27 rat strains that served as popular disease models of hypertension, diabetes, and insulin resistance resulted in the identification of a number of genomic variants and coevolved gene clusters [16]. The researchers produced the sequence data with 20-fold coverage on average for all strains except for BBDP/Wor and WKY/NHsd rat strains that reached approximately a tenfold coverage level. Variant data from each of these large-scale strain-specific sequencing projects are available at the Rat Genome Database (*see* below for details).

1.3 Reported Reference Genome Errors

The regions that posed special problems to complete genome assembly were regions with unusual repeat structures, polymorphisms, possible BAC rearrangements, and low sequence coverage. The Rnor3.4 assembly contained many gaps, inconsistencies, and sequence errors due to the relatively low coverage and errors associated with capillary technology. Genetic single nucleotide polymorphism (SNP) mapping by the STAR consortium in 2008 identified discrepancies between the genetic map and the draft genome: a p11-centromeric segment of chromosome 1 was wrongly inserted into the p14-telomeric region of chromosome 17, intra- and inter-chromosomal relocations were observed in regions of chromosomes 2, 4, 11, 12, 14, and 17 [16]. The relocation in the p14 region of chromosome 17 and one conflict on chromosome 9 were discovered during the revision of differences between BCM and Celera rat genome assemblies [9]. In the study of the rat genomic variation in complex traits, four pairs of regions on chromosomes 1, 4 (2 regions), 9, 12, 14, 17, and 19 showed high inter-chromosomal linkage disequilibrium, due to mis-assembly of the Rnor3.4 reference sequence [15]. Analyses at the Rat Genome Database of changes in NCBI (National Center for Biotechnology Information) gene position annotations between rat genome assemblies showed a number of co-localized genes that in upgraded reference versions were re-annotated, frequently together, to different chromosomes (*see* Table 3). We observed 25 out of 49 genes that occupied an 8.3 Mbp region of chromosome 1 in the reference version 3.4 were relocated to two different genomic regions in v5.0: 17 genes moved to chromosome 8, while 6 genes moved to chromosome 9. All together, we found 7 clusters of 19 to 96 co-localized genes, that span 1.2 to 8.3 Mb regions in chromosomes 1, 4, 7, 8, 17 and chromosome X of the assembly v3.4 that changed genomic position in the v5.0 (*see* Fig. 1). The changes were less profound between assemblies v5.0 and v6.0. Four clusters of 10 to 93 genes spanning 1.5 to 6 Mbp regions of chromosomes 1, 3, and X changed genomic position in the v6.0 assembly with reference to v5.0. However, there were also numerous smaller changes in other chromosomes in both transitions: 435 genes in

Table 3
Annotation changes between different rat genome assemblies

Compared Asemblies	Chromosome OLD position	Number of genes[a]/ total	Mbp	Genomic position (bp) on older assembly	Number of genes[a]/ total	Chromosome new position
Rnor3.4	1	25/49	8.3	58877734-67150234	17/49	8
and					6/49	9
Rnor5.0	4	19/28	4.6	99068636–103646030	17/28	3
	7	49/69	1.7	137288635–1389558414	49/69	X
					9/69	6
	8	40/41	1.2	40493892–41711691	18/41	15
					16/41	4
	17	27/33	5.3	44656–5318612	28/33	1
	X	96/132	7	153730373–160683450	83/132	1
		23/132	6.5	122692432–129236338	17/132	3
Rnor5.0	1	93/113	6	147946356–153934661	95/113	X
and		10/113	6.5	64184207–70664968	6/113	11
Rnor6.0	3	20/23	1.5	52231912–53715589	20/23	X
	X	60/63	1.6	114700497–116300972	60/63	7

[a]Including protein-coding genes, noncoding genes, pseudogenes, and genes under revision

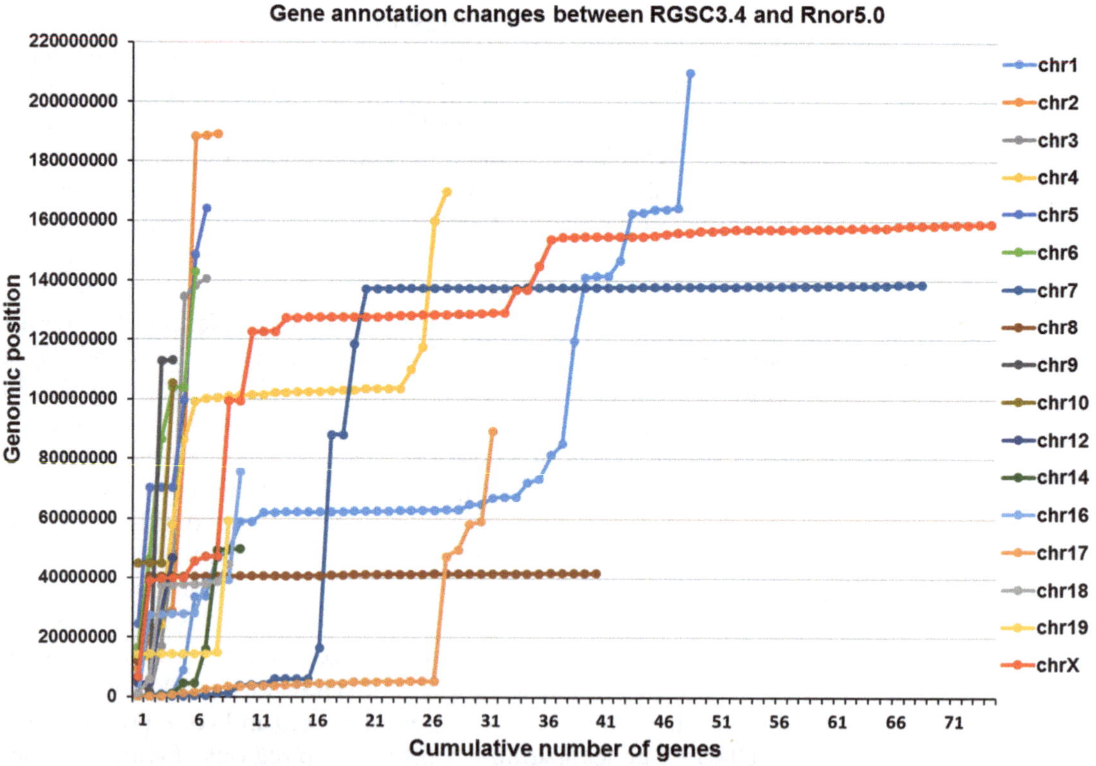

Fig. 1 Number of genes re-annotated from chromosomes of the assembly Rnor3.4 to different genomic position in Rnor5.0; Number of genes limited to 100 for better visualization

total changed chromosome position between assemblies v3.4 and v5.0, 249 genes between assemblies v5.0 and v6.0.

In particular, repetitive regions of the genome are often misassembled. Researchers reported that a duplication of the *Fcgr3* locus associated with autoimmune nephritis in a rat model and in human was not represented in the Rnor3.4 genome assembly [5, 17]. BAC and genomic Southern blots and clonotype analysis suggest that *Fcgr3* underwent at least two duplications during the time of divergence between the mouse and rat lineages, and the rat has at least three expressed genes. Because of the similarity in the sequence between the duplicated genes and the presence of a SINE repeat element, this region presents a particular challenge for assembly. Other regions reported as improperly assembled included the *Ttn* gene with highly complex alternative splicing, a region of duplication around the *Cd36* gene, and a 5 Mbp region with repeat sequences on chromosome 1 which contains a number of genes studied in stroke, hypertension, and metabolic syndrome (*P2ry2, P2ry6, Pde2a,* and *Slco2b1*) [5].

2 Rat Genome Assembly Annotations

2.1 Gene Model Prediction

In 2004, the Ensembl gene prediction pipeline predicted 20,973 genes with 28,516 transcripts and 205,623 exons for the Rnor3.1 assembly [1]. The improvement provided by reassembly of the reference sequence in general, and by the Rnor 6.0 assembly in particular, had a positive impact on the assembly annotation. Gene model predictions consider known protein and transcript data for rat, as well as homology to other sequences, including rodent proteins, non-rodent vertebrate proteins, rat cDNA data from RefSeq and EMBL, and mouse cDNAs from Riken, RefSeq, and EMBL. The statistics depend on the quality of the genome sequence, the gene prediction method, the alignment criteria, and the amount of expressed sequence evidence. Table 4 lists the current number of gene model predictions provided by NCBI for the v3.4, v5.0, and v6.0 rat genome assemblies. There is an increase in the numbers of genes, protein coding sequences (CDS), and defined noncoding 5′ and 3′ untranslated regions (UTRs). Genome annotations and prediction accuracy benefit from the addition of different evidence resources, such as the use of RNA-Seq data, and new methodologies. This is clearly demonstrated by the substantial increase in the number of predicted transcripts and transcript features for the v5.0 and v6.0 assemblies where the incorporation of RNA-seq transcriptomic data improved the identification of isoforms, UTRs, exon boundaries and transcripts with only low expression. Worth noting is that the number of noncoding genes doubles from v3.4 to v5.0 and more than triples from v3.4 to v6.0.

Table 4
Number of genomic features for the rat assemblies

NCBI annotations RGD QC	Rnor3.4	Rnor5.0	Rnor6.0
Genes	**24,949**	**33,330**	**41,517**
Protein-coding genes	19,623	22,480	23,485
Non-protein-coding genes	**5235**	**10,694**	**17,932**
Transcripts	**16,973**	**38,453**	**71,613**
Exons	**147,831**	**373,816**	**726,638**
5'UTR	19,006	46,936	88,311
3'UTR	14,814	32,999	56,722
CDS	139,669	353,905	685,260
QTLs	**2272**	**2294**	**2266**
Genetic markers	**43,306**	**45,663**	**44,828**

2.2 NCBI and Ensembl Gene Annotation Models

Genome annotations, i.e., the prediction and localization of genes and other genomic elements on a genome sequence, differ between NCBI and Ensembl because of variations in annotation strategies, algorithms, and input data. The NCBI Eukaryotic Genome Annotation Pipeline [18] utilizes a suite of informatic tools that include the alignment programs Splign and ProSplign, and the gene prediction program Gnomon to generate sets of genes with their associated transcripts and proteins. The annotation process relies heavily on the availability of transcript or protein evidence for the species. Originally implemented in 2000 as a semi-manual process to align Genbank and RefSeq transcripts to the genome using the BLAST algorithm then supplementing these with ab initio gene predictions using GenomeScan [19], NCBI's pipeline has gone through a number of substantial improvements. These include the addition of EST and protein data as input and the development of the splicing-aware aligners Splign for transcripts and ProSplign for proteins. Addition of RNA-Seq data improved the quality of the annotations, particularly for organisms that have little or no experimental mRNA or EST data available. Reengineering the pipeline using a new framework for parallel execution enhanced its extensibility, robustness and reproducibility, as well as improving tracking, all of which were necessary to keep pace with both annotation of new genomic sequences and reannotation of improved genome assemblies.

The current Eukaryotic Genome Annotation Pipeline takes as input same-species transcripts, proteins and RNA-seq reads, and where necessary, transcripts and proteins from closely related species. Input transcripts include known coding and noncoding

RefSeq transcripts (i.e., those with NM_ or NR_ prefixes), long transcripts and ESTs from NCBI's Nucleotide database. Both short and long RNA-Seq sequences are utilized from the Sequence Read Archive (SRA). In addition, the following proteins are aligned: known RefSeq (NP_) proteins and proteins derived from transcripts by the International Nucleotide Sequence Database Collaboration (INSDC—includes DDBJ, ENA and GenBank collaborators). Curated RefSeq genomic sequences (i.e., those with NG_ prefixes, representing non-transcribed pseudogenes and manually annotated gene clusters) can be used, if available. Aligned sequences are then submitted to the Gnomon tool. Gnomon uses a two-step gene prediction program that assembles overlapping alignments into "chains" and extends these chains into complete models in an ab initio prediction step, using a Hidden Markov Model (HMM). Predicted gene models are aligned against proteins from the curated Swiss-Prot database (a subset of the UniProtKB knowledgebase) to confirm and refine the predictions, and the best model for each gene is selected.

The Ensembl gene annotation process starts with a model-building phase that involves the alignment of protein, cDNA, EST, and RNA-Seq sequences to the genome assembly [20]. The methods used in this phase depend on the input data available at the time of annotation, with the same-species data preferred over data from other species, and with annotated sequences preferred over computed sequences. The Targeted Pipeline uses same-species protein sequences to identify the genomic location of protein-coding genes, and then to produce coding models using GeneWise [21]. Only high-confidence same-species protein sequences are downloaded from Swiss-Prot and TrEMBL sequences (labeled as PE level 1 or 2) and RefSeq (annotated sequences with NP_ and AP_ accessions). The pipeline finds the genomic location of transcripts by aligning protein sequences to the genome, and both DNA and protein sequences for this region are passed to GeneWise. The software uses a splice-aware algorithm and generates a protein-coding transcript model as an output. In addition, cDNA data together with its annotated coding sequence (CDS) from INSDC are used to complement protein-coding gene models with the addition of untranslated regions (UTRs).

The Similarity Pipeline uses as an input UniProtKB proteins from a wide range of species. The output of the Similarity Pipeline is a set of models that supplements the Targeted Pipeline models. This approach is especially useful for species that do not have many same-species proteins. Increases in the amount of available RNA-Seq sequence data prompted the addition of an RNA-Seq pipeline, which uses these data to produce both protein-coding and noncoding transcript models in the gene annotation process. The model-building steps of the pipeline are followed by a filtering step, that selects models with the highest confidence at each location (generates a hierarchy of models). Protein-coding models that overlap with RNA-Seq, cDNA, and/or EST models are ranked as top priority and the output are protein-coding transcript models that are further extended to include UTR regions. Selected models are then passed to the GeneBuilder module which removes redundant transcript models in the process of clustering protein-coding models into multi-transcript gene structures. In addition to the modules which produce protein-coding gene predictions, the Ensembl GeneBuild generates annotations of pseudogenes, short noncoding RNAs such as microRNAs and transfer RNAs, and long intergenic noncoding RNAs (lincRNAs). Output from all of these pipelines is integrated to produce the final Ensembl gene set [20].

2.3 Gene Annotation Differences

The genome size for human is 3.257Gb (GRCh38), slightly larger than for rat—2.870Gb (Rnor6.0) or mouse—2.819Gb (GRCm38), but there are substantial differences in the number of annotated proteins and transcripts between the three of them. Currently the amount of expressed sequence evidence is much more abundant for human and mouse than for rat. Table 5 presents data collected from annotation pages for individual species that are available in the NCBI and Ensembl databases [22, 23]. There are more than 2 times the number of transcripts and 3–4 times more EST data used for the human and mouse gene prediction models than for the rat in both Ensembl and NCBI. There are 4 times more protein sequences used for the human model, and 2 times more protein data for the mouse in the NCBI prediction. We compared the number of genes between NCBI and Ensembl for human, mouse, and rat (*see* Fig. 2). There are 19,633 rat genes shared by the two gene models, 25,372 human genes and 24,637 mouse genes. Even though the amount of evidence for both human and mouse is much higher than for rat, the proportion of overlapping predicted genes is low. It suggests that the results of the prediction do not depend on the number of provided evidences but depend on a design of the prediction strategy. We counted how many of NCBI's and Ensembl's predicted genes have been assigned the same genomic position in the genome reference (gene boundaries—from start to stop position). We found that an exact match of position applies to only a limited number of genes: 15% of human genes (9595), 12% of rat (3902), and 10% of mouse (5637) genes. Examples of differences in rat gene model prediction between Ensembl and NCBI are shown in Fig. 3. Some genes are predicted to be in the same genomic location but differ in length, number of exons or the exons' positions. In some cases, exons have the same predicted positions but different genes are assigned to that position by two models. There are examples of single genes in one database that are split into two genes in another one. In some locations one model predicts the presence of genes whereas the other one does not. Number and length of predicted transcripts also differ.

Table 5
Resources for generating gene models

Alignment type	Organism / Database	Human: NCBI—GRCh38.p7	Human: ENSEMBL—GRCh38	Mouse: NCBI—GRCm38.p4	Mouse: ENSEMBL—GRCm38	Rat: NCBI—Rnor6.0	Rat: ENSEMBL—Rnor6.0
Transcript	RefSeq(NM_/NR_)	55,892	159,081	34,827	213,883	17,901	[a]65,192
	GenBank	178,172		223,325		81,472	
	EST	4,398,669	3,976,554	3,110,250	3,094,335	818,919	994,706
Protein	GenBank/ENA/DDBJ/RefSeq	119,177	108,207	60,386	56,044	17,558	[a]65,192
	RefSeq (NP_)/UniProt	42,610	74,356	29,728	30,746	19,226	30,721
	Other resources	—	—	Human, rat RefSeq	Mammalia, vertebrate UniProt	Mouse, human RefSeq	Mouse UniProt/Refseq, HI-KNAW RNASeq

[a]RefSeq cDNAs with accession prefix "NM_" matching RefSeq proteins with "NP_" prefix

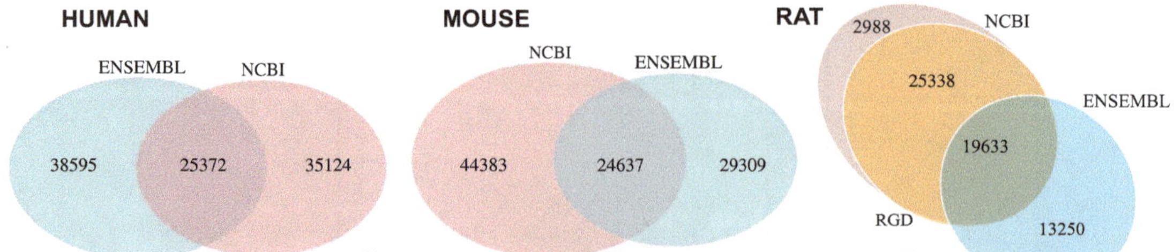

		Human	Mouse	Rat
Total Gene Number (IDs)	**NCBI**	60496	69020	47959
	ENSEMBL	63967	53946	32883
Gene overlap (by position)	**Exact match**	9594	5637	3902
	No match	15457	8134	5027

Fig. 2 The comparison of gene annotations for human, rat, and mouse. Number of genes with unique ID shared between NCBI and Ensembl

Kumar et al. proposed to improve the annotation of the rat genome by utilizing transcriptomics and proteomics data together [25]. The researchers built the reference-based transcriptome assembly from RNA-Seq reads and analyzed publicly available RNA-Seq and mass spectrometry (MS)-based proteomics data. They discovered hundreds of novel peptides in rat brain microglia that were expressed by 249 genes. The evidence helped to identify unknown exons, pseudogenes, and splice variants for various loci, many of which have important disease associations. Wu et al. showed that the genome annotation choice has a significant influence on human RNA-Seq expression analysis outcome [26]. These examples highlight a need to improve existing methods of genome annotation and utilize all available resources in the annotation process. The Rat Genome Database (RGD) is in the process of incorporation of Ensembl annotations in addition to NCBI annotations that currently are prioritized in our database (Fig. 2).

3 Rat Variants and Genomic Markers: RGD Repository

3.1 Genomic Markers

There are over 700 inbred strains of rats, and the history of their generation and evolution is not always well known. Markers are important in research studies involving cross-breeding between different rat strains, and essential in finding the level of polymorphism and genetic homogeneity between them (inter-strain and intra-strain differences).

Years before the sequencing of the rat genome was possible, there was an intensive search for novel markers that could be integrated into rat genetic and radiation hybrid maps [27–29]. The inbred mouse and rat strains were known by coat colors and MHC until the 1970s, when biochemical markers

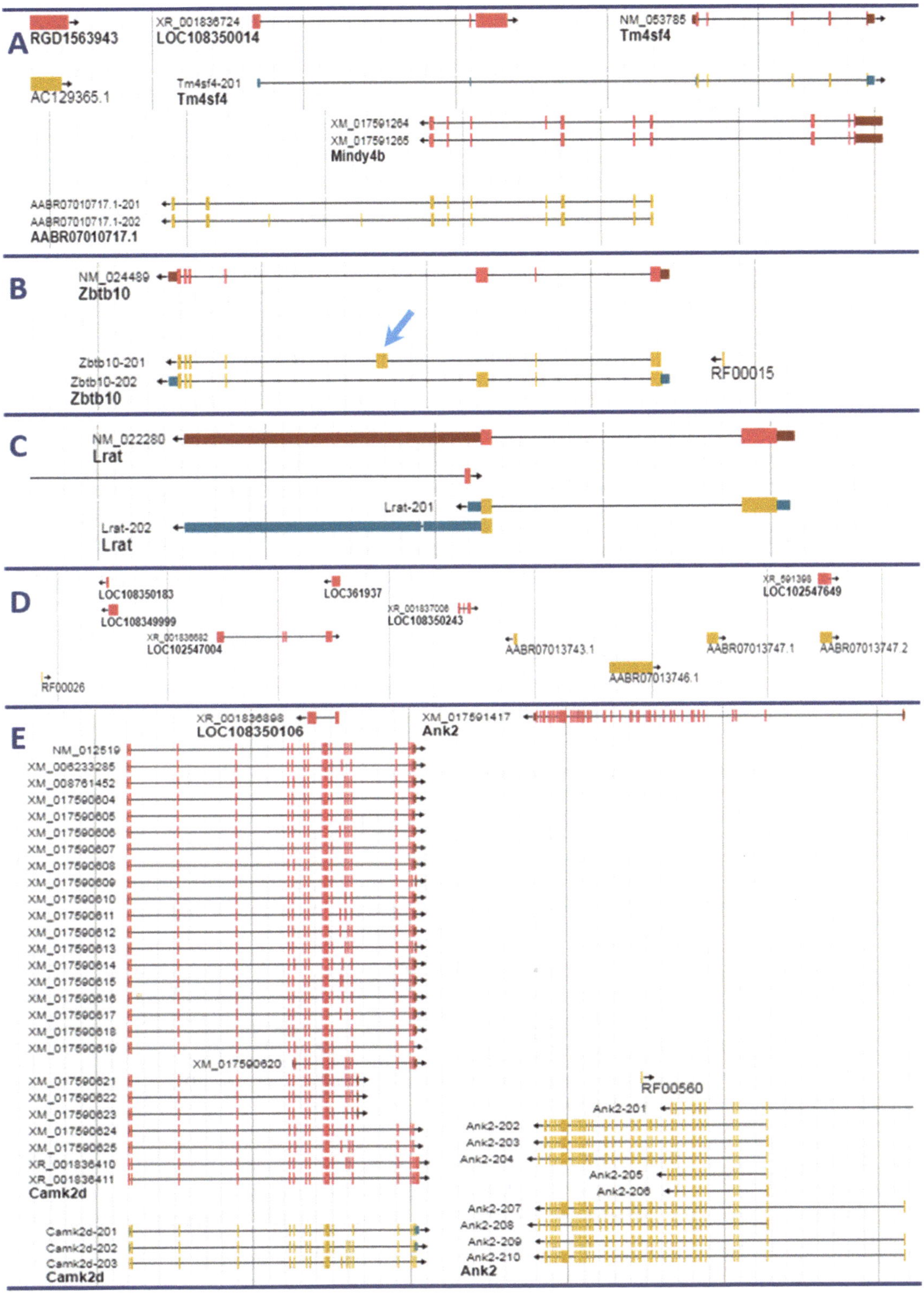

Fig. 3 Examples of differences in gene model prediction between ENSEMBL (yellow-blue) and NCBI (red-brown) using the JBrowse genome viewer [24]: (**a**) genes in the same genomic location differ in length and number of exons or other genes are assigned to that position; (**b**) the same genes differ in the exons' position; (**c**) one gene in NCBI is split into two genes in ENSEMBL; (**d**) in some locations one model predicts the presence of genes whereas the other one does not; (**e**) number and length of predicted transcripts differ

became popular [30, 31] and a decade later polymorphic DNA markers were introduced [32]. DNA fingerprinting originally relied on the use of restriction fragment length polymorphisms (RFLP) that with the discovery of PCR were replaced by amplification of simple sequence repeats (SSRs) also known as simple sequence length polymorphisms (SSLPs) [33]. SSLPs (also called microsatellites) are DNA regions containing di-, tri-, or tetranucleotide repeats that are found randomly and abundantly throughout the genome. The number of these repeats is highly polymorphic between inbred rats, allowing identification of different strains [34]. In 1991, Jacob et al. developed a set of 112 SSRs that were found to be polymorphic in length between the SHRSP and WKY strains and used them to discover a region of chromosome 10 linked to hypertension in this model [35]. In 1995, the same group expanded this set of markers to 432 SSLPs and used them to construct a more complete linkage map for the rat [28]. Linkage maps are created using various types of polymorphic markers and calculated based on recombination rates between these markers. They show the linear position of genes or markers on a chromosome, provide information on recombination rates, give insight into intra- and interspecies gene rearrangements, and help to identify Mendelian loci and Quantitative Trait Loci (QTL). In the 1990s many reports were published on construction of rat linkage maps that incorporated increasing numbers of markers and facilitated the localization of disease loci (*see* Table 6 and [27–40]). Current genetic monitoring and linkage mapping relies on single nucleotide polymorphisms (SNPs). The identification of SNPs has advanced rapidly and is routinely used in linkage and haplotype mapping, association studies, pharmacogenetics, and forensics. The most complete rat linkage map was published recently with a distinction of the variation between male and female recombination rates [40]. Littrell et al. constructed the refined genetic map (870 meioses; 95,769 markers), comparable with the high-resolution human map (104,246 meioses; 833,754 markers) [41] and mouse map (15,832 meioses; 120,789 markers) [42]. Increased accuracy of marker placement in high-density genetic maps is essential for QTL localization and subsequent fine gene mapping.

3.2 Genetic Variations

For personal genomics it is important to collect all information about genetic variations in order to establish the extent to which a gene's function can be compromised. Disease phenotypes are often determined by many genetic and environmental interactions and described as a multifactorial or complex genetic trait as opposed to a single-locus Mendelian trait. Genome-wide association studies (GWAS) have discovered thousands of associations between SNPs and complex traits [43, 44]. Most complex traits result from the combination of many genetic variants with small

Table 6
Rat linkage maps

Publication year	1992	1995	1997	1998	1999	2004	2018
First author last name	Serikawa	Jacob	Bihoreau	Brown	Steen	Jensen-Seaman	Littrell
Number of markers	125	432	767	678	4736	2305	95,769
Number of crosses (meiosis)	9	1	3	4	2	2 (90)	528 (870)
Map length (cM)	~2400	1509	1998	1749	1503	1542	1708
Average intermarker interval or (Map resolution) cM	~10	3.7	4.7	2.43	0.403	(1.1)	(<0.02)
Software or (Mapping function)	GeneLink	MapMaker	JoinMap, GMS, Multimap	MapMaker, Multimap	MapMaker	(Kosambi function)	Lep-MAP3
Reference (PubMed PMID)	27 (1628813)	28 (7704027)	37 (1472068)	38 (9657848)	29 (10400928)	39 (15059993)	40 (29760201)

effects, that together could account for a considerable portion of the variation in disease risk [43]. Mapping causal variants in model organisms requires time-consuming and costly breeding and/or mutation procedures. It also requires development of statistical algorithms for genomic predictions to estimate the distribution of SNP effects and distinguish true causal mutations from neutral variants.

RGD collects information on genetic variation from the worldwide community of rat researchers (*see* Table 7) and provides tools for searching and retrieving these data (*see* Chapter 3). Currently, we show details about almost 605 million variants (SNVs and small insertions and deletions) and about the studies that have identified these variants using different genome reference assemblies and methods in our Variant Visualizer [45].

The STAR project greatly contributed to increase the dbSNP repository [46]. The STAR Consortium identified *three million* new SNPs and evaluated a subset of *20,238* SNPs across 167 distinct inbred rat strains and two rat recombinant inbred (RI) panels [9]. High-density maps were constructed for mapping disease genes and identification of functional effects of 325,788 SNPs. 56 nonsynonymous coding SNPs were predicted to affect protein function, and seven of these SNPs were located in genes involved in hereditary diseases and cancer (*Aldh2*—acute alcohol intolerance, *Pccb*—propionic acidemia, *Aff4*—leukemia). In addition, 250 significant QTL were identified in the panel of 33 RI strains, that could be applicable in the detection of traits and risk factors for complex diseases [9]. Atanur et al. identified *3.6 million* high-quality SNPs and *343,243* short indels ($\leq$15 bp) in the SHR/OlaIpcv rat strain [16]. These variants produced 161 gains or losses of stop codons, but most importantly the large deletions resulted in complete or partial absence of 107 genes in the SHR/OlaIpcv genome. They found an intronic deletion in the SHR *Echdc2* gene (beta-oxidation pathway of fatty acids), a deletion in an intergenic region near the *Cyp4a8* gene (hypertension candidate gene), and confirmed the presence of a CNV in the *Cd36* gene (insulin resistance, dyslipidemia, and hypertension). The regions with cis-regulated expression quantitative trait loci (eQTL) were enriched with SNPs, short indels, and larger deletions [16]. Simonis et al. identified *3.2 million* single nucleotide variants, and *425,924* small insertions and deletions in the SHR/OlaIpcv and BN-Lx/Cub strains, the founders of an RI panel (HXB/BXH) [11]. The distribution of SNVs was not random. They found eight regions with a high SNV density, with a combined length of 51 Mb, together holding 97% of the SNVs. Nine substrains of the BN rat strain from different institutes had the same genotype in the seven regions as BN-Lx and there were no SNV high-density regions in the SHR strain. RNA sequencing of the liver tissues in both strains identified 532 differentially expressed genes. 1% of the SNVs and

Table 7
Rat variants resources in RGD

Assembly	SNVs total	Number of rat strains	Secondary analysis provider	Primary data source	PMID	Publication data provider	Sequence platform	dbSNP	dbSNP build
3.4	205,581,620	28	ICL	ICL, MCW, WTSI, UI, MDC, KNAW, ERIBA-UMCG, CHPM-UT, UMMC, UniSR, ICAMS, ILA KyotoU, EMBL-EBI, DZHK, INSERM; **Simonis et al. 2012**	23890820, 24628878, 20430781	Atanur et al. 2013, Ma et al. 2014, Atanur et al. 2010	Illumina HiSeq 2000, SOLiD 2,3 and 4 (2 strains)	**4,877,558**	**dbSnp136**
		12	HI-KNAW	**RGSMC**[1]: WTCHG, HI-KNAW, ERIBA-UMCG, KI-CNS, ICAMS, IUSM, INSERM, KI-MBB, MDC, INc-UAB, CEA-CNG, MPIMG, ECRC, HGSC-BCM, WTSI, ICL, EMBL-EBI, DZHK	23708188	Baud et al. 2013	SOLiD 4 and 5500, Illumina HiSeq 2000 (1 strain)		
		2	HI-KNAW	HI-KNAW, ICL, BCGSC, IPHYS-CAS	22541052	Simonis et al. 2012	SOLiD 2,3 and 4 (2 strains)		
		2	ICAHN	BGI, FIMR, UESTC	23695301	Guo et al. 2013	Illumina HiSeq 2000		
		4	UMich	UMich	PhD thesis[2]	Dr. Jun Z. Li group	Illumina HiSeq 2000		
		7	MCW	MCW (1 strain SMPH-UW)	NA	Dr. Howard Jacob group	Illumina HiSeq 2000		

(continued)

Table 7 (continued)

Assembly	SNVs total	Number of rat strains	Secondary analysis provider	Primary data source	PMID	Publication data provider	Sequence platform	dbSNP	dbSNP build
		3	MDC	MDC	NA	Dr. Norbert Huebner group	Illumina/Solexa, Genome Analyzer II		
5.0	261,827,016	42	HI-KNAW	HI-KNAW, MDC, UCSM, MRC-ICS, ERIBA - UMCG; Ma et al. 2014, Guo et al. 2013, Atanur et al. 2013, Baud et al. 2013, Simonis et al. 2012	25943489	Hermsen et al. 2015	Illumina HiSeq 2000	**4,806,887**	**dbSnp138**
		20	MiB-KyushuU	MiB-KyushuU, KDRI, ILA-KyotoU	27882299	Yoshihara et al. 2016	Illumina NextSeq 500		
		3	UDEL	Nemours, UDEL, Penn Med	26502805	Barthold et al. 2016	Illumina HiSeq 2500		
		9	MCW	MCW (1 strain SMPH-UW)	NA	Dr. Howard Jacob group	Illumina HiSeq 2000		
6.0	137,495,090	25	RGD	**Atanur et al. 2013**	NA	in preparation	Illumina HiSeq 2000	**4,726,744**	**dbSnp146**
		9	MCW	MCW (1 strain SMPH-UW)	NA	Dr. Howard Jacob group	Illumina HiSeq 2000		
		8	MCW (lift-over results)	HI-KNAW	NA	Dr. Michael Flister group	SOLiD 4 and 5500		

1. Rat Genome Sequencing and Mapping Consortium, 2. Ref. [57]. BCM-HGSC Human Genome Sequencing Center, Baylor College of Medicine, USA, BGI BGI-Shenzhen, China, CBMR The Novo Nordisk Foundation Center for Basic Metabolic Research, University of Copenhagen, Denmark, CEA-CNG Commissariat a l'Energie Atomique, Centre National de Génotypage, France, DZHK German Centre for Cardiovascular Research, Germany, ECRC Experimental and Clinical Research Center, Charité Universittsmedizin Berlin, Germany, ERIBA UMCG European Research Institute for the Biology of Ageing, University Medical Center Groningen, Netherlands, EMBL-EBI European Molecular Biology Laboratory, European Bioinformatics Institute, Wellcome Trust Genome Campus, UK, FIMR Laboratory of Experimental Rheumatology, Feinstein Institute for Medical Research, USA, HGSC-BCM Human Genome Sequencing Center, Baylor College of Medicine, USA, HI-KNAW Hubrecht Institute, Royal Netherland Academy of Arts and Sciences, Netherlands, ICL Imperial College London, UK, ICAHN The Icahn School of Medicine at Mount Sinai, USA, ICAMS The Institute of Cardiovascular and Medical Sciences,

University of Glasgow, UK, ILA-KyotoU Institute of Laboratory Animals, Kyoto University, Japan, INc-UAB Institute of Neurosciences, Universitat Autònoma de Barcelona, Spain, INSERM INSERM UMR-S872, Cordeliers Research Centre & INSERM U698, Hôpital Bichat, France, IPHYS-CAS Institute of Physiology, Czech Academy of Sciences, Czech Republic, IUSM Department of Medical and Molecular Genetics, Indiana University School of Medicine, USA, KDRI Department of Technology Development, Kazusa DNA Research Institute, Japan, KI-CNS Department of Clinical Neuroscience, Karolinska Institute, Sweden, KI-MBB Department of Medical Biochemistry and Biophysics, Karolinska Institute, Sweden, MiB-KyushuU Medical Institute of Bioregulation, Kyushu University, Japan, MDC Max Delbruck Center for Molecular Medicine, Germany, MCW Medical College of Wisconsin, USA, MPIMG Department of Computational Biology, Max Planck Institute for Molecular Genetics, Germany, MRC-ICS Physiological Genomic and Medicine Group, Institute of Clinical Sciences, UK, Nemours Nemours Alfred I. duPont Hospital for Children, USA, Penn Med Perelman School of Medicine, University of Pennsylvania, USA, SMPH-UW School of Medicine and Public Health, University of Wisconsin, USA, UCSF University of California, San Francisco, USA, UCSM Department of Pharmacology, University of Colorado School of Medicine, USA, UDEL Center for Bioinformatics and Computational Biology, University of Delaware, USA, UESTC School of Life Science and Technology, University of Electronic Science and Technology of China, China, UMich Department of Human Genetics, University of Michigan, USA, UMMC University of Mississippi Medical Center, USA, UniSR The Vita-Salute San Raffaele University, Italy, WTCHG Wellcome Trust Centre for Human Genetics, UK, WTSI The Wellcome Trust Sanger Institute, UK

indels were located in coding regions or splice sites, and 489 genes were affected by variant changes. Only 19 of 193 genes expressed in liver were differentially expressed between SHR and BN-Lx. The variant predictions were correct but the unaffected genes produced transcripts without affected exons. The duplicated gene *Mx2* showed heterozygous SNVs and the expression was also heterozygous in the RNA-seq data. In BN-Lx, which does not carry the duplications, all positions were homozygous at the DNA and RNA levels. The results implied that many of the analyzed liver transcripts were not spliced according to the available annotations. Changes in transcript structure rarely overlapped with genomic variants [11].

The Rat Genome Sequencing and Mapping Consortium analyzed 160 phenotypes from an outbred rat heterogeneous stock (NIH-HS) and showed the presence of segregating variation in commonly used laboratory rats [15]. NIH-HS rats were phenotyped for six disease models (anxiety, diabetes, hypertension, aortic elastic lamina ruptures, multiple sclerosis, and osteoporosis) and several related risk factors (lipid and cholesterol levels, cardiac hypertrophy, etc.). The researchers identified 355 QTL for 122 phenotypes using *265,551* polymorphic high-quality SNPs. They investigated the extent to which the variants would identify genes and causative mutations. 212 QTL (62%) had no candidate variant. The median proportion of heritability explained by QTL in rats and mice was above 30% and differed substantially from the mean proportion of heritability in human, which was less than 10%. The important observation was that genetic variants present in both inbred rat strains and inbred mouse strains rarely contributed to the same phenotype [15].

Atanur et al. studied coevolved gene clusters they named "putative artificial selective sweep (PASS)" regions and defined by the presence of many fixed rare variants and at least one variant that contributed to selection [16]. PASS regions co-localized with QTL, indicating an increased genetic variation in these regions and therefore an enrichment of variation in genes associated with disease phenotypes for which the strains were selected. They identified *9,665,340* SNVs and *3,502,117* short indels, and *29,131* SNVs were nonsynonymous coding (NSC) variants across 27 rat strains, including 11 models of hypertension, diabetes, and insulin resistance. Half of all single strain SNVs resided within the 189 segments that occupied only 0.8% of the genome, so private SNVs were concentrated in a small number of discrete regions of the genome. They identified clusters of coevolved transcripts that were unique for each disease model. In the Milan hypertensive rat strain (MHS), a cluster of 65 transcripts (47 genes) was found containing NSC sequence variants. The NSC cluster included the *Add1* gene that is known to cause hypertension in MHS rats due to amino acid substitution (F316Y). *ADD1* is also associated with human hypertension and responsiveness to antihypertensive medications. Unfortunately, they could not find clusters with shared functional mutations between strains representing the same disease model, like hypertensive rats (FHH, MHS, SHR, SHRSP, and LH). Hermsen et al. reanalyzed primary rat sequence data of 40 rat strains using an upgraded rat reference Rnor5.0 version and identified over *12 million* genomic variants (SNVs, indels, and structural variants) [47]. They found 601 SNVs that have a deleterious effect on gene function including stop_gained variants (SnpEff consequences type; SO:0001587) and alterations of splice sites, but in the case of 60 SNVs (10%), a neighboring single variant or indel restored the open reading frame. Genes affected by the remaining 541 high impact SNVs were expressed at lower levels. The biological relevance of the high impact SNVs was limited due to exon skipping or the canceling effect of adjacent variants. They observed more non-synonymous SNVs in the substrain variants compared to the control set. They identified 3006 protein-coding genes that contain 6 or more SNVs in the coding region and analyzed 909 genes that were under positive selection. Functional annotation showed that this set of genes is enriched for genes related to the immune and olfactory system. The WKY strain had the highest degree of substrain variation, partly because of a geographical relocation before the inbreeding was completed.

Evidently, causal variants are not easy to identify. An accurate and complete set of called variants together with gene information and QTL data, available for many rat strains for a broad range of complex traits, may assist in identifying the potential causal variants. Recently She and Jarosz proposed a novel computational strategy for an optimal crossing scheme in yeast that enables a single-nucleotide resolution mapping of the genetic variation and identification of the causal variants from adjacent passenger mutations [48]. This new strategy could yield a breakthrough in the search for disease susceptibility markers.

In addition to rat data, 420,850 human variants from ClinVar along with their predicted effects are available in the Variant Visualizer tool [45]. As the rat is used to study gene-phenotype associations or specific diseases, researchers need access to all information necessary to connect genes to phenotypes or diseases of interest. RGD currently integrates multi-organism data and has improved multilevel navigation to allow finding this information.

3.3 Phylogeny

Researchers interested in the relationships between rat strains used different types of markers to produce phylogenetic trees. Early researchers, using 28 biochemical marker loci, distinguished 52 genetically different rat strains that grouped in three clusters [49]. In later studies, highly variable microsatellites and high-density SNPs were used as genetic markers to construct phylogenetic trees, establish the relatedness of organisms, and predict

Table 8
Comparison of phylogeny studies of laboratory rats

Publication year	1984	1995	1997	2003	2005	2006	2008	2008	2013	2015	2015	2017
First author	Festing	Canzian	Canzian	Thomas	Smits[a]	Mashimo	Nijman	STAR	Atanur	Hermsen[a]	Battula	Puckett
Number of strains	52	13	63 (214 sub)	48	39	93	37	167	28	41	51	326 (29 inb)
Marker type	Bioch	RAPD	Bioch, SSLP	SSLP	SNP	SSLP	SNP	SNP	SNP	SNP	SSLP	SNP
Number of markers	28	264	995	>4800	861	357	820	20,283	9.6 Mln	~ 9 Mln	76	32,127
Brown Norway	BN											
Long Evans	5	1	14	8	1	6	1	10	1	3	1	3
Sprague Dawley 1	4	4 1	7	2 3	3	4	2	5	1	6	8	1
Sprague Dawley 2	2 2 2 2	1 1	3 2 6	2 5 3	5	5	3 1	10	2	3	2 3	3 2
Wistar - WKY	4 4	1 1	7 2	5	1	3	7 1	22	2	2	2	2
Wistar - BB	2	1	5	2	2	5 4	4 2	7	1	5	2	3 1
F344			3	8	2 2	33	1 1	3	6	4	1 4	5
Sabra & Cohen			7		3		4	7	6	2	1	3
PVG			3					11	6	5 2	5 2	
BD			3 4 3					7			6	
ACI			4 2 5					4				
			5 2									

[a]Phylogeny studies that did not explore phylogeny distances

linkage between genomic regions and various phenotypes and diseases (*see* Table 8). In 1995, Canzian et al. published a phylogenetic tree for 13 commonly used rat strains with construction based on 264 genome-wide loci [50]. In 1997 Dr. Canzian built a tree for 63 inbred strains and 214 substrains using 995 microsatellite and biochemical markers [51]. The average polymorphism for pairwise comparisons of rat strains derived using both biochemical and microsatellite markers was 53%, whereas the average increased to 64% when only microsatellite markers were included in the analysis, possibly due to differences in selective pressures between biochemical and genetic markers and/or the fact that a single biochemical type can result from more than a single genetic polymorphism. Thomas et al. used 48 substrains from 46 distinct strains and over 4800 microsatellite markers in their population study [52]. Mashimo et al. constructed a phylogenetic tree for 93 rat strains from a 357 SSLP marker set [53]. Nijman et al. used a selected genome-wide set of 820 SNPs on 38 rats of 34 different strains and 3 wild rat strains [54]. The STAR Consortium presented a phylogenetic network using 20,283 SNPs from 167 inbred strains [9]. Atanur et al. constructed a tree using 9.6 million SNVs [16]. In 2015, two studies reported rat strain phylogenic analyses: Hermsen et al. presented a population structure of 40 rat strains using almost nine million SNVs and Battula et al. reported the genetic relatedness of WNIN and WNIN/Ob rats with other strains using 76 unlinked microsatellite markers [47, 55].

All published phylogenetic studies for rat strains showed consistently that the Brown Norway strain was the most distant from all the other strains. The STAR Consortium found 10 clusters: two Wistar (WKY cluster—with SHR, SHRSP, WKY, GK strains; BB—with MHS, MNS, Lew, BB, LUDW strains), two Sprague-Dawley subtrees (SD1—with DRH, NAR strains; SD2—with SS, SR, LN, LH strains), Cohen with Sabra rats (CD—SBH, SBN, CDR, CDS strains), a Fisher 344 group (F344—with F344, BUF, MES strains), rats grouped with the Piebald Virol Glaxo strain (PVG—

WAG, BS, GH, LOU strains), a cluster of Berlin Druckrey rats (BD—with BDIX, BH, E3), the furthest from the BN root—the cluster containing the August Copenhagen Irish strain (ACI—with ACI, DA, COP), and the nearest to the BN was the Long Evans cluster (LE—with KDP, FHH, LE, R33 strains) [13]. We used their cluster structure to compare other published trees in Table 7 that shows the numbers of strains that were included in 10 clusters. Two studies of Hermsen et al. and Smits et al. identify the number of populations (clusters) and explore similarities between them instead of constructing the phylogenetic trees [47, 56]. Smits et al. presented a phylogenetic network that explores alternative evolutionary paths along the network [56] and Hermsen et al. investigated a "population" structure, appropriate for studying ancestry of samples genotyped at a large number of genetic markers and with a complex evolutionary history [47].

Nijman et al. determined seven very similar clusters, but the WKY cluster was further in distance from the BN strain, the PVG strain was separated with the LOU and WAG strains (a similar separation was present in the Canzian 1997 tree), the ACI cluster was closer to the BN root and the most distant was the BD group. All Wistar strains (WKY and BB clusters) were also separated from the group of BN in the Atanur et al. 2013 study. That was in contrast to three other phylogenic analyses, where the WKY group was in close proximity to the BN root, the Canzian 1997, Thomas 2003, and Battula 2015 trees. The LE group strains were the closest branch to BN in most of the trees. The ACI cluster was close to the F344 group in the Thomas, STAR Consortium, and Atanur trees. The F344 strain was not included in the Nijman tree analysis, but the Buffalo strain was placed next to the ACI cluster. SS, SR, SHR, and MHS strains that represent models of hypertension remain close in distance in most of the trees (SD2 and BB clusters). The MNS and MHS strains were bred together and they are tightly related, but in the Canzian 1997 tree they were placed separately. Two inbred WNIN strains in the Batutula analysis were placed as an individual cluster between the F344 group and the LE strain. Some substrains showed greater interstrain genetic differences than others [51]. Variation between substrains were observed in LE (29% in the pairwise substrain comparison), WKY (up to 19%), LEW (13%), SHR (11%), BB (10%), PKD (5%) and, to a lesser degree, in GK (1%) and BN (0.6%) inbred strains [51]. These observations are important for research strategy design, as the results and reproducibility depend on the choice of particular substrains.

4 Conclusion

In summary, RGD provides a resource for rat genetic markers that includes simple sequence length polymorphisms (SSLP), copy number variations (CNV), deletions, insertions, single nucleotide variants (SNV), and QTL. The data are available for three rat genome assemblies for a range of commonly used laboratory rat strains. This repository is valuable for researchers that use rats in medical research but also for those who do comparative analysis using other organisms. RGD's major goal is to present rat genomic and phenotypic data making it easy to interpret, to assist in experimental design and in the aftermath to facilitate rat research and interspecies comparison. RGD's resources may improve the reproducibility of scientific research between laboratories and thus ensure the overall quality of biomedical animal research.

References

1. Gibbs RA, Weinstock GM, Metzker ML, Muzny DM, Sodergren EJ, Scherer S et al (2004) Genome sequence of the Brown Norway rat yields insights into mammalian evolution. Nature 428(6982):493–521
2. Havlak P, Chen R, Durbin KJ, Egan A, Ren Y, Song XZ et al (2004) The Atlas genome assembly system. Genome Res 14(4):721–732
3. Krzywinski M, Wallis J, Gösele C, Bosdet I, Chiu R, Graves T et al (2004) Integrated and sequence-ordered BAC- and YAC-based physical maps for the rat genome. Genome Res 14 (4):766–779
4. Kren V, Qi N, Krenova D, Zidek V, Sladká M, Jáchymová M, Míková B et al (2001) Y-chromosome transfer induces changes in blood pressure and blood lipids in SHR. Hypertension 37(4):1147–1152
5. Gibbs R, Weinstock G (2005) Upgrading the DNA sequence of the rat genome. White paper available at https://www.genome.gov/pages/research/sequencing/seqproposals/ratupgradeseq.pdf
6. van Boxtel R, Cuppen E (2010) Rat traps: filling the toolbox for manipulating the rat genome. Genome Biol 11(9):217. https://doi.org/10.1186/gb-2010-11-9-217
7. Prokop JW, Underwood AC, Turner ME, Miller N, Pietrzak D, Scott S et al (2013) Analysis of Sry duplications on the *Rattus norvegicus* Y-chromosome. BMC Genomics 14:792. https://doi.org/10.1186/1471-2164-14-792
8. Rozen S, Warren WC, Weinstock G, O'Brien SJ, Gibbs RA, Richard K et al (2006) Sequencing and annotating new mammalian Y chromosomes. White paper available at https://www.genome.gov/pages/research/sequencing/seqproposals/ychromosomewp.pdf
9. STAR Consortium, Saar K, Beck A, Bihoreau MT, Birney E, Brocklebank D et al (2008) SNP and haplotype mapping for genetic analysis in the rat. Nat Genet 40(5):560–566. https://doi.org/10.1038/ng.124
10. Atanur SS, Birol I, Guryev V, Hirst M, Hummel O, Morrissey C et al (2010) The genome sequence of the spontaneously hypertensive rat: analysis and functional significance. Genome Res 20(6):791–803. https://doi.org/10.1101/gr.103499.109
11. Simonis M, Atanur SS, Linsen S, Guryev V, Ruzius FP, Game L et al (2012) Genetic basis of transcriptome differences between the founder strains of the rat HXB/BXH recombinant inbred panel. Genome Biol 13(4):r31. https://doi.org/10.1186/gb-2012-13-4-r31
12. Guo X, Brenner M, Zhang X, Laragione T, Tai S, Li Y et al (2013) Whole-genome sequences of DA and F344 rats with different susceptibilities to arthritis, autoimmunity, inflammation and cancer. Genetics 194 (4):1017–1028. https://doi.org/10.1534/genetics.113.153049
13. Li R, Yu C, Li Y, Lam TW, Yiu SM, Kristiansen K et al (2009) SOAP2: an improved ultrafast tool for short read alignment. Bioinformatics 25(15):1966–1967. https://doi.org/10.1093/bioinformatics/btp336
14. Li R, Zhu H, Ruan J, Qian W, Fang X, Shi Z et al (2010) De novo assembly of human genomes with massively parallel short read sequencing. Genome Res 20(2):265–272. https://doi.org/10.1101/gr.097261.109
15. Rat Genome Sequencing and Mapping Consortium, Baud A, Hermsen R, Guryev V, Stridh P, Graham D et al (2013) Combined sequence-based and genetic mapping analysis of complex traits in outbred rats. Nat Genet 45 (7):767–775. https://doi.org/10.1038/ng.2644
16. Atanur SS, Diaz AG, Maratou K, Sarkis A, Rotival M, Game L et al (2013) Genome sequencing reveals loci under artificial selection

that underlie disease phenotypes in the laboratory rat. Cell 154(3):691–703. https://doi.org/10.1016/j.cell.2013.06.040

17. Aitman TJ, Dong R, Vyse TJ, Norsworthy PJ, Johnson MD, Smith J et al (2006) Copy number polymorphism in Fcgr3 predisposes to glomerulonephritis in rats and humans. Nature 439(7078):851–855
18. Thibaud-Nissen F, Souvorov A, Murphy T, DiCuccio M, Kitts P (2013) Eukaryotic genome annotation pipeline. In: The NCBI handbook, 2nd edn. National Center for Biotechnology Information, Bethesda. https://www.ncbi.nlm.nih.gov/books/NBK169439/
19. Yeh RF, Lim LP, Burge CB (2001) Computational inference of homologous gene structures in the human genome. Genome Res 11 (5):803–816
20. Aken BL, Ayling S, Barrell D, Clarke L, Curwen V, Fairley S et al (2016) The Ensembl gene annotation system. Database (Oxford) 2016:baw093. https://doi.org/10.1093/database/baw093
21. Birney E, Clamp M, Durbin R (2004) GeneWise and Genomewise. Genome Res 14 (5):988–995
22. National Center for Biotechnology Information (2005) US National Library of Medicine, Bethesda. http://www.ncbi.nlm.nih.gov. Accessed 1 Feb 2015
23. Yates A, Akanni W, Amode MR, Barrell D, Billis K, Carvalho-Silva D et al (2016) Ensembl 2016. Nucleic Acids Res 44:D710–D716. https://doi.org/10.1093/nar/gkv1157
24. Buels R, Yao E, Diesh CM, Hayes RD, Munoz-Torres M, Helt G et al (2016) JBrowse: a dynamic web platform for genome visualization and analysis. Genome Biol 17:66. https://doi.org/10.1186/s13059-016-0924-1
25. Kumar D, Yadav AK, Jia X, Mulvenna J, Dash D (2015) Integrated transcriptomic-proteomic analysis using a proteogenomic workflow refines rat genome annotation. Mol Cell Proteomics 15(1):329–339. https://doi.org/10.1074/mcp.M114.047126
26. Wu PY, Phan JH, Wang MD (2013) Assessing the impact of human genome annotation choice on RNA-seq expression estimates. BMC Bioinformatics 11:S8. https://doi.org/10.1186/1471-2105-14-S11-S8
27. Serikawa T, Kuramoto T, Hilbert P, Mori M, Yamada J, Dubay CJ et al (1992) Rat gene mapping using PCR-analyzed microsatellites. Genetics 131(3):701–721
28. Jacob HJ, Brown DM, Bunker RK, Daly MJ, Dzau VJ, Goodman A et al (1995) A genetic linkage map of the laboratory rat, *Rattus norvegicus*. Nat Genet 9(1):63–69
29. Steen RG, Kwitek-Black AE, Glenn C, Gullings-Handley J, Van Etten W, Atkinson OS et al (1999) A high-density integrated genetic linkage and radiation hybrid map of the laboratory rat. Genome Res 9(6): AP1–AP8. Erratum in: Genome Res 1999 9 (8):793
30. Hutton JJ, Roderick TH (1970) Linkage analyses using biochemical variants in mice. 3. Linkage relationships of eleven biochemical markers. Biochem Genet 4(2):339–350
31. Moutier R, Toyama K, Charrier MF (1973) Biochemical polymorphism in the rat, *Rattus norvegicus*: genetic study of four markers. Biochem Genet 8(3):321–328
32. Botstein D, White RL, Skolnick M, Davis RW (1980) Construction of a genetic linkage map in man using restriction fragment length polymorphisms. Am J Hum Genet 32(3):314–331
33. Williams JG, Kubelik AR, Livak KJ, Rafalski JA, Tingey SV (1990) DNA polymorphisms amplified by arbitrary primers are useful as genetic markers. Nucleic Acids Res 18:6531–6535
34. Bryda EC, Riley LK (2008) Multiplex microsatellite marker panels for genetic monitoring of common rat strains. J Am Assoc Lab Anim Sci 47(3):37–41
35. Jacob HJ, Lindpaintner K, Lincoln SE, Kusumi K, Bunker RK, Mao YP et al (1991) Genetic mapping of a gene causing hypertension in the stroke-prone spontaneously hypertensive rat. Cell 67(1):213–224
36. Levan G, Szpirer J, Szpirer C, Klinga K, Hanson C, Islam MQ (1991) The gene map of the Norway rat (*Rattus norvegicus*) and comparative mapping with mouse and man. Genomics 10:699–718
37. Bihoreau M-T, Sebag-Montefiore L, Godfrey RF, Wallis RH, Brown JH, Danoy PA et al (1997) A high-resolution consensus linkage map of the rat, integrating radiation hybrid and genetic maps. Genomics 75:57–69
38. Brown DM, Matise TC, Koike G, Simon JS, Winer ES, Zangen S et al (1998) An integrated genetic linkage map of the laboratory rat. Mamm Genome 9(7):521–530
39. Jensen-Seaman MI, Furey TS, Payseur BA, Lu Y, Roskin KM, Chen CF et al (2004) Comparative recombination rates in the rat, mouse, and human genomes. Genome Res 14 (4):528–538
40. Littrell J, Tsaih SW, Baud A, Rastas P, Solberg-Woods L, Flister MJ (2018) A high-resolution genetic map for the laboratory rat. G3 (Bethesda) 8(7):2241–2248

41. Bhérer C, Campbell CL, Auton A (2017) Refined genetic maps reveal sexual dimorphism in human meiotic recombination at multiple scales. Nat Commun 8:14994
42. Morgan AP, Gatti DM, Najarian ML, Keane TM, Galante RJ, Pack AI et al (2017) Structural variation shapes the landscape of recombination in mouse. Genetics 206:603–619
43. Ulirsch JC, Nandakumar SK, Wang L, Giani FC, Zhang X, Rogov P et al (2016) Systematic functional dissection of common genetic variation affecting red blood cell traits. Cell 165 (6):1530–1545. https://doi.org/10.1016/j.cell.2016.04.048
44. Wood AR, Esko T, Yang J, Vedantam S, Pers TH, Gustafsson S et al (2014) Defining the role of common variation in the genomic and biological architecture of adult human height. Nat Genet 46(11):1173–1186. https://doi.org/10.1038/ng.3097
45. Shimoyama M, De Pons J, Hayman GT, Laulederkind SJ, Liu W, Nigam R et al (2015) The Rat Genome Database 2015: genomic, phenotypic and environmental variations and disease. Nucleic Acids Res 43(Database issue): D743–D750
46. Twigger SN, Pruitt KD, Fernández-Suárez XM, Karolchik D, Worley KC, Maglott DR et al (2008) What everybody should know about the rat genome and its online resources. Nat Genet 40(5):523–527. https://doi.org/10.1038/ng0508-523
47. Hermsen R, de Ligt J, Spee W, Blokzijl F, Schäfer S, Adami E et al (2015) Genomic landscape of rat strain and substrain variation. BMC Genomics 16:357. https://doi.org/10.1186/s12864-015-1594-1
48. She R, Jarosz DF (2018) Mapping causal variants with single-nucleotide resolution reveals biochemical drivers of phenotypic change. Cell 172(3):478–490. https://doi.org/10.1016/j.cell.2017.12.015
49. Festing MF, Bender K (1984) Genetic relationships between inbred strains of rats. An analysis based on genetic markers at 28 biochemical loci. Genet Res 44(3):271–281
50. Canzian F, Ushijima T, Pascale R, Sugimura T, Dragani TA, Nagao M (1995) Construction of a phylogenetic tree for inbred strains of rat by arbitrarily primed polymerase chain reaction (AP-PCR). Mamm Genome 6(4):231–235
51. Canzian F (1997) Phylogenetics of the laboratory rat *Rattus norvegicus*. Genome Res 7 (3):262–267
52. Thomas MA, Chen CF, Jensen-Seaman MI, Tonellato PJ, Twigger SN (2003) Phylogenetics of rat inbred strains. Mamm Genome 14(1):61–64
53. Mashimo T, Voigt B, Tsurumi T, Naoi K, Nakanishi S, Yamasaki K et al (2006) A set of highly informative rat simple sequence length polymorphism (SSLP) markers and genetically defined rat strains. BMC Genet 7:19
54. Nijman IJ, Kuipers S, Verheul M, Guryev V, Cuppen E (2008) A genome-wide SNP panel for mapping and association studies in the rat. BMC Genomics 9:95. https://doi.org/10.1186/1471-2164-9-95
55. Battula KK, Nappanveettil G, Nakanishi S, Kuramoto T, Friedman JM, Kalashikam RR (2015) Genetic relatedness of WNIN and WNIN/Ob with major rat strains in biomedical research. Biochem Genet 53 (4–6):132–140. https://doi.org/10.1007/s10528-015-9679-8
56. Smits BM, Guryev V, Zeegers D, Wedekind D, Hedrich HJ, Cuppen E (2005) Efficient single nucleotide polymorphism discovery in laboratory rat strains using wild rat-derived SNP candidates. BMC Genomics 6:170
57. Ren Y (2016) Multi-omics analysis of a rat model of aerobic exercise capacity and metabolic fitness. PhD dissertation, University of Michigan, Michigan

Chapter 3

Rat Genome Databases, Repositories, and Tools

Stanley J. F. Laulederkind, G. Thomas Hayman, Shur-Jen Wang, Matthew J. Hoffman, Jennifer R. Smith, Elizabeth R. Bolton, Jeff De Pons, Marek A. Tutaj, Monika Tutaj, Jyothi Thota, Melinda R. Dwinell, and Mary Shimoyama

Abstract

Resources for rat researchers are extensive, including strain repositories and databases all around the world. The Rat Genome Database (RGD) serves as the primary rat data repository, providing both manual and computationally collected data from other databases.

Key words Database, Genomics, Analysis, Visualization, Disease, Phenotype, Pathway, Gene, Annotation, Model organism

1 Introduction

The laboratory rat (*Rattus norvegicus*) has been used as an animal model for physiology, pharmacology, toxicology, nutrition, behavior, immunology, and disease for over 150 years [1]. It was the first animal to be domesticated for use by scientists [2]. The rat's value continues to grow as indicated by the more than 1.5 million publications in PubMed, with about 40,000 being added every year. Advanced sequencing technologies, genome modification techniques, and the development of embryonic stem cell protocols ensure that the rat remains an important mammalian model for disease studies. The 2004 release of the reference genome has been followed by the sequencing of genomes for more than two dozen individual strains utilizing NextGen sequencing technologies. These analyses have identified over 50 million variants [3, 4].

This explosion of genomic data has been accompanied by the ability to selectively edit the rat genome, leading to hundreds of new strains through technologies using the CRISPR/Cas9 system [5], zinc finger nucleases [6], transcription activator-like effector nucleases [7], transposons [8], and meganucleases [9]. A number

G. Thomas Hayman et al. (eds.), *Rat Genomics*, Methods in Molecular Biology, vol. 2018,
https://doi.org/10.1007/978-1-4939-9581-3_3,

of resources have been developed to provide investigators access to precision rat models, comprehensive datasets, and sophisticated software tools necessary for their research. Those include the Rat Genome Database (RGD), Gene Editing Rat Resource Center (GERRC), Rat Resource and Research Center (RRRC), the National BioResource Project-Rat (NBRP-Rat), PhenoGen, and more, as detailed later.

2 Rat Strain Repositories

Hundreds of rat strains have been developed during the past 100 years, including inbred, consomic, congenic, ENU-mutants, and, more recently, genetically engineered mutant strains. This large number of strains is managed mainly by two rat resource centers, the Rat Resource and Research Center (RRRC) in the United States and the National Bio Resource Project-Rat (NBRP-Rat) in Japan. These resource centers collect, maintain, and distribute rat strains as animals or cryopreserved embryos and spermatozoa. These two centers also perform phenotypic and genetic characterization of the specimen with dissemination of that information through their respective, publically accessible databases.

2.1 RRRC

Many important rat strains for life science research have been maintained by scientists in individual laboratories. This type of resource propagation is inefficient and susceptible to changes in funding or local interest. The NIH rat model repository workshop was held in 1998, with scientists from around the world discussing the needs, opportunities, and parameters for optimal standardization, maintenance, and distribution of genetically defined rat strains. Those scientists strongly encouraged the NIH to establish a national rat genetics resource center, and as a result, the Rat Resource Research Center (RRRC) was established in 2001.

The service functions of the RRRC (https://www.rrrc.us/) involve the procurement of non-commercial rat lines, sperm and embryo cryopreservation, cryo-resuscitation or rederivation with pathogen and genotype quality control, genotyping and cytogenetic services, gut microbiome characterization, and distribution of rats, cell lines, and tissues to biomedical investigators. RRRC also performs research to make improvements in rat model development and enhancement.

2.2 NBRP-Rat

The National BioResource Project-Rat (NBRP-Rat) (https://www.anim.med.kyoto-u.ac.jp/nbr/) was initiated in 2002 to establish a system to facilitate the systematic collection, preservation, and provision of laboratory rats. It is the world's largest rat repository with specimens kept as live animals and cryopreserved embryos or sperm. Hundreds of laboratories across Japan and the world have been supplied with rat strains or rat DNA from NBRP-Rat. Protocols for cryopreservation and rederivation techniques have also been supplied by NBRP-Rat to the research community [10].

NBRP-Rat's Phenome Project was a reevaluation of more than 150 strains based on over 100 phenotypic parameters in seven general categories [11]. A major benefit of all these phenotypic measurements is the generation of biological ranges of various parameters, which allows visualization of normal and abnormal values for the different rat strains examined. The data can be visualized on the NBRP-Rat web site or in the RGD PhenoMiner tool [12].

More than 700 rat strains have been deposited at NBRP-Rat, with most of those available as cryopreserved sperm or embryos and the remaining available as live animals. All of the deposited strains can be obtained by interested researchers. The Kyoto University Rat Mutant Archive (KURMA) was added to NBRP-Rat to provide ENU mutant strains, which provide many models for biomedical research. More than 150 strains have been genotyped by NBRP-Rat with more than 300 microsatellite markers [13, 14]. These genotyped rats have provided data to create phylogenetic charts and SSLP charts, which allow a visual approximation of the genetic distance between different strains. There are also various other tools which allow public access to data at NBRP-Rat.

2.3 Gene Editing Rat Resource Center

The Gene Editing Rat Resource Center (GERRC; http://rgd.mcw.edu/wg/gerrc) at the Medical College of Wisconsin committed in 2013 to produce about 200 genetically modified rat strains over a five-year period for use by researchers. These selected strains have specific genes knocked out using several different gene editing technologies. There were two application rounds each year, during which researchers requested genes to be knocked out in a specific strain, with up to two applications allowed per laboratory. Applications were reviewed by an external advisory board to determine which models, up to 25, were to be made. After the strains were created, usually 9 to 12 months after application, the requesting investigator received the first breeder pair. Any other breeder pairs are available to other investigators on a first come, first served basis. An annotated list of all the mutant strains generated by the project is available on the GERRC web site.

3 Rat-Specific Data Resources

3.1 Rat Genome Project

The original 2004 release of the reference genome for the rat [15] was done by the Rat Genome Sequencing Consortium (RGSC) led by the Human Genome Sequencing Center at Baylor College of Medicine (BCM-HGSC). Access to the original data and assembly updates (including Rnor 6.0) is available on the BCM-HGSC web site (https://www.hgsc.bcm.edu/other-mammals/rat-genome-project) and at the National Center for Biotechnology Information (NCBI) (https://www.ncbi.nlm.nih.gov/genome/73).

3.2 Rat Genome Database

The Rat Genome Database (https://rgd.mcw.edu/) was established in 1999 as a resource to support the emerging genomic data for the rat. This role has continued to expand with continuing work on the rat reference genome sequence (the current assembly is Rnor_6.0 - RGSC Genome Assembly v6.0), strain-specific DNA sequencing [16], expanded SNP discovery, and large-scale phenotyping projects such as the PhysGen project (http://pga.mcw.edu) and NBRP [10], all needing to be integrated with existing and newly published research data. As the amount of data has grown, so has the challenge of mining relevant information and defining its meaning in the broader context of biomedical science. With this in mind, much effort has gone into the development and incorporation of biomedical ontologies such as the Gene Ontology [17], the Mammalian Phenotype Ontology [18], the Pathway Ontology [19], and others [20]. These are incorporated into the search and analysis tools, greatly facilitating the discovery of information and interpretation of its meaning.

Many researchers using the rat as a model system are ultimately studying a specific phenotype or disease with the goal of applying this knowledge to humans. To meet this need, RGD has developed "Disease Portals" that present RGD data and tools from the perspective of a particular disease. The Disease Portals allow researchers to visit a single page that is focused on a single disease area like cardiovascular, neurological, etc. These disease categories are being expanded in an ongoing process of targeted curation to create more portals devoted to particular disease areas that will cater directly to researchers working in those areas. The rest of RGD is accessible via these portals, but researchers will find the items of their greatest interest first, reducing the challenge of finding the data and interpreting its meaning. Similarly, the Phenotypes & Models Portal and the Pathways Portal focus on specific areas of research, which allows easier access to targeted searches for relevant data.

In addition to the portal style of data organization, the access to different software tools at RGD is an important part of the database. Ranging from annotation-based analysis to sequence-based analysis, the options are extensive to manipulate both RGD data and user-uploaded data. Further analysis may be done with downloaded data via the FTP site or the REST API.

3.2.1 RGD Data Objects

RGD stores data about various "objects," including genes, quantitative trait loci (QTLs), markers, references, strains, and cell lines. Report pages for these objects are presented in a similar format. The most data-rich report pages are gene, QTL, and strain pages. Disease data for genes, QTLs, and strains can also be accessed through various RGD Disease Portals. Pathway data for genes can be accessed through the RGD Pathway Portal. Physiological data for strains is accessible through the Phenotypes and Models Portal.

Genes

To allow comparative investigation, RGD has historically provided gene report pages for rat, mouse, and human, with additional species added recently. A typical rat gene report page is shown in Fig. 1. The top section contains the gene name and annotation-based description, links to orthologs and external information/analysis sites, and map information. Next are data sections (*see* Fig. 1B), the first being the "Annotation" section containing both RGD manual and imported annotations using various ontologies: Disease Ontology (DO), Gene Ontology (GO), Pathway Ontology (PW) and phenotype ontologies (Mammalian Phenotype Ontology (MP) and Human Phenotype Ontology (HP). Each annotation has a term, an evidence code, reference, and source information. Sections on Genomics, Sequence (DNA and protein), Strain (Sequence) Variation and Additional Information are found below the Annotation section.

Quantitative Trait Loci

Another type of RGD object presented on report pages is the Quantitative Trait Locus (QTL), which is a large region of DNA associated with a physiological or pathological phenotype. RGD has data on a large variety of QTLs (rat, mouse, and human) describing physiologic and anatomic traits, like blood pressure and organ weight, to disease traits for cancer, diabetes, and other pathological conditions.

The top section of a QTL report page provides the QTL name, trait and measurement type. Significance scores, map information, and strains crossed to derive the QTL are also provided. The Annotation section contains disease annotations with DO terms, phenotype annotations with MP (rat, mouse) or HP (human) terms, and experimental data annotations, which use the following ontologies: the Vertebrate Trait Ontology (VT), the Clinical Measurement Ontology (CMO), the Measurement Method Ontology (MMO), and the Experimental Condition Ontology (XCO). The CMO, MMO, and XCO are RGD-produced and maintained ontologies. References and disease portal links are also provided in the Annotation section of the QTL report page. The "Region" section provides position markers for the QTL, and genes, markers, and overlapping QTLs in the region.

Strains

Rat strains in RGD are named according to the official nomenclature rules (https://rgd.mcw.edu/nomen/rules-for-nomen.shtml) and are organized hierarchically by the Rat Strain (RS) ontology [21] to facilitate access to curated strain data. The RS is available in the RGD ontology browser [20] for easy navigation through the

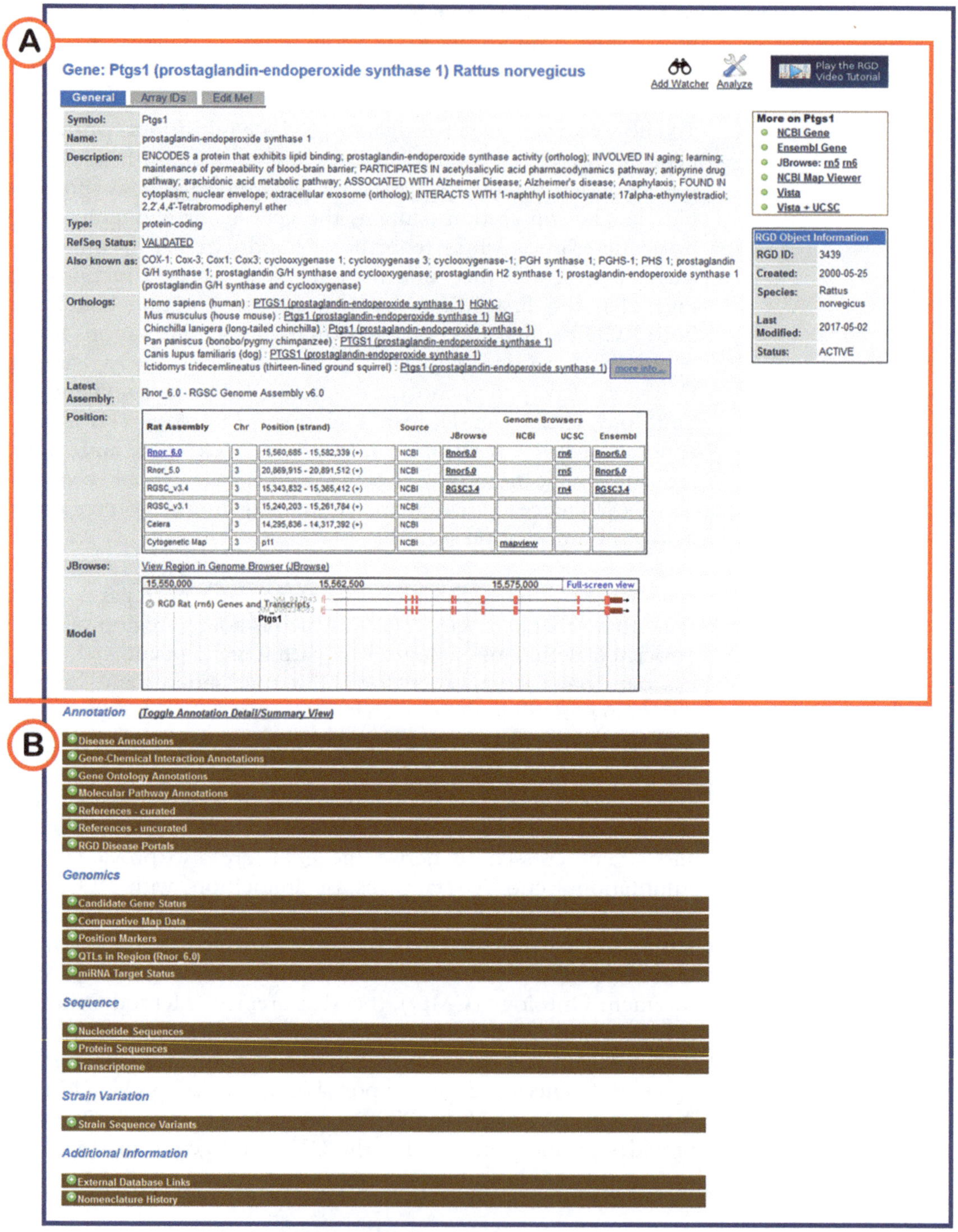

Fig. 1 Gene Report Page for Rat Ptgs1. (*A*) The top half of the page contains general information, ortholog assignments, genomic positions, JBrowse model, and links to external sites. (*B*) The bottom half of the page has annotations in various categories, genomic information, sequence information, and more, all in expandable, labeled bars

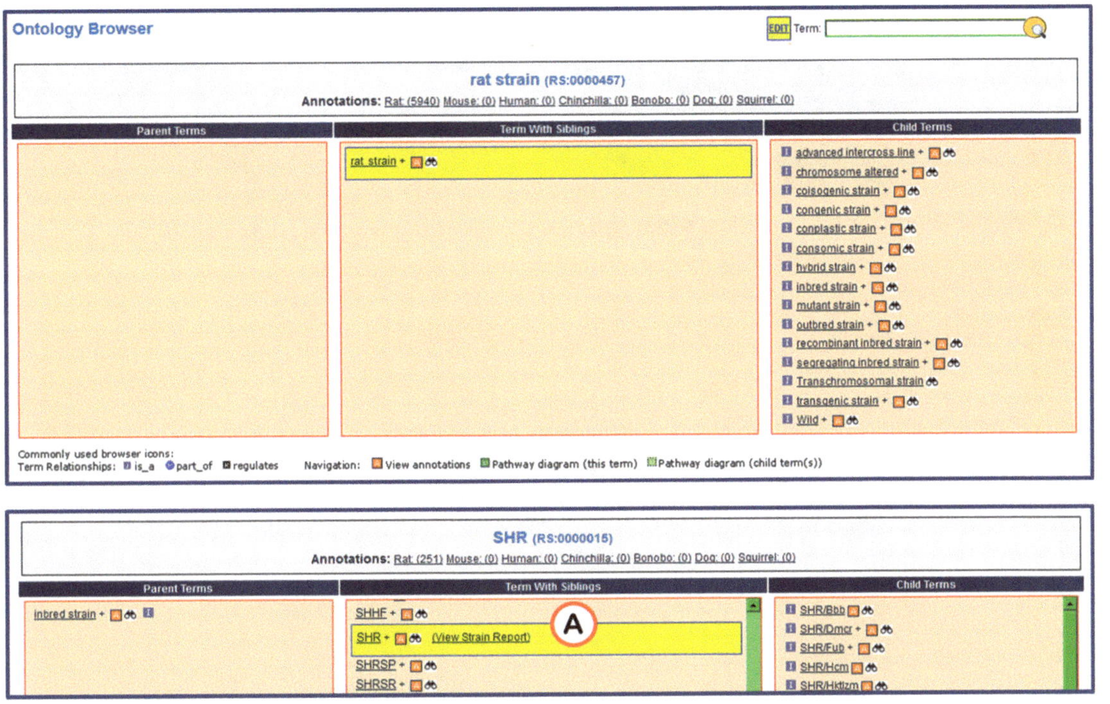

Fig. 2 Ontology Term Browser. The RGD term browser with "rat strain" (RS) selected in the top panel. (*A*) Selection "SHR" in the bottom panel has an accompanying "View Strain Report" link

rat strain nomenclature to find strains and sub-strains of interest. Users can access strain report pages through the general RGD search at the top of most RGD pages, through the strain search page, or the link next to the selected strain (*see* Fig. 2A) in the ontology browser. Each registered strain has a report page that includes information on the source and availability of the strain, and the type of manipulation used to derive the strain. On the rat strain report pages the Annotation section has Disease, Phenotype, Experimental Data, and Phenotype Values via Phenominer subsections containing annotations from curation of the scientific literature or user submission data. The Experimental Data annotations and Phenotype Values via Phenominer annotations relate to quantitative physiological data that is available in the RGD PhenoMiner tool (*see* RGD Data Analysis and Visualization Tools section below).

3.2.2 Portal Access to RGD Data

Disease Portals

RGD currently has 12 disease portals encompassing many disease areas from developmental and age-related to cardiovascular and neurological. Each portal is an entry point where investigators can access data and tools relevant to their research area. One can access rat, mouse, and human genes and QTLs, and rat strains annotated to a selected disease category or subcategory (*see* Fig. 3A). Annotations for a disease-related phenotype, biological process, or

Fig. 3 Hematologic Disease Portal Home Page. (*A*) Drop down menus for selection of disease category and specific disease. (*B*) Numerical summary of results for the selected disease category/disease. (*C*) GViewer display of results with approximate positions of all disease genes, QTLs, and strains. (*D*) Lists of genes, QTLs, and strains annotated to selected disease category/disease. (*E*) Graphs showing Gene Ontology annotations for all selected disease-annotated genes, using GO slim (subset) representations of the three GO aspects

pathway can also be accessed through a tab selected at the top of the portal homepage. A summary box of the number of objects annotated to the selected category is shown (*see* Fig. 3B). These objects are presented in a genomic context via an instance of the Genome Viewer below the summary box. The Genome Viewer can be set to rat, mouse, or human, and to synteny views of the unselected species. Beneath the Genome Viewer, the genes, QTLs, and strains

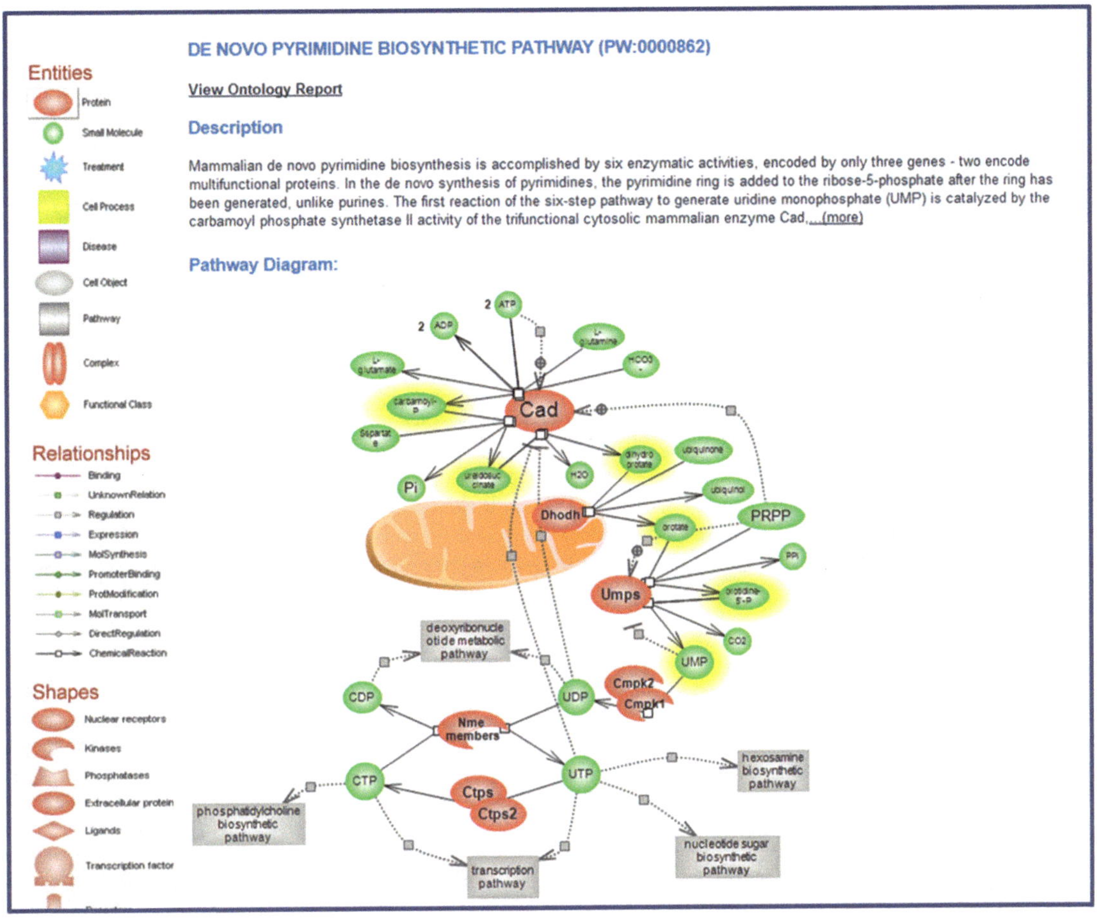

Fig. 4 De Novo Pyrimidine Biosynthetic Pathway Diagram. The diagram is accompanied by a text description above it and a key to the left of it

are listed, linked to their respective report pages. The bottom of the page shows graphs displaying GO annotation enrichment data.

Pathway Portal

The RGD Pathway Portal presently contains 200 interactive pathway diagram pages organized into five branches, based on the five branches of the Pathway Ontology, which was developed at RGD. Some pathway pages are organized into suites of related pathways, and suite networks—higher order organizations of suites. The molecular pathway diagrams (*see* Fig. 4) are designed with Elsevier's Pathway Studio software (https://support.pathwaystudio.com/) and feature hyperlinks from most of the objects in the diagram to RGD pages representing the respective term, gene, chemical, or associated secondary pathway. Beneath the diagram is a downloadable list of genes in the pathway (*see* Fig. 5A), with tabs for rat, human, mouse, and other species. Below the gene lists are tables of additional elements in the pathway (*see* Fig. 5B), disease

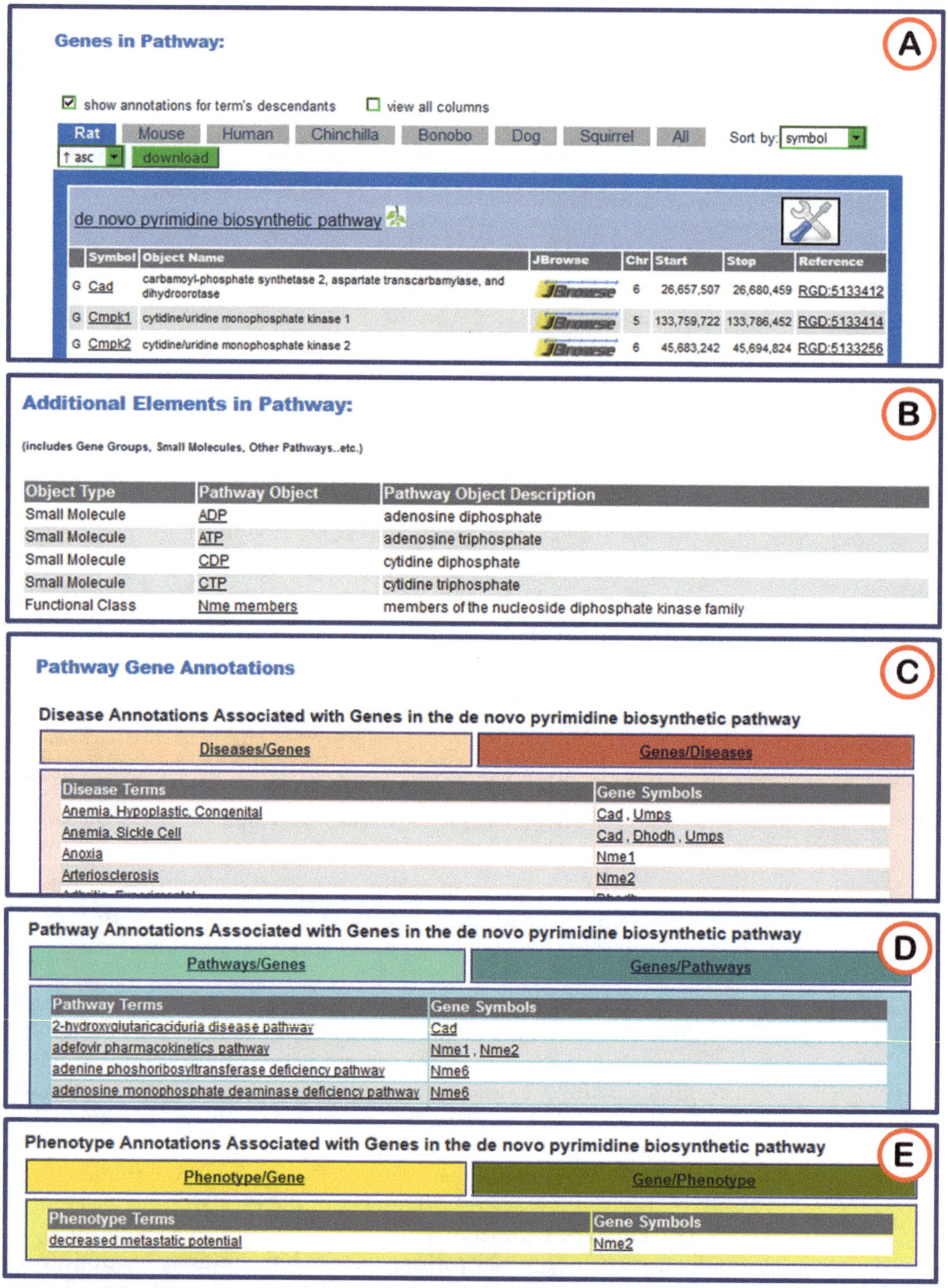

Fig. 5 Pathway Gene/Element Lists. A number of gene lists are found on pathway diagram pages below the diagram. (*A*) A list of genes annotated to de novo pyrimidine biosynthetic pathway and its children terms. The

annotations to genes in the pathway (*see* Fig. 5C), additional pathway annotations to genes in the diagrammed pathway (*see* Fig. 5D), and, when available, phenotype annotations to the genes in the pathway (*see* Fig. 5E). These tables toggle from annotation/gene to gene/annotation displays, with all objects linked to report pages. Below the gene lists there is a reference list of publications associated with the diagrammed pathway. Lastly, below the references is an ontology graph that shows the diagrammed term and all its ancestor terms up to the root term.

Phenotypes & Models Portal

This portal contains data related to rat strains and phenotypes, as well as essential information for conducting physiological research, identifying disease models, and comparative analysis of strain-centered data. Comparative analysis is the main focus of the PhenoMiner tool, which is described below.

3.2.3 RGD Data Analysis and Visualization Tools

Overview

Some of the data analysis tools at RGD are database-specific instances of freely available software. These include JBrowse, RatMine, and InterViewer (Cytoscape). The remaining analysis/visualization tools described in this section were developed at RGD: Gene Annotator, GViewer (Genome Viewer), OLGA (Object List Generator & Analyzer), PhenoMiner, and Variant Visualizer. They all provide different views or different types of analysis of the data in RGD. All of the tools may be accessed by the "Analysis & Visualization" icon in the middle of the RGD home page or the tab near the top of most RGD pages.

InterViewer

InterViewer, RGD's Cytoscape-based (https://www.cytoscape.org/) [22] protein-protein interaction viewer, takes one or more gene symbols, RGD gene IDs, or UniProtKB protein IDs for rat, mouse, human, and/or dog (*see* Fig. 6) and displays pairwise protein interactions for them, with information about the types of interactions and links to the associated genes in RGD, proteins in UniProt (http://www.uniprot.org/), and the originating interaction records at IMEX (International Molecular Exchange (IMEx) consortium) [23, 24].

JBrowse

The JBrowse genome browser [25, 26] from the Generic Model Organism Database project (http://www.gmod.org) is an interactive tool which allows researchers to visualize a variety of genetic and phenotypic data types in their genomic context. Virtually all of

Fig. 5 (continued) list includes links to RGD gene report pages, JBrowse, and reference pages. (*B*) A list of additional elements in the pathway. (*C*) A list of disease ontology terms/genes that can be toggled by the title bar to genes/disease terms. All the disease terms link to ontology report pages and the gene symbols link to gene report pages. (*D*) A list of additional pathways associated with genes annotated to the diagrammed pathway. (*E*) A list of phenotypes associated with the genes annotated to the diagrammed pathway

Fig. 6 Interviewer Search/Results. The target protein (rat Grb2) that initiated the search is shown in the center of the graphic display. Individual proteins are indicated by color-coded circles (red—rat, green—mouse, blue—human). The types of interactions are designated by color-coded lines between the circles

the data within the Rat Genome Database has been associated with the genome sequence in one way or another. As fundamental datasets such as genes, quantitative trait loci, microsatellite and SNP markers, and sequence resources such as ESTs, are aligned with the genome sequence, they bring with them phenotypic and other information. This information includes gene-chemical interaction data, genetic associations with disease, RNA-Seq data, synteny views of rat, mouse, and human genomes, and many types of variant/mutation data. Any or all of these can be accessed via the JBrowse genome browser and their relationship to the genomic sequence explored.

RatMine

RatMine integrates data on function, disease, phenotype, variation, and comparative genomics from RGD, UniProtKB, Ensembl (https://www.ensembl.org), NCBI (https://www.ncbi.nlm.nih.gov/), PubMed (https://www.ncbi.nlm.nih.gov/pubmed), and KEGG (https://www.genome.jp/kegg/) to form a web-based data warehousing, mining, and analysis tool tailored to the needs of rat researchers. Datasets derived from querying this data or from uploading researchers' own data can be saved, manipulated, and/or downloaded for use in other applications.

RatMine also has interaction datasets imported from BioGrid (Biological General Repository for Interaction Datasets) (https://thebiogrid.org) [27] and IntAct (https://www.ebi.ac.uk/intact) [23]. The BioGRID database manually curates the biomedical literature for genetic, protein, and chemical interaction data for major model organisms and humans. IntAct is a molecular interaction database that provides data derived from literature curation or direct user submissions to IntAct.

A key component of RatMine and of InterMine instances in general is the "MyMine" feature. Logging in as a specific user allows one to keep object lists (genes, etc.), user-created queries, and a history of activity. An API (application program interface) allows queries to run in RatMine from various web-based programs (Perl, Python, Ruby, or Java).

Gene Annotator

The Gene Annotator (GA) takes a list of gene symbols, RGD IDs, GenBank accession numbers, Ensembl identifiers, and/or a chromosomal region, and retrieves annotation data from RGD. The tool will retrieve annotations from most ontologies used at RGD for genes and their orthologs, as well as links to additional information at other databases. The entry page (*see* Fig. 7A) is very similar to the InterViewer entry page.

The first GA page after a search is an annotation/external link/species selection page where everything is selected by default (*see* Fig. 7B). Clicking the submit button returns a page with all annotations for the first gene (and selected orthologs) in the list. The lists include links to RGD gene pages, ontology term pages, annotation pages, and external data pages (*see* Fig. 7C).

A list of links at the top of the page allows the user to pick a particular type of analysis to view (Annotation Distribution or Comparison Heat Map) or to send the gene list to another tool by selecting the "All Analysis Tools" link.

On the "Annotation Distribution" page (*see* Fig. 7D) there are enrichment lists of terms by category, which rank the terms according to how many of the searched genes are annotated to those particular terms. Each entry in the list can be opened to see which genes and which specific terms are in the annotations. Subsets of annotations can be displayed by selecting at least two of the check boxes which appear to the right of every term in the lists.

Fig. 7 The Gene Annotator (GA) Tool. (*A*) The start page of the GA tool with large text box for entry of gene lists or object identifiers. (*B*) The results selection page where category of ontology annotations, external links, and orthologs may be chosen. (*C*) Results page listing annotations and external links for the rat, human, and mouse A2M genes. (*D*) Annotation distribution shows what percentage of the gene list is annotated to lists of terms in various ontologies/vocabularies. (*E*) The Comparison Heat Map shows an interactive matrix comparing numbers of annotated genes between terms of the disease vocabulary and the pathway ontology

The "Comparison Heat Map" (*see* Fig. 7E) compares the number of genes annotated to selected terms of two different ontologies by displaying how many genes are annotated to specific terms in one ontology while also being annotated to the selection of terms in the other ontology. The default view shows annotations to disease terms versus annotations to pathway terms. The dropdown menus on the left side of the page allow the ontology to be changed on either axis of the heat map. Also, clicking on any of the hyperlinked terms on either axis of the heat map will reset that respective axis to include just child terms of the selected term. By clicking on any non-zero numbered square in the heat map, a list is returned of all genes annotated to both terms which cross at that spot in the matrix.

To download annotations retrieved by the tool at any stage in viewing results, the "This Gene" or "All Genes" links are available at the upper right corner of the results pages.

GViewer

The Genome Viewer (GViewer) provides users with a complete genome view of genes, QTLs, and congenic strains annotated to a molecular function, biological process, cellular component, phenotype, disease, or pathway. The tool will search for terms from the Gene Ontology, Mammalian Phenotype Ontology, RGD Disease Ontology, or Pathway Ontology. The search page for GViewer (*see* Fig. 8A) features an ontology selection section where any or all available ontologies may be chosen. Complex searches may be made by clicking "Add Search Term" on the right side of the page. The returned results include all annotations to the chosen term and its children terms.

The main feature of the GViewer results page is an ideogrammic view of chromosomes with genes, QTLs, and congenic strains marked by color-coded bars in their approximate positions according to genome coordinates (*see* Fig. 8B). This GViewer graphic is used on every ontology term report page to visualize annotated objects. Below the chromosome view in the GViewer tool is a list of returned objects (listed alphabetically under type of object) and the searched ontology term(s), all of which link to the appropriate RGD report page. Clicking on any chromosome opens the zoom pane (*see* Fig. 9) that is displayed beneath the chromosome view. The zoom pane features a horizontal view of the chosen locus with a closer view featuring labeled genes, QTLs, and congenic strains. Alternate views of the data and other options are available on the bottom menu of the chromosome display.

OLGA—Object List Generator & Analyzer

OLGA, the Object List Generator & Analyzer tool, is a list builder for rat, mouse, human, and other species genes, rat, mouse, or human QTLs, or rat strains, using any of a variety of querying options. RGD objects may be searched using ontologies (*see* Fig. 10) or genomic positions. Options for using the generated

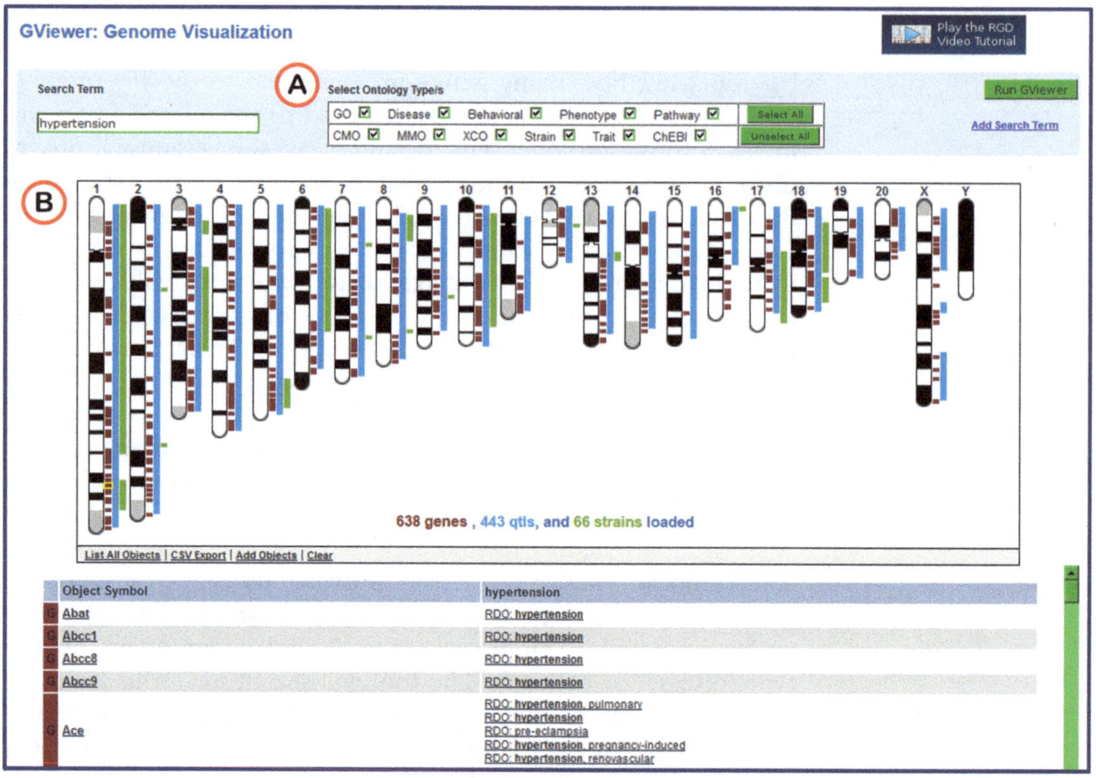

Fig. 8 GViewer Search Page. (*A*) The GViewer search page has one text box and optional additional text boxes ("Add Search Term" link) with multiple ontology check boxes to restrict the search. A link to a tutorial video is available on the upper right side of the page. (*B*) Ideogrammic view of chromosomes with returned genes, QTLs, and congenic strains marked by color-coded bars in their approximate positions according to genome coordinates

list are sending the list to a variety of RGD's analysis tools or downloading.

OLGA can generate multiple lists and integrate them in a number of ways. OLGA gives the user a choice of union, intersection, or subtraction for combining a second list with a previously generated list. After two or more lists have been integrated, the result list may be transferred to another RGD tool or downloaded.

PhenoMiner

The purpose of the PhenoMiner tool is to integrate phenotypic data from different rat strains, collected by a variety of measurement methods under various experimental conditions. The data in PhenoMiner comprises results from the PhysGen Program for Genomic Applications (http://pga.mcw.edu) [28], the National BioResource Project – Rat (http://www.anim.med.kyoto-u.ac.jp/nbr/) (*see* section above) [10], and manual annotation from the rat physiological literature.

The PhenoMiner start page (*see* Fig. 11) features a choice of rat strains, clinical measurements, measurement methods, or

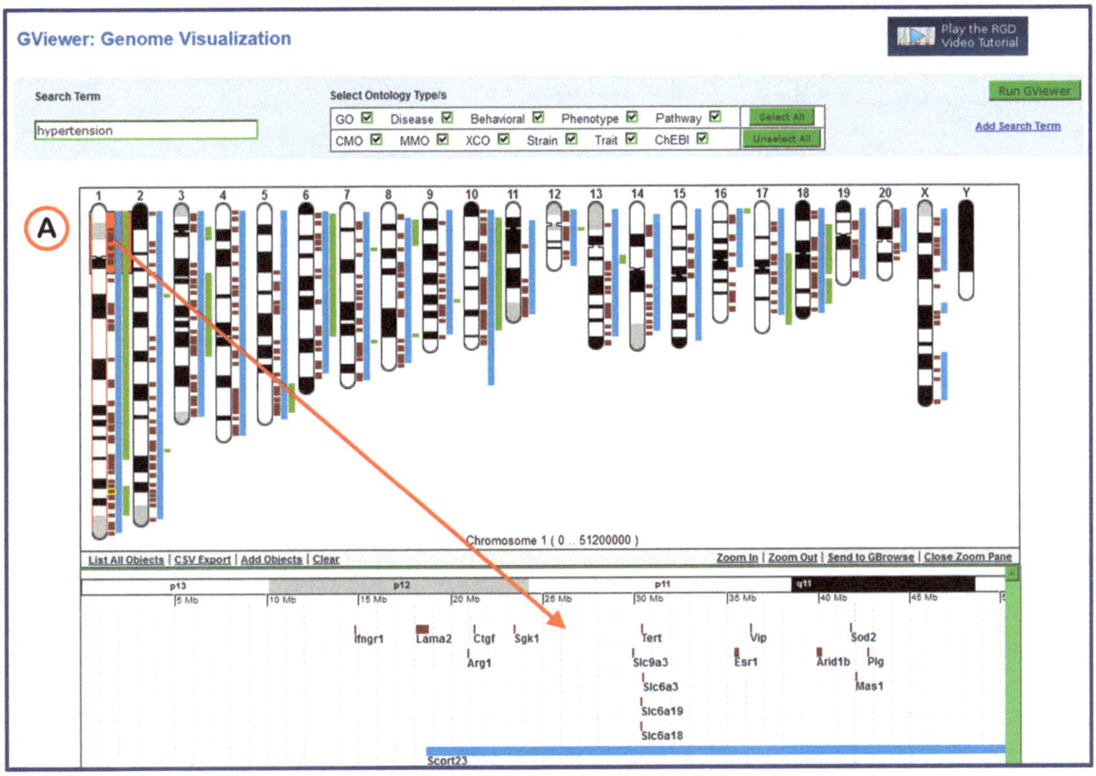

Fig. 9 GViewer Genome Visualization. The ideogrammic view of chromosomes with the location of returned genes marked in brown, QTLs in blue, and strains in green. (*A*) *A* Clicking on the end of chromosome 1 opens the zoom pane to show a closer view of genes and QTLs in the locus

experimental conditions to begin a search for quantitative phenotypic data. After a category and a specific selection are made, a tally of results is made (*see* Fig. 12A1) and the other three categories remain as options. Each consecutive selection limits the remaining selections based on what strains were measured for what parameter, by which method, and under what condition. A report may be generated at any point after at least one category has been selected.

All the data available for the selected parameters is displayed in a graph or made accessible on the PhenoMiner report page (*see* Fig. 12B). The presentation of the data can be manipulated by altering the choices in the drop-down menus or the check boxes at the top of the report page (*see* Fig. 12B1). The data is also presented in table form below the bar graph.

Variant Visualizer

Variant Visualizer is a viewing and analysis tool for rat strain-specific sequence polymorphisms and human ClinVar variants. The selection options include choices of rat strain or human assembly, type of variant, genomic feature, potential effect of variant upon protein

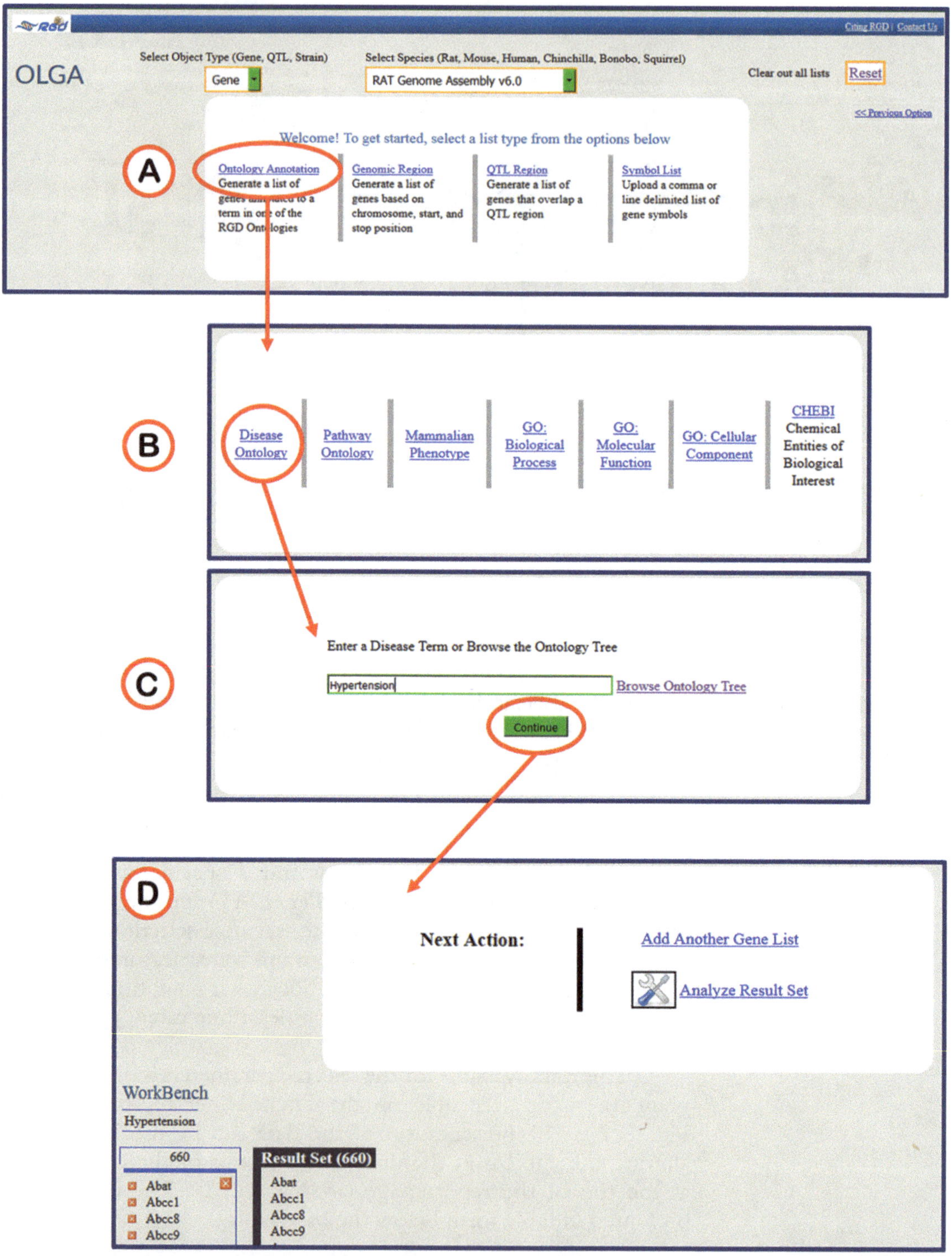

Fig. 10 Object List Generator & Analyzer (OLGA) tool. (*A*) OLGA selection screen for object, species/assembly, ontology, and genomic region. (*B*) Ontology/vocabulary selection screen. (*C*) Autocomplete text entry search box. (*D*) Results page with links to "Add Another Gene List" (which adds another *A*, *B*, and *C* to the search process) and the "Analyze Result Set" pop-up window

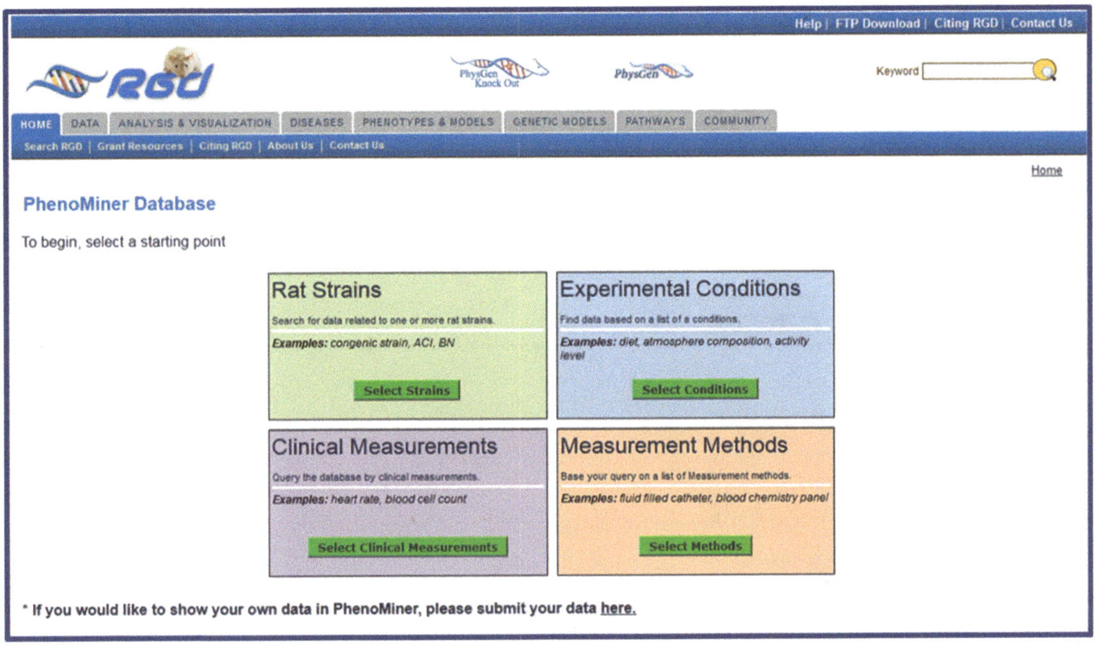

Fig. 11 PhenoMiner Start Page. Selection of "Rat Strains," "Experimental Conditions," "Clinical Measurements," or "Measurement Methods" begins the quantitative physiology data search process

product, and statistics involved in the variant calls. The graphic display of results shows a horizontal view of DNA sequence of the strain/assembly compared to the reference sequence (*see* Fig. 13A, B). Variants are identified by chromosome coordinate and base designation, and are visible in a scrollable display frame. From the sequence display page optional views and links to further analysis are available from a pop-up selection box accessed by a link in the upper right corner of the display page (*see* Fig. 13D).

VCMap

The Virtual Comparative Map (VCMap) tool was originally developed at RGD to explore the syntenic relationships between rat, mouse, and human genomes. A newer version of VCMap has been developed by a collaboration of Iowa State University, University of Iowa, Medical College of Wisconsin, and RGD. The current VCMap expands both the versatility and utility of this valuable tool by incorporating pig, chicken, cow, and horse genomes.

After an anchor species is chosen, additional species may be successively chosen for comparison. The chromosomes are displayed vertically, with the coordinates of the anchor chromosome increasing from top to bottom. The chromosome sections of the other species are arranged such that orthologs and syntenic regions line up, as much as possible, along the horizontal axis.

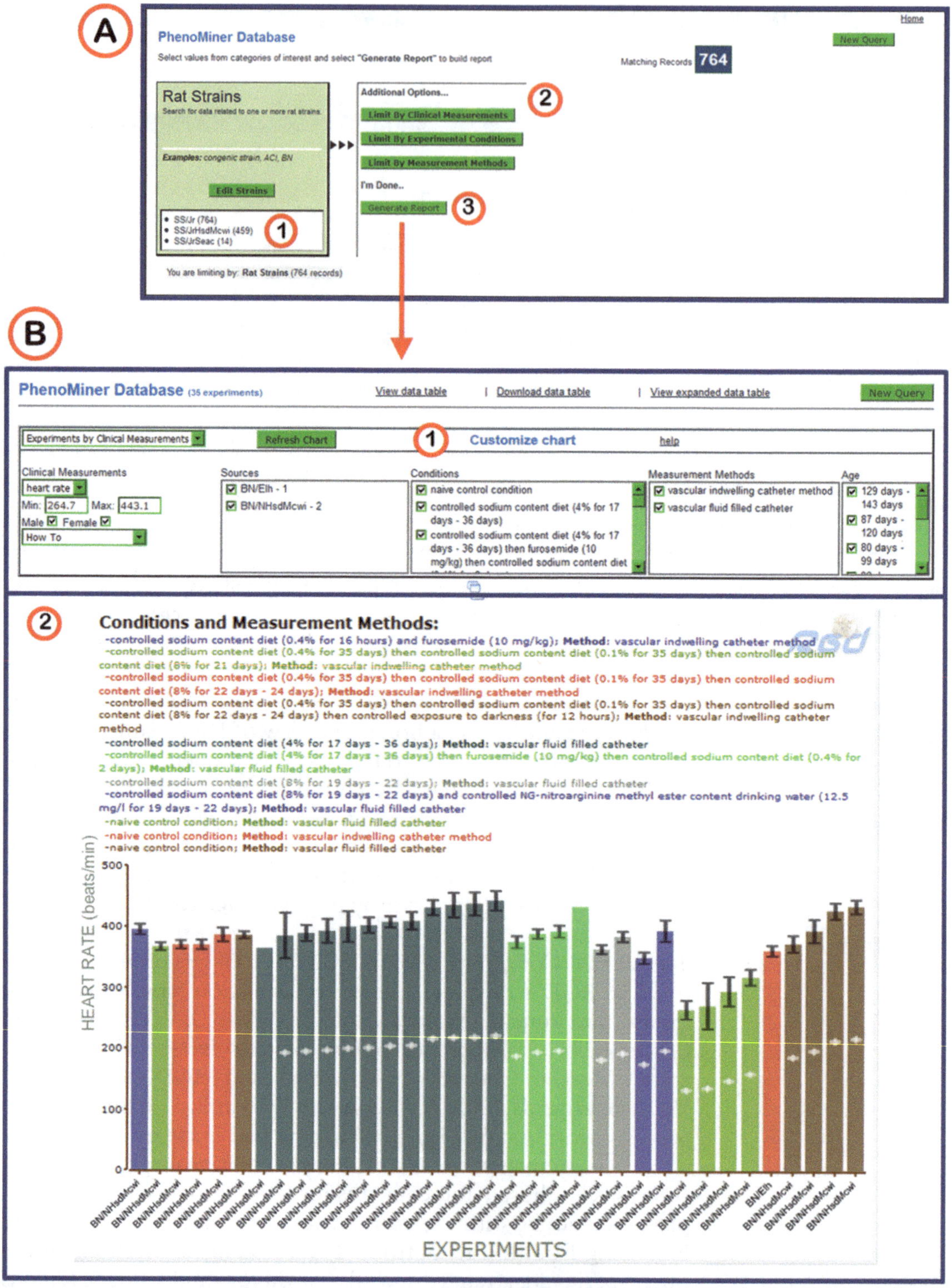

Fig. 12 PhenoMiner Selection Process and Report Page. (*A*) Intermediate selection page with scoreboard (*1*) of selected strains, more limiting options (*2*), and a "Generate Report" link (*3*). (*B*) PhenoMiner Report Page with customizable result options (*1*), featuring a color-coded bar graph (*2*) with all selected strains, methods, and conditions

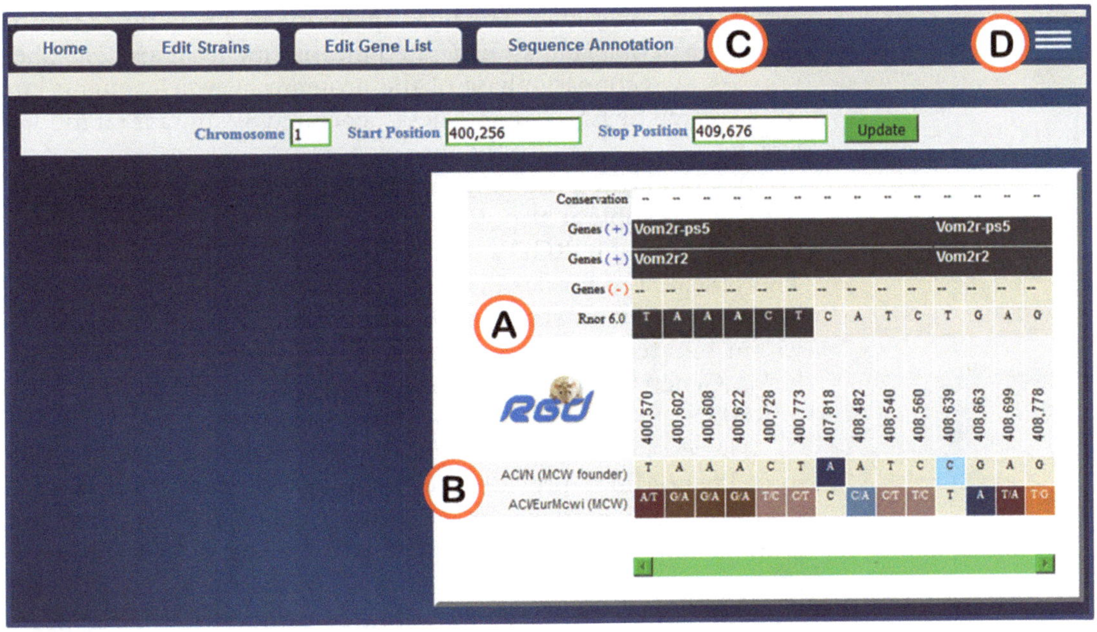

Fig. 13 Variant Visualizer Results Display. (*A*) The reference sequence of selected region. (*B*) Selected Strains. (*C*) Options of editing prior selections of strain, gene, annotation, or coordinates. (*D*) Pop-up links to more tool options

Automated Access to RGD Data

RGD has facilitated bulk download and other automated access to curated and other data. Data is accessible by both FTP download and REST API. The RGD FTP site (ftp://rgd.mcw.edu/pub/) maintains regularly updated files of all RGD data that can be downloaded and used in subsequent studies. These include the curated gene, QTL, strain and marker datasets, mapping information, genome annotation (in GFF format), sequence files for RGD data, and RGD-developed ontologies/vocabularies. The FTP site can be reached by clicking the "FTP Download" link found in the menu bar on the upper right of most RGD web pages.

3.3 PhysGen

The Programs for Genomic Applications (PGA) were designed by the National Heart, Lung and Blood Institute (NHLBI) to utilize the tools and information designed and generated by the Human Genome Project (HGP) for identification of genomic elements associated with cardiovascular, pulmonary, hematological, and sleep traits. The PhysGen program (Physiogenomics of Stressors in Derived Consomic Rats) at the Medical College of Wisconsin was one of eleven PGA projects at major research centers across the United States. In addition to phenotyping 11 inbred strains, the program generated and phenotyped reciprocal consomic rat strains using SS, BN, and FHH genetic backgrounds. The data generated through the phenotyping program was placed into a repository that has served as a freely accessible resource to the scientific community (http://pga.mcw.edu). The initial phenotypic screening of consomic and parental animals generated measurements across 213 phenotypes. A secondary phase of the program in which mutant rats were created by ENU mutagenesis generated 39 mutant rat models which were screened for 99 phenotypes.

Although the program officially ended in July 2009, the data remains accessible at http://pga.mcw.edu. The PhysGen data is also available through RGD's PhenoMiner tool as described above. This has allowed for flexible, user-defined queries of parental, consomic, and mutant strains with automated statistical comparisons within selected phenotypes, sexes, and environmental conditions [12]. Genetic information for each PGA strain can be found in RGD.

The evolution of the PhysGen dataset and its incorporation into PhenoMiner provide the scientific community a valuable resource for identifying quantitative phenotype ranges from an array of sequenced inbred and mutant rat strains. The utility of this information maintains relevance as investigators seek to identify refined disease models to capture pathophysiological traits observed across human populations.

3.4 The Virtual Physiological Rat Project

The Virtual Physiological Rat Project (VPR) (http://www.virtualrat.org) [29] focuses on a systems biology approach to study physiological and pathophysiological phenotypes emerging from various levels of interactions and manifesting at the organismal level. Its research areas are Cardiovascular Systems Dynamics, Hypertensive Heart Disease, Exercise & Metabolism, and Systems Pharmacology. These research projects are supported by two research cores. First, the Physiome Knowledgebase Dissemination core provides services for annotating, archiving, and distributing models and data within the VPR and larger scientific community. Second, the Animal Phenotyping and Functional Assessment core provides cohesive and easy accessible phenotyping service to all project investigators. All the computational resources (software and models) used and produced by VPR can be accessed from the VPR web site (http://virtualrat.org/resources).

3.5 Rat BodyMap

The Rat BodyMap (http://pgx.fudan.edu.cn/ratbodymap/index.html) [30] is an RNA-Seq expression data resource that includes data sets from 320 individual RNA samples derived from Fischer 344 rats. Samples were taken from eleven organs at four developmental stages for both sexes. Each sample was aligned to the genome and was annotated with its specific expression patterns by comparing among organs and developmental stages. The menu bar across the home page provides several convenient ways to use the resources. The Browse function allows viewing whole data sets or through filters such as gene symbols, genome positions, or expression patterns. The Search function allows retrieval of the expression data of a single gene or a defined chromosome region. The retrieved data can be filtered by gene symbols and expression patterns as in the Browse function. The complete database can be downloaded in full or in parts.

4 General Resources with Rat Data

Many genomic and proteomic databases have data on a large number of species. Rat is always included in these databases as a model organism for human physiology and disease. The following is a sample of the more important general databases for rat data.

4.1 Alliance of Genome Resources

The Alliance of Genome Resources (AGR) (https://www.alliancegenome.org/) is a group of six model organism databases (FlyBase, Mouse Genome Database (MGD), Saccharomyces Genome Database (SGD), Rat Genome Database (RGD), WormBase, and the Zebrafish Information Network (ZFIN)) and the Gene Ontology Consortium (GOC) which was formed in 2016 to create a collaborative, sustainable genome information resource that facilitates the use of diverse model organisms in understanding the genetic and genomic basis of human biology and disease. Currently, the AGR web site provides (1) gene pages with a sequence feature viewer, GO annotations, orthology designations, and disease associations (using Disease Ontology terms) and (2) disease pages with definition, synonyms, cross-references, and gene associations. The gene pages link to internal ortholog and disease pages, originating database sources, and literature references. The disease pages make reciprocal links to internal gene pages, originating database sources, and literature references. Future plans include the addition of other data objects such as alleles, genotypes, and strains, annotated to disease and phenotype terms.

4.2 Ensembl—European Bioinformatics Institute (EMBL-EBI) and Welcome Trust Sanger Institute

The Ensembl database (http://www.ensembl.org) and genome browser have genomic data for dozens of vertebrate species, including rat. The Ensembl database includes tools, APIs, and web interfaces for querying the data for access and analysis [31]. The rat gene set includes manual annotation from HAVANA (Human And Vertebrate Analysis aNd Annotation) [32].

4.3 Gene Ontology Consortium (GOC)

The GOC web site (http://www.geneontology.org) is a bioinformatics resource that provides data about gene products using ontologies to characterize biological knowledge. The GOC web site provides a term browser to find ontology terms across the three aspects of GO (cellular component, biological process, molecular function) and data across many model organisms, including rat. The majority of the rat annotations at the GOC web site originate at RGD, with additional annotations provided by UniProt-Gene Ontology Annotation (UniProt-GOA) and others.

4.4 Kyoto Encyclopedia of Genes and Genomes (KEGG)

KEGG is a bioinformatics resource that provides genomic, chemical, and network information. The genomic information covers genes, ligands, and orthology. The information is accessed via many different entry points including species, disease, and drugs. The database covers many organisms, with over 50 mammals including rat [33].

4.5 National Center for Biotechnology Information (NCBI)

NCBI (https://www.ncbi.nlm.nih.gov/) represents a large collection of databases and software tools, which provide information and data analysis across many areas of biomedical research and species, including rat. With everything from GenBank to PubChem to Taxonomy, NCBI covers much biological and chemical information related to rat research.

4.6 Phenogen

PhenoGen (https://phenogen.ucdenver.edu/PhenoGen/) [34] is an informatics web site that provides tools and data sets to explore DNA variants, RNA expression, and QTLs. It currently has DNA microarray data, markers, and expression QTLs for rat and mouse, RNA-Seq and strain-specific genomic data for rat, and some human genotyping data for download.

The DNA microarray expression output includes data from brain, heart, liver, and brown adipose tissue of HXB/BXH recombinant inbred rats. The RNA-Seq data is from brain, liver and heart of various rat strains. Both protein-coding transcripts and non-coding RNAs are included in the data sets. The rat strain-specific genomic sequences are mapped to the Rn5 and/or Rn6 reference assembly(ies). The Genome/Transcriptome Data Browser is the point of entry for data visualization. The tool provides customizable tracks that can display DNA and RNA data from several public repositories, including Ensembl transcripts, RefSeq transcripts, and QTLs from RGD.

4.7 Reactome

The Reactome Knowledgebase (https://reactome.org) provides molecular details of numerous cellular processes as networks of molecular interactions. Manual annotation of gene products for Reactome pathways is made for human. Homologous pathways in multiple other species (including rat) are presented based on orthology of interacting components. Reactome serves as both a database of interactions and a resource for data analysis.

4.8 UCSC Genome Browser

The UCSC Genome Browser (https://genome.ucsc.edu) provides a platform for investigating annotated genome assemblies of many different species, including all of the different genome assemblies for rat. In addition to basic gene/transcript annotation, the UCSC browser provides data tracks for interactions, expression, and CRISPR/Cas9 binding sites to allow for more complex analysis.

4.9 UniProt

UniProt (https://www.uniprot.org) is a large database of protein sequence data and associated functional annotation. UniProt contains mostly automated annotation (UniProtKB/TrEMBL), but also includes manual annotation (UniProtKB/Swiss-Prot). The number of reference eukaryotic proteomes represented in UniProt is in excess of 1000. Data is available by browsing the web site or using the REST API or SPARQL endpoint accessible via the UniProt web site.

Acknowledgements

RGD is supported by the National Heart, Lung, and Blood Institute on behalf of the National Institutes of Health [HL64541].

References

1. Worthey EA, Stoddard AJ, Jacob HJ (2010) Sequencing of the rat genome and databases. Methods Mol Biol 597:33–53. https://doi.org/10.1007/978-1-60327-389-3_3
2. Richter CP (1954) The effects of domestication and selection on the behavior of the Norway rat. J Natl Cancer Inst 15(3):727–738
3. Atanur SS, Diaz AG, Maratou K, Sarkis A, Rotival M, Game L et al (2013) Genome sequencing reveals loci under artificial selection that underlie disease phenotypes in the laboratory rat. Cell 154(3):691–703. https://doi.org/10.1016/j.cell.2013.06.040
4. Baud A, Hermsen R, Guryev V, Stridh P, Graham D, McBride MW et al (2013) Combined sequence-based and genetic mapping analysis of complex traits in outbred rats. Nat Genet 45(7):767–775. https://doi.org/10.1038/ng.2644
5. Li D, Qiu Z, Shao Y, Chen Y, Guan Y, Liu M et al (2013) Heritable gene targeting in the mouse and rat using a CRISPR-Cas system. Nat Biotechnol 31(8):681–683. https://doi.org/10.1038/nbt.2661
6. Geurts AM, Moreno C (2010) Zinc-finger nucleases: new strategies to target the rat genome. Clin Sci (Lond) 119(8):303–311. https://doi.org/10.1042/cs20100201
7. Tong C, Huang G, Ashton C, Wu H, Yan H, Ying QL (2012) Rapid and cost-effective gene targeting in rat embryonic stem cells by TALENs. J Genet Genomics 39(6):275–280. https://doi.org/10.1016/j.jgg.2012.04.004
8. Carlson DF, Geurts AM, Garbe JR, Park CW, Rangel-Filho A, O'Grady SM et al (2011) Efficient mammalian germline transgenesis by cis-enhanced Sleeping Beauty transposition. Transgenic Res 20(1):29–45. https://doi.org/10.1007/s11248-010-9386-5
9. Menoret S, Fontaniere S, Jantz D, Tesson L, Thinard R, Remy S et al (2013) Generation of Rag1-knockout immunodeficient rats and mice using engineered meganucleases. FASEB J 27(2):703–711. https://doi.org/10.1096/fj.12-219907
10. Serikawa T, Mashimo T, Takizawa A, Okajima R, Maedomari N, Kumafuji K et al (2009) National BioResource Project-Rat and related activities. Exp Anim 58(4):333–341
11. Mashimo T, Voigt B, Kuramoto T, Serikawa T (2005) Rat Phenome Project: the untapped potential of existing rat strains. J Appl Physiol 98(1):371–379. https://doi.org/10.1152/japplphysiol.01006.2004
12. Laulederkind SJ, Liu W, Smith JR, Hayman GT, Wang SJ, Nigam R et al (2013) PhenoMiner: quantitative phenotype curation at the rat genome database. Database 2013: bat015. https://doi.org/10.1093/database/bat015
13. Mashimo T, Voigt B, Tsurumi T, Naoi K, Nakanishi S, Yamasaki K et al (2006) A set of highly informative rat simple sequence length polymorphism (SSLP) markers and genetically defined rat strains. BMC Genet 7:19. https://doi.org/10.1186/1471-2156-7-19
14. Saar K, Beck A, Bihoreau MT, Birney E, Brocklebank D, Chen Y et al (2008) SNP and haplotype mapping for genetic analysis in the rat. Nat Genet 40(5):560–566. https://doi.org/10.1038/ng.124
15. Gibbs RA, Weinstock GM, Metzker ML, Muzny DM, Sodergren EJ, Scherer S et al (2004) Genome sequence of the Brown

Norway rat yields insights into mammalian evolution. Nature 428(6982):493–521. https://doi.org/10.1038/nature02426

16. Hermsen R, de Ligt J, Spee W, Blokzijl F, Schafer S, Adami E et al (2015) Genomic landscape of rat strain and substrain variation. BMC Genomics 16:357. https://doi.org/10.1186/s12864-015-1594-1
17. Ashburner M, Ball CA, Blake JA, Botstein D, Butler H, Cherry JM et al (2000) Gene ontology: tool for the unification of biology. The Gene Ontology Consortium. Nat Genet 25 (1):25–29. https://doi.org/10.1038/75556
18. Smith CL, Goldsmith CA, Eppig JT (2005) The Mammalian Phenotype Ontology as a tool for annotating, analyzing and comparing phenotypic information. Genome Biol 6(1): R7. https://doi.org/10.1186/gb-2004-6-1-r7
19. Petri V, Jayaraman P, Tutaj M, Hayman GT, Smith JR, De Pons J et al (2014) The pathway ontology – updates and applications. J Biomed Semantics 5(1):7. https://doi.org/10.1186/2041-1480-5-7
20. Laulederkind SJ, Tutaj M, Shimoyama M, Hayman GT, Lowry TF, Nigam R et al (2012) Ontology searching and browsing at the Rat Genome Database. Database 2012:bas016. https://doi.org/10.1093/database/bas016
21. Nigam R, Munzenmaier DH, Worthey EA, Dwinell MR, Shimoyama M, Jacob HJ (2013) Rat Strain Ontology: structured controlled vocabulary designed to facilitate access to strain data at RGD. J Biomed Semantics 4(1):36. https://doi.org/10.1186/2041-1480-4-36
22. Shannon P, Markiel A, Ozier O, Baliga NS, Wang JT, Ramage D et al (2003) Cytoscape: a software environment for integrated models of biomolecular interaction networks. Genome Res 13(11):2498–2504. https://doi.org/10.1101/gr.1239303
23. Orchard S, Ammari M, Aranda B, Breuza L, Briganti L, Broackes-Carter F et al (2014) The MIntAct project--IntAct as a common curation platform for 11 molecular interaction databases. Nucleic Acids Res 42(Database issue): D358–D363. https://doi.org/10.1093/nar/gkt1115
24. Orchard S, Kerrien S, Abbani S, Aranda B, Bhate J, Bidwell S et al (2012) Protein interaction data curation: the International Molecular Exchange (IMEx) consortium. Nat Methods 9 (4):345–350. https://doi.org/10.1038/nmeth.1931
25. Skinner ME, Holmes IH (2010) Setting up the JBrowse genome browser. Current protocols in bioinformatics/editoral board, Andreas D Baxevanis, et al. Chapter 9:Unit 9.13. https://doi.org/10.1002/0471250953.bi0913s32
26. Westesson O, Skinner M, Holmes I (2013) Visualizing next-generation sequencing data with JBrowse. Brief Bioinform 14 (2):172–177. https://doi.org/10.1093/bib/bbr078
27. Chatr-Aryamontri A, Oughtred R, Boucher L, Rust J, Chang C, Kolas NK et al (2017) The BioGRID interaction database: 2017 update. Nucleic Acids Res 45(D1):D369–d379. https://doi.org/10.1093/nar/gkw1102
28. Dwinell MR (2010) Online tools for understanding rat physiology. Brief Bioinform 11 (4):431–439. https://doi.org/10.1093/bib/bbp069
29. Beard DA, Neal ML, Tabesh-Saleki N, Thompson CT, Bassingthwaighte JB, Shimoyama M et al (2012) Multiscale modeling and data integration in the virtual physiological rat project. Ann Biomed Eng 40(11):2365–2378. https://doi.org/10.1007/s10439-012-0611-7
30. Yu Y, Fuscoe JC, Zhao C, Guo C, Jia M, Qing T et al (2014) A rat RNA-Seq transcriptomic BodyMap across 11 organs and 4 developmental stages. Nat Commun 5:3230. https://doi.org/10.1038/ncomms4230
31. Aken BL, Achuthan P, Akanni W, Amode MR, Bernsdorff F, Bhai J et al (2017) Ensembl 2017. Nucleic Acids Res 45(D1):D635–d642. https://doi.org/10.1093/nar/gkw1104
32. Harrow JL, Steward CA, Frankish A, Gilbert JG, Gonzalez JM, Loveland JE et al (2014) The Vertebrate Genome Annotation browser 10 years on. Nucleic Acids Res 42(Database issue):D771–D779. https://doi.org/10.1093/nar/gkt1241
33. Tanabe M, Kanehisa M (2012) Using the KEGG database resource. Current protocols in bioinformatics/editoral board, Andreas D Baxevanis. et al. Chapter 1:Unit1.12. https://doi.org/10.1002/0471250953.bi0112s38
34. Bhave SV, Hornbaker C, Phang TL, Saba L, Lapadat R, Kechris K et al (2007) The PhenoGen informatics website: tools for analyses of complex traits. BMC Genet 8:59. https://doi.org/10.1186/1471-2156-8-59

Chapter 4

Next Generation Transgenic Rat Model Production

Wanda E. Filipiak, Elizabeth D. Hughes, Galina B. Gavrilina, Anna K. LaForest, and Thomas L. Saunders

Abstract

The next generation of new genetically engineered rat models by microinjection is described. Genome editors such as CRISPR/Cas9 have greatly increased the efficiency with which the rat genome can be modified to generate research models for biomedical research. Pronuclear microinjection of transgene DNA into rat zygotes results in random multicopy transgene integration events that use exogenous promoters to drive expression. Best practices in transgenic animal design indicate the use of precise single copy transgene integration in the genome. This ideal can be achieved by repair of CRISPR/Cas9 chromosome breaks by homology directed repair. The most effective way to achieve this type of transgenic rat model is to deliver genome modification reagents to rat zygotes by pronuclear microinjection. The keys to success in this process are to obtain fertilized eggs (zygotes) from the rat strain of choice, to purify the microinjection reagents, to deliver the reagents to the eggs by pronuclear microinjection, to use the surgical transfer of microinjected eggs to pseudopregnant rats to obtain G0 founder animals that carry the novel genetic modification. Ultimately the success of new rat models is measured by changes in gene expression as in the expression of a new reporter protein such as eGFP, Cre recombinase, or other protein of interest.

Key words Transgenic rats, CRISPR/Cas9, Gene editing, DNA, Transgenes, Gene knockout techniques

1 Introduction

Rat models of human disease are actively investigated to help understand the basis of human health [1–3]. Genetically engineered rats have proven to be valuable research tools to understand disease processes [4–6]. Traditional methods of transgenic rat production rely on the pronuclear microinjection of DNA and random integration of transgene DNA in the genome [7, 8]. In this type of experiment, the transgene DNA is designed to overexpress proteins under the control of a tissue specific promoter [9, 10]. The random integration of DNA transgenes in the genome results in damage to the chromosome at the integration site and also fragmentation of the transgene, producing a structure that resembles chromothripsis [11, 12]. Thus, each transgenic founder bears a unique transgene

G. Thomas Hayman et al. (eds.), *Rat Genomics*, Methods in Molecular Biology, vol. 2018,
https://doi.org/10.1007/978-1-4939-9581-3_4, © Springer Science+Business Media, LLC, part of Springer Nature 2019

integration site and a unique multicopy transgene array. The genomic region of the integration site can influence the expression pattern of the transgene, and transgene expression does not correlate with copy number unless each inserted gene includes a large (100 kb or more) region of genomic DNA to direct its expression [13–15]. Because of these limitations, it has been recognized that the best practices in transgenic animal model construction call for single copy transgene integration in a specific targeted site in the genome [16].

The use of CRISPR/Cas9 technology to direct site-specific single copy transgene integration is revolutionizing transgenic rat model designs. The basic principle is that Cas9-specific chromosome breaks stimulate the cellular DNA repair machinery in rat zygotes to insert transgenes in specific locations by homology directed repair. The requirements for this application are to identify a specific sgRNA that will induce a chromosome break in the desired location and then to provide the zygote with a template for HDR that will insert the transgene precisely. This approach to transgenic rats was first demonstrated to be feasible with zinc finger nuclease technology [17]. Subsequent work demonstrated that CRISPR/Cas9 could also be used to achieve similar results [18, 19, unpublished results]. The development of this category of transgenic rat model requires the use of pronuclear microinjection.

The principle alternative method to microinjection for delivery of CRISPR/Cas9 reagents to rat zygotes is electroporation. This method has been demonstrated to successfully deliver reagents that produce gene knockouts and small knockins such as point mutations or peptide tags [20, 21]. Electroporation offers important advantages over microinjection such as the substitution of an electroporator for a micromanipulation workstation. Much less training and practice is needed to master the electroporation method than to acquire skill in micromanipulator operation. Electroporation also results in higher viability and birth rates because of the absence of physical damage due to microinjection needle insertion. Importantly, attempts to introduce large transgenes by electroporation have not been successful in rat zygotes [21], thus microinjection remains a vitally important technique for the generation of single copy transgene integrations in defined sites in the rat genome.

We describe the process of microinjecting rat zygotes with transgene DNA and with CRISPR/Cas9 reagents, with or without DNA donors, for the purpose of generating transgenic rat models. To be successful each step in the process is optimized because each one compounds the efficiency of subsequent steps. Suboptimal performance in any one stage significantly reduces the efficiency of the overall process. The first step is to obtain highly purified reagents for microinjection, including the microinjection buffer. If desired, these reagents can be first microinjected into rat or mouse zygotes to test for overt toxicity evidenced by the failure of the zygotes to develop normally in vitro or by outright death of the zygotes. We routinely screen all CRISPR/Cas9 reagents in this way and have observed toxicity in some preparations. Concentrations of microinjected materials need to be monitored. High concentrations of linear DNA transgenes are toxic to developing zygotes and will result in extremely low birth rates [22, 23]. The size and shape of microinjection needles is important for effective penetration of zygotes without causing lysis of the cell. A rat colony that includes stud males, egg donors, and pseudopregnant recipients needs to provide plentiful zygotes for microinjection and numerous pseudopregnant recipients. Proper rat husbandry practices and diet are important for successful pregnancies and high reproductive performance. Superovulation treatments with hormone preparations that are quality controlled are essential for high responses to treatments. Incubators and zygote culture medium that will support the normal development of rat zygotes are vital. Microscopes, micromanipulation equipment and skilled operators need to perform at a high level for high productivity. Under optimal conditions the efficiency of transgenic rat production is equivalent to that of mouse transgenic production.

2 Materials

2.1 Buffers and Solutions

Agarose, Low Electroendosmosis Agarose Gel Loading Buffer.

Agarose Gel Running Buffer: 10× TBE: 1 M Tris base, 1 M Boric acid, 20 mM EDTA.

Ethidium Bromide: 10 mg/mL in water.

Approved Anesthetic Agent.

Approved Analgesic Agent.

eSpCas9(1.1) Protein: Sigma-Aldrich cat. no. ESPCASPRO.

sgRNA or crRNA/tracrRNA Duplex (*see* **Note 1**).

Embryo Tested Mineral Oil.

Epinephrine Injection, USP: final concentration in sterile PBS 0.01 mg/mL.

M2 Medium.

KSOMaa Medium.

Bovine Hyaluronidase Stock: 10 mg/mL in sterile water.

Hyaluronidase Working Solution: 1 mg/mL in M2 Medium.

Pregnant Mare's Serum Gonadotropin (PMSG): 50 IU/mL in PBS.

Human Pregnancy Urine Chorionic Gonadotropin (HCG): 50 IU/mL in PBS.

LHRH Agonist: [des-Gly10,D-Ala6]-LH-RH ethylamide (LHRHa) 200 μg/mL in PBS.

Microinjection Buffer: 10 mM Tris–HCl, pH 7.4, 0.1 mM EDTA, Anotop filtered.

Prepare 1 L of buffer with molecular biology grade water, 1 M Tris–HCl, pH 7.4, 0.5 M.

EDTA stock solutions. Use disposable sterile tissue culture pipets to add stock solutions to 1 L disposable sterile tissue culture plastic bottle. Filter 500 mL through a 500 mL filter unit to wash membrane. Discard filtrate. Filter remaining 500 mL of microinjection buffer and place in disposable sterile plastic tissue culture bottle. Prepare 1 mL aliquots of 0.02 μM Anotop filtered microinjection buffer in microtubes.

RNAse-Free Microinjection Buffer: prepare the buffer with RNAse-free stock reagents:

Para lube Ophthalmic Ointment.

Phosphate Buffered Saline (PBS).

Restriction Enzyme Digested Transgene DNA.

Sigmacote: Sigma-Aldrich cat. no. SL2.

2.2 Equipment

4-0 Suture.

Autoclip Wound Clip Applier.

Betadine Solution Swabsticks.

Cauterizing Tool: Bovie high temperature cauteries.

Dumont #5 Forceps.

Bulldog Serrefine Clamp.

Iris Curved Dissection Scissors.

Oster Animal Clipper with size 40 blade set or razor.

Slide Warmer or other gentle heat source.

Size 1 Monel Ear Tags and Applicator: National Band and Tag.

Tissue Forceps.

1.5 mL Sterile, RNAse-Free Microcentrifuge Tubes.

2.0 mL Sterile, RNAse-Free Microcentrifuge Tubes.

50 °C Water Bath.

Centrifuge for 1.5 and 2.0 mL microtubes.

Anotop Syringe Filters: 0.02 μM pore size.

4 mm Syringe Filters: 0.2 μM pore size, cellulose acetate membrane.

Disposable vacuum filter units: 500 mL, surfactant-free cellulose acetate membrane, 0.2 μM pore size.

Embryo Transfer Pipets: Origio.com cat. no. MXL3-175.

Insulin Syringes: 1 mL for rat injections.

Dow Corning High Vacuum Grease.

Glass Capillaries with Filament: World Precision Instruments. Cat. No. TW100F-4.

Glass Cover Slip: 9 × 22 mm: Electron Microscopy Sciences Cat. No. 72190-22.

Glass Microscope Slides.

Holding Pipets: BioMedical Instruments or Origio or other supplier.

Horizontal agarose gel electrophoresis apparatus.

Luer Tip Syringes: 3 mL for use with Anotop syringes.

Mouth Pipettor: Sigma –Aldrich Aspirator tube assembly.

NucleoSpin Extract II Kit.

Pipettor such as Gilson Pipetman.

Pipettor Tips, RNAse-Free, Sterile.

Plexiglas Spacer: 3 mm × 3 mm × 10 mm or glass blocks.

Petri Dish: 35 mm petri dish Falcon 353,001.

Petri Dish: 60 mm petri dish (2) Falcon 351,007.

Petri Dish: 15 cm sterile tissue culture petri dishes.

Razor Blades.

Transilluminator.

Millipore dialysis filter (Millipore #VMWP02500, pore size 0.05 μm).

2.3 Animals

Animal Room: 0600–1800 lights on.

Female Rats: Egg donors 4–5 weeks of age.

Female Rats: Pseudopregnant recipients 10–12 weeks of age.

Male Rats: Intact Studs 10 weeks or more housed one per cage.

Male Rats: Vasectomized Studs 10 weeks or more housed one per cage.

Love Mash Rodent Reproductive Diet: Bio-Serv.com

Rat Diet: including 6–6.5% crude fat content.

2.4 Micromanipulation Work Station

De Fonbrune Type Microforge or Narishige MF-900 Microforge.

Horizontal Micropipette Puller such as Sutter P-97.

Inverted Brightfield Tissue Culture Microscope: 4×, 10×, 20× objectives with 10× eyepieces.

Inverted Research Microscope: 4× brightfield, 10×, 20×, and 40× differential interference contrast objectives and 10× eyepieces.

Microinjector such as Eppendorf Femtojet, Tritech, picospritzer, or other.

Micromanipulators Mounted on the Research Microscope: such as Eppendorf Transferman or Narishige Three-axis Hanging Joystick Oil Hydraulic Micromanipulator or Leica Mechanical Micromanipulator.

Humidified Tissue Culture Incubator: 37 °C, 5% CO_2.

Microscope Vibration Isolation Platform: Vistek or other.

Piston Syringe Holding Pipette Controller such as Eppendorf Cell Tram Air or other.

Stereomicroscope with fiber optic ring illumination for embryo transfer surgery.

Stereomicroscope with transmitted illumination for embryo collection.

3 Methods

3.1 CRISPR/Cas9 Reagents for Microinjection

1. Resuspend Cas9 protein at 100 ng/μL in buffer supplied with lyophilized protein.
2. Resuspend synthetic sgRNA at 60 ng/μL in RNAse-free microinjection buffer (*see* **Note 2**). If using crRNA and tracrRNA, then mix together an equimolar concentration of crRNA and tracrRNA and anneal the crRNA and tracrRNA by incubating the mixture for 5 min at 95 °C and then placing the mixture on ice for 20 min.
3. Combine one volume of Cas9 protein and one volume of sgRNA. Incubate at room temperature for 20–30 min to allow complex formation, then place the mixture on ice until microinjection.
4. Microinjection final concentration is 50 ng/μL Cas9; 30 ng/μL sgRNA.

3.2 CRISPR/Cas9 Reagents with Oligonucleotide for Microinjection

1. Resuspend oligonucleotide in microinjection buffer.
2. Place 200 mL of microinjection buffer in a sterile 15 cm petri dish.
3. Carefully place a Millipore dialysis filter so that it floats onto the surface of the buffer (place the filter shiny side up). Use Millipore filter forceps to manipulate the filter.
4. Carefully spot the resuspended DNA solution on the center of the filter with a pipettor. Replace the Petri dish lid. Dialyze for 2–3 h (*see* **Note 3**).

5. After 2 h remove the solution with a pipettor. Place the pipet tip in the middle of the DNA droplet and carefully aspirate as much as possible without stopping. Transfer the DNA solution into a sterile microtube.
6. Measure the DNA concentration. Recoveries between 50% and 70% of the original volume are normal. The rest remains attached as a very thin liquid layer onto the surface of the filter and it is difficult to pipette it off.
7. Prior to microinjection prepare a tube containing 20 ng/μL of oligonucleotide DNA in RNAse-free microinjection buffer.
8. Resuspend Cas9 protein at 200 ng/μL in buffer supplied with lyophilized protein.
9. Resuspend synthetic sgRNA at 120 ng/μL in RNAse-free microinjection buffer. If using crRNA and tracrRNA, then anneal them as described in Subheading 3.1.
10. Combine one volume of Cas9 protein and one volume of sgRNA. Incubate at room temperature for 20–30 min to allow complex formation, then place the mixture on ice until microinjection.
11. Immediately before microinjection combine one volume of the oligonucleotide solution with one volume of the Cas9 ribonucleoprotein mixture.
12. Microinjection final concentration is 50 ng/μL Cas9, 30 ng/μ L sgRNA, 10 ng/μL DNA.

3.3 CRISPR/Cas9 Reagents with ds DNA Plasmid Donor for Microinjection

1. Use an endotoxin-free plasmid DNA kit to purify the ds plasmid DNA.
2. Resuspend the plasmid DNA in microinjection buffer instead of the provided kit buffer.
3. Measure the DNA concentration.
4. Prior to microinjection prepare a tube containing 20 ng/μL of ds DNA donor in RNAse-free microinjection buffer.
5. Resuspend Cas9 protein at 200 ng/μL in buffer supplied with lyophilized protein.
6. Resuspend synthetic sgRNA at 120 ng/μL in RNAse-free microinjection buffer. If using crRNA and tracrRNA, then anneal them as described in Subheading 3.1.
7. Combine one volume of Cas9 protein and one volume of sgRNA. Incubate at room temperature for 20–30 min to allow complex formation, then place the mixture on ice until microinjection.
8. Immediately before microinjection combine one volume of circular ds DNA plasmid solution with one volume of the Cas9 ribonucleoprotein mixture.

9. Microinjection final concentration is 50 ng/μL Cas9, 30 ng/μL sgRNA, 10 ng/μL DNA.

3.4 Microinjection DNA Purification

1. Perform restriction digest to liberate transgene from plasmid vector sequences. Final yield should be 50 μg of transgene insert.
2. Separate restriction digest products on 1% agarose gel made from TBE.
3. Place gel on transilluminator. Cut out band(s) of interest with a clean razor blade. Remove as much excess agarose as possible. Minimize DNA exposure to UV light to prevent photochemical damage (less than 1 min).
4. Transfer agarose slice(s) to a preweighed tube. Reweigh tube to determine weight of agarose in tube.
5. For each 100 mg of gel, add 200 μL Nucleospin buffer NT. If agarose concentration is greater than 1%, add proportionately more buffer. For example, if a 2% agarose gel is used, add 400 μL buffer NT for each 100 mg of gel.
6. Place at 50 °C for 10 min to dissolve agarose, vigorously vortexing every 2–3 min until the agarose is completely dissolved.
7. Place a NucleoSpin cartridge in a 2 mL microtube and load 750 μL dissolved gel slice onto the cartridge.
8. Spin at 11,000 × *g* for 60 s in a microcentrifuge.
9. Discard the flow-through. The cartridge has a capacity of 15 μg DNA, so you can run several 750 μL loads of dissolved gel slice through a single cartridge.
10. Add 750 μL of buffer NT3 to the cartridge and spin at 11,000 × *g* for 60 s in a microcentrifuge. Discard the flow-through.
11. Replace tube with a fresh microtube. Spin the empty cartridge at 11,000 × *g* for 60 s in a microcentrifuge to completely remove buffer NT3.
12. To elute the DNA from the cartridge, replace tube with a fresh microtube. Add 50 μL of preheated (60 °C) elution buffer to cartridge and incubate 1 min. Spin at 11,000 × *g* for 60 s in a microcentrifuge (*see* **Note 4**).
13. If desired, repeat **step 12** to increase yield. We find 90% of the DNA in the first elution.
14. Quantitate DNA solution.
15. Verify size and intact condition of DNA on mini-gel.
16. Store eluted DNA at −20 °C.

17. Prior to microinjection adjust the DNA concentration to 1 ng/μL with microinjection buffer.

3.5 Vasectomy

1. Adult males are anesthetized with an approved anesthetic (*see* **Note 5**).
2. Preoperative analgesia is administered and analgesia is provided for 48 h postoperatively (*see* **Note 6**).
3. Hair is removed from the surgical site by clipping (Oster animal clipper with size 40 blade set) or shaving with razors.
4. Skin is disinfected three times by swabbing with betadine alternated with 70% ethanol or 3–6% hydrogen peroxide.
5. Make a 1 to 1.5 cm transverse incision through the skin. Make a second incision through the peritoneal wall at the level of the knees.
6. The testicular fat pad is grasped with tissue forceps and both the fat pad and testes are exteriorized.
7. The vas deferens is grasped with tissue forceps and Dumont forceps are used to separate it from connective tissue by blunt dissection.
8. The vas deferens is raised so as to produce a loop. A cauterizing tool is used to cut the vas deferens twice and the loop of tissue is discarded.
9. The testis is returned to the abdominal cavity and the procedure is repeated on the vas deferens of the second testis.
10. The peritoneal wall is closed with two sutures.
11. The skin is closed with wound clips that are removed seven to ten days after surgery.
12. The animal is placed in a cage kept on a 37 °C slide warmer until alert and returned to sternal recumbency.

3.6 Estrus Synchronization and Superovulation

1. On Day 1: time 1100. Use insulin syringe to give a subcutaneous injection of 0.2 mL LHRHa (40 μg) to adult female rats used as pseudopregnant recipients.
2. On Day 3: time 1100. Use insulin syringe to give an intraperitoneal injection of 0.4 mL PMSG (20 IU) to each juvenile female egg donor.
3. On Day 5: time 1100. Use insulin syringe to give an intraperitoneal injection of 1.0 mL HCG (50 IU) to each juvenile female egg donor. Place one egg donor with one intact male. Usually five outbred (Sprague Dawley) or eight inbred (F344) females are superovulated for one microinjection session.
4. On Day 5: time 1500–1700. Place one LHRHa treated female with one vasectomized male. Usually six females are treated and mated for each microinjection session.

5. On Day 6: early morning, check LHRHa treated females and separate those with copulation plugs from those that are not plugged. Leave LHRHa treated females in the housing room (*see* **Note 7**).
6. On Day 6. Separate superovulated females from males and bring them to the microinjection laboratory after checking for copulation plugs (*see* **Note 7**).

3.7 Preparation of Petri Dishes for Egg Collection

1. Equilibrate KSOMaa in petri dishes in incubator: add 0.3 mL KSOM to one 35 mm dish and cover medium with oil, then place in incubator.
2. In a second 35 mm dish place four 75 μL drops of KSOM, cover with oil, then place in incubator.
3. Place 0.2 mL M2 medium on the center of a 60 mm dish lid. Ring this drop with six drops of 0.1 mL hyaluronidase. Do this just before Subheading 3.4.
4. On a second 60 mm dish lid put up to ten drops of 0.1 mL hyaluronidase. Do this just before Subheading 3.4.

3.8 Dissection of Oviducts

1. Euthanize egg donors (*see* **Note 8**).
2. Place rat on its back, wet fur with 70% ethanol, then make a transverse cut in the skin over the abdomen.
3. Pinch the skin on both sides of the cut and peel the skin toward the head and tail.
4. Make a transverse cut in the peritoneal wall near the groin, then make two lateral cuts in the peritoneal wall, from the groin to the rib cage.
5. Lift the resulting flap and push the intestines up and out of way.
6. Grasp the uterus with blunt forceps and insert the tip of the scissors between the uterus and the mesometrium.
7. Run the scissors along the uterus to free it from the mesometrium.
8. Grasp the uterotubal junction with watchmaker's forceps and use the scissors to cut between the ovary and the oviduct, and then cut through the uterus below the forceps to free the oviduct.
9. Place the dissected oviduct in a drop of hyaluronidase. Remove the second oviduct in the same way and place it in a second drop of hyaluronidase.
10. Continue until all of the oviducts are in their own drops of hyaluronidase.

1. Place petri dish on a stereomicroscope at 10–15× magnification.

3.9 Ampullae Microdissection and Release of Eggs

2. Each oviduct contains a swollen ampulla in which is a clump of eggs surrounded by cumulus cells. Isolate the loop of the oviduct that contains the ampulla with watchmaker's forceps and tear it free from the rest of the oviduct.
3. Place the ampulla in a hyaluronidase drop ringing the central M2 drop.
4. Continue until all of the ampullae are in hyaluronidase drops ringing the M2 drop.
5. Grasp ampulla on either side of eggs with watchmaker's forceps and pull it apart. The eggs will come out. If they do not, continue tearing apart the ampulla. Take care not to squeeze the eggs between in the jaws of the forceps. Continue until all of the ampullae are torn open.
6. After a few minutes the hyaluronidase will release most of the cumulus cells from the eggs.
7. Pick up the eggs with a transfer pipet and move them into the M2 drop to wash away the hyaluronidase.
8. After all the eggs are in the M2 drop, place them in the 0.3 mL drop of KSOMaa in the incubator.

3.10 Hanging Drop Microinjection Chamber Preparation

1. Use a small amount of Dow Corning high vacuum grease to tack two Plexiglas spacers on the slide center so that the spacers are perpendicular to the long axis of the slide (Fig. 1).
2. Put a small amount of vacuum grease on top of the spacers. Place a 70 μL drop of M2 between the spacers.
3. Place a 50 μL drop of M2 on a 9 × 22 mm coverslip. Invert the coverslip so that the drop hangs down.
4. Place the coverslip on the spacers so that the two M2 drops fuse; gently press on the coverslip ends so that the vacuum grease holds the coverslip in place (*see* **Note 9**).
5. Use a polyethylene pipet to surround the M2 drop with mineral oil to prevent M2 evaporation during the microinjection session.

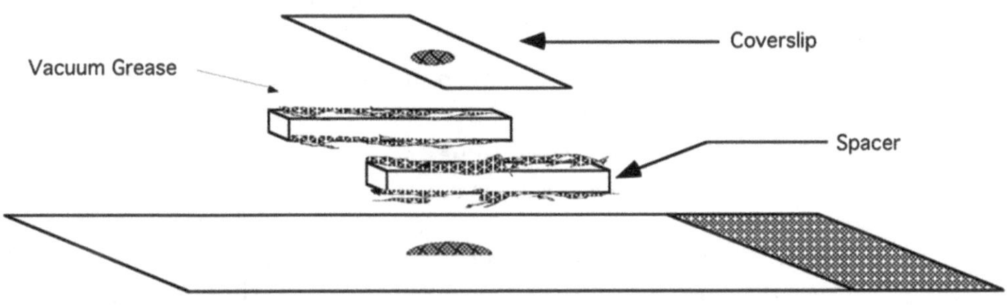

Fig. 1 Diagram of the assembly of the microinjection chamber from a glass slide, spacers, silicone grease, and M2 medium as detailed in the text

6. Place the assembled hanging drop slide chamber on the microscope stage.
7. Place a holding pipet on the glass capillary holder attached to an air-filled piston syringe such as the Eppendorf CellTram Air.
8. Attach the holder to the micromanipulator on the left side of the microscope if you are right-handed.
9. Position the tip of the holding pipet in the center of the field of view in the M2 medium droplet (*see* **Note 10**).
10. Microinjection needles are pulled with the pipet puller and placed in microtubes containing 25–50 μL of microinjection solution (*see* **Note 11**).
11. The butt end of the needle is placed in the solution. Do not contaminate the glass capillaries by careless handling. Use glass handling forceps or use forceps with silicone tubing covered tips to avoid skin oil contamination or dust particles from gloves [24]. After a few minutes the solution will wick to the tip of the needle, as shown by the appearance of a liquid meniscus near the needle's shoulder. The filled needle is then placed on the capillary holder attached to the microinjector and the holder is placed on the micromanipulator on the side opposite of the holding pipet.

3.11 Zygote Microinjection

1. Place the dish of collected eggs on the inverted tissue culture microscope and examine the eggs at 100× magnification. The fertilized eggs will contain two visible pronuclei, one from each gamete.
2. Aspirate as many eggs as you can inject in an hour with the embryo transfer pipet. Set the microinjection microscope magnification to 40× and arrange the eggs in a column as shown in Fig. 2.
3. Position the holding pipette so its tip is in the same plane of focus as the embryos and center the holding pipette tip in the field of view. Repeat this process with the microinjection needle.
4. Switch the microinjection microscope to 100× magnification and re-center the needle and holding pipette, if necessary.
5. Switch to 400× magnification and re-center the needle and holding pipette, if necessary.
6. Check the injection needle to test whether it is open by placing the needle tip at the equator of a free egg and ejecting DNA. If the needle tip is open, the DNA solution will flow out of the needle and move the egg. If the needle tip is closed, the egg will not move and the needle tip is fused shut. Open the needle tip by breaking it on the holding pipette. This can be done by brushing or tapping the tip of the needle against the holding

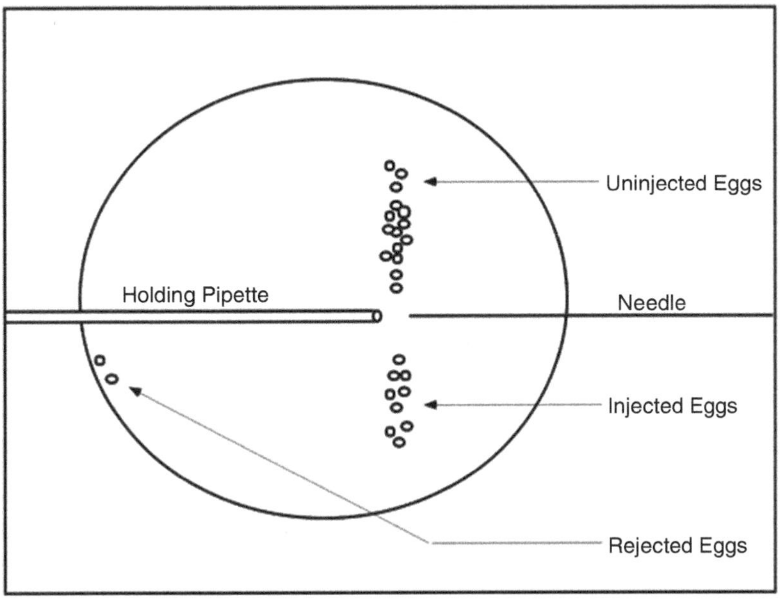

Fig. 2 Rat zygotes are placed in a column to track the eggs as they are microinjected. Microinject the egg at the bottom of the column and use the microscope stage movement control to move stage toward you and the uninjected eggs toward the micromanipulators. Unfertilized eggs can be moved to the far left of the chamber to separate them from the injected eggs

pipette until it breaks. Test to see that the needle is open by moving an egg with the DNA stream from the needle (*see* **Note 12**).

7. Aspirate an egg into the holding pipette. Raise or lower the egg to bring the male pronuclear membrane into focus. The larger of the two pronuclei farthest from the polar body is the male pronucleus. The pronucleus should be in line with the central axis of the holding pipette. If necessary, reposition the egg until the pronucleus is centered on the holding pipet axis.
8. Bring the injection needle into the same plane of focus by raising/lowering it with the micromanipulator.
9. Guide the needle into the egg and into the pronucleus without touching the sticky nucleoli within the pronucleus. The pronuclear member is very flexible, thus it is often necessary to push the needle tip through the entire egg before the pronuclear membrane is punctured. Apply pressure to the needle with the microinjector to eject DNA into the pronucleus (*see* **Note 13**). A successful injection is indicated by swelling of the pronucleus. If a small "droplet" forms at the needle tip as you inject and the pronucleus does not swell, the pronuclear membrane was not pierced and the ejected DNA solution is inflating a vesicle made up of the egg's cell membrane. If you do not observe swelling,

a vesicle, or cytoplasmic swirling, then the needle is blocked by particulates contaminating the injection solution (*see* **Note 14**).

10. Remove the needle with a smooth motion, the cytoplasm should not extrude into the perivitelline space between the egg and the zona pellucida. If a trail of nuclear contents follows the needle out of the egg discard the needle because the tip of the needle will become sticky and damage subsequently injected eggs.
11. Move the injected egg so that a clear boundary exists between injected and uninjected eggs (*see* Fig. 2).
12. After all of the eggs in the chamber are microinjected, use the embryo transfer pipet to wash them through the four drops of 75 μL KSOM in the 35 mm dish in the incubator.
13. Keep the eggs in the incubator until all of the usable eggs are microinjected and it is time to transfer them to pseudopregnant recipients.

3.12 Oviduct Transfer of Rat Eggs to Pseudopregnant Females

1. Pseudopregnant females are anesthetized with an approved anesthetic (*see* **Note 5**).
2. Preoperative analgesia is administered and analgesia is provided for 48 h postoperatively (*see* **Note 6**).
3. Hair is removed from the surgical site by clipping (Oster animal clipper with size 40 blade set) or shaving with razors.
4. Skin is disinfected three times by swabbing with betadine alternated with 70% ethanol or sterile saline or 3–6% hydrogen peroxide.
5. Make a 1 to 1.5 cm transverse incision through the skin over the ovary. Make a second incision through the peritoneal wall and exteriorize the reproductive tract by grasping the fat pad attached to the ovary with tissue forceps.
6. Place a serrefine clamp on the fat pad to hold the ovary in place.
7. A few drops of epinephrine are applied to vasoconstrict the blood vessels running through the bursa membrane that covers the ovary and oviduct.
8. Use Dumont forceps to tear the bursa open where the oviduct approximates the ovary. Alternatively, microscissors can be used to cut the bursa open.
9. The opening of the infundibulum is palpated with the tips of a Dumont forceps and the eggs are placed into the infundibulum of the oviduct with a sterile transfer pipet that has a 200 μm opening.
10. The peritoneal wall is closed with two sutures.

11. The skin is closed with wound clips that are removed seven to ten days after surgery.
12. A single dose of ampicillin (50 mg) is delivered intraperitoneally to improve pregnancy rates [25].
13. The animal is placed in a cage on a 37 °C slide warmer until alert and returned to sternal recumbency.
14. Rat diet is supplemented with Love Mash Rodent Reproductive Diet until pups are weaned.

4 Notes

1. When working with RNA-containing solutions remember to use proper RNA handling techniques, including wear gloves; tubes and tips and solutions should be sterile and RNAse free. Glass capillaries that will be used for microinjection should be dry heat sterilized at 200 °C for 2 h. After heat treatment, handle the capillaries with dry heat treated forceps. Do not use DEPC treated water or buffers because DEPC will kill the zygotes.
2. Synthetic sgRNA is preferred to in vitro transcribed sgRNA because of toxicity from the in vitro process [26]. Chemically modified sgRNA is preferred because the phosphorothioate groups increase the stability of sgRNA molecules [27].
3. Up to 200 μL of solution can be placed on the filter without losing it to the buffer. Let the dialysis proceed quietly without any shaking or movement. Do not let the dialysis go more than 3 h; otherwise the drop might begin to evaporate.
4. Check the pH of the elution buffer just before you use it: best yields are obtained at a pH of 8.5 or greater. If you have problems with particulates plugging the microinjection needles, prefilter the wash and elution buffers with Anotop 0.02 μM syringe filters. Do not filter the DNA through the filters—the small pore size will trap DNA molecules.
5. Consult with your local veterinary authority to identify an approved anesthesia such as 50 mg/kg and 10 mg xylazine/kg [28].
6. Consult with your local veterinary authority to identify an approved analgesic agent such as Carprofen 5 mg/kg [29].
7. Copulation plugs occur when the male ejaculate coagulates. The hard waxy plug may be difficult to see in the rat vagina without using a speculum and a blunt dissecting probe. It is possible to replace vasectomized males with the use of mechanical stimulation to induce the lordosis response in LHRHa treated female rats [30].

8. Euthanasia should be carried out according to an approved method in consultation with your local authority (AVMA Guidelines for the Euthanasia of Animals. American Veterinary Medical Association Schaumburg, IL).
9. The key to success in assembling the hanging drop chamber is to treat the glass slide with a siliconizing agent such as Sigma-cote siliconizing reagent. This will cause the M2 droplet to form a tall dome instead of spreading out over the surface of the slide. We have found that Rain-X Original Glass Water Repellent treatment also works well.
10. An alternative to the purchase of embryo holding pipets and embryo transfer pipettes is to fabricate them by hand. To make embryo holding pipets, take a microinjection needle from the pipette puller. Place it on the microforge. Using the reticle in the eyepiece, position the needle so that it contacts the glass anvil on the heating filament at an outer diameter of 100 μm. Apply heat until the anvil and needle fuse. Shut off the heat and the filament will cool and contract, making a flush break in the needle. Move the needle away from the filament. Apply heat to the filament and move the flush opening of the needle to the filament until it melts down to an opening of 10–20 μm. Embryo transfer pipets are hand pulled from a 9″ Pasteur pipet or other glass capillary over a small flame. A diamond pencil is used to score the glass so as to produce a flush break with an 180–200 μm diameter opening. The size of the opening can be measured with a microforge reticle and the sharp edges of the glass smoothed by heat polishing with the microforge filament.
11. When RNA or ribonucleoprotein will be microinjected the glass capillaries are baked for 2 h at 200 °C in dry heat to inactivate RNAse. A pair of forceps is similarly treated and used to handle the glass capillaries. The operator wears gloves while handling these materials and placing the needles in the microinjection solution.
12. When needle tip opening is less than 1 μm in diameter it cannot be resolved at 400× magnification. If the opening at the tip the needle is visible, then it is too large and must be discarded. At larger diameters the needle will not pierce the rat egg's membrane easily and survival will be extremely low.
13. A simple microinjector can be assembled from a 50 ml glass syringe connected to a Leitz instrument tube [31] by a length of Tygon tubing. The syringe and tubing are filled with air, the needle is placed in the holder and injection is accomplished by pushing the syringe piston. If desired the holding pipet can be controlled by attaching the holding pipet tubing to an aspirator assembly instead of a piston syringe and using mouth suction to control the holding pipet.

14. When it happens that injection solutions are contaminated by particulate matter that plug up the needles, then the solution can be passed through 4 mm 0.2 μm pore filters or by centrifugation at high speed in a benchtop microtube centrifuge. CRISPR/Cas9 will retain its DNA cleavage activity after passing through a 0.2 μm filter (unpublished results). In practice this pore size does not always trap all of the contaminating debris and the centrifugation does not always pellet all of the debris. Do not use the Anotop 0.02 μm pore size to filter microinjection solutions because this small pore size will trap the nucleic acids that need to be delivered to the egg. Plasmid DNA purification with resin-based kits can introduce particulates that plug microinjection needles. This can be avoided by the use of CsCl gradient centrifugation [32].

References

1. de Jesus Perez VA (2016) Molecular pathogenesis and current pathology of pulmonary hypertension. Heart Fail Rev 21:239–572
2. Nicolini C, Fahnestock M (2018) The valproic acid-induced rodent model of autism. Exp Neurol 299:217–227
3. Shull JD, Dennison KL, Chack AC, Trentham-Dietz A (2018) Rat models of 17β-estradiol-induced mammary cancer reveal novel insights into breast cancer etiology and prevention. Physiol Genomics 50:215–234
4. Creed RB, Goldberg MS (2018) New developments in genetic rat models of Parkinson's disease. Mov Disord 33(5):717–729. https://doi.org/10.1002/mds.27296
5. Haller ST, Kumarasamy S, Folt DA, Wuescher LM, Stepkowski S, Karamchandani M et al (2017) Targeted disruption of Cd40 in a genetically hypertensive rat model attenuates renal fibrosis and proteinuria, independent of blood pressure. Kidney Int 91:365–374
6. Kikuchi M, Wickman L, Hodgin JB, Wiggins RC (2015) Podometrics as a potential clinical tool for glomerular disease management. Semin Nephrol 35:245–255
7. Charreau B, Tesson L, Soulillou JP, Pourcel C, Anegon I (1996) Transgenesis in rats: technical aspects and models. Transgenic Res 5:223–234
8. Filipiak WE, Saunders TL (2006) Advances in transgenic rat production. Transgenic Res 15:673–686
9. Wharram BL, Goyal M, Wiggins JE, Sanden SK, Hussain S, Filipiak WE et al (2005) Podocyte depletion causes glomerulosclerosis: diphtheria toxin-induced podocyte depletion in rats expressing human diphtheria toxin receptor transgene. J Am Soc Nephrol 16:2941–2952
10. Witten IB, Steinberg EE, Lee SY, Davidson TJ, Zalocusky KA, Brodsky M et al (2011) Recombinase-driver rat lines: tools, techniques, and optogenetic application to dopamine-mediated reinforcement. Neuron 72:721–733
11. Chiang C, Jacobsen JC, Ernst C, Hanscom C, Heilbut A, Blumenthal I et al (2012) Complex reorganization and predominant non-homologous repair following chromosomal breakage in karyotypically balanced germline rearrangements and transgenic integration. Nat Genet 44:390–397
12. Dubose AJ, Lichtenstein ST, Narisu N, Bonnycastle LL, Swift AJ, Chines PS et al (2013) Use of microarray hybrid capture and next-generation sequencing to identify the anatomy of a transgene. Nucleic Acids Res 41:e70
13. al-Shawi R, Kinnaird J, Burke J, Bishop JO (1990) Expression of a foreign gene in a line of transgenic mice is modulated by a chromosomal position effect. Mol Cell Biol 10:1192–1198
14. Giraldo P, Montoliu L (2001) Size matters: use of YACs, BACs and PACs in transgenic animals. Transgenic Res 10:83–103
15. Palmiter RD, Brinster RL (1986) Germ-line transformation of mice. Annu Rev Genet 20:465–499
16. Rossant J, Nutter LM, Gertsenstein M (2011) Engineering the embryo. Proc Natl Acad Sci U S A 108:7659–7660
17. Brown AJ, Fisher DA, Kouranova E, McCoy A, Forbes K, Wu Y et al (2013) Whole-rat conditional gene knockout via genome editing. Nat Methods 10:638–640

18. Ma Y, Yu L, Pan S, Gao S, Chen W, Zhang X et al (2017) CRISPR/Cas9-mediated targeting of the Rosa26 locus produces Cre reporter rat strains for monitoring Cre-loxP-mediated lineage tracing. FEBS J 284:3262–3277
19. Wang L, Shao Y, Guan Y, Li L, Wu L, Chen F et al (2015) Large genomic fragment deletion and functional gene cassette knock-in via Cas9 protein mediated genome editing in one-cell rodent embryos. Sci Rep 5:17517
20. Kaneko T, Mashimo T (2015) Simple genome editing of rodent intact embryos by electroporation. PLoS One 10:e0142755
21. Remy S, Chenouard V, Tesson L, Usal C, Ménoret S, Brusselle L et al (2017) Generation of gene-edited rats by delivery of CRISPR/Cas9 protein and donor DNA into intact zygotes using electroporation. Sci Rep 7:16554
22. Brinster RL, Chen HY, Trumbauer ME, Yagle MK, Palmiter RD (1985) Factors affecting the efficiency of introducing foreign DNA into mice by microinjecting eggs. Proc Natl Acad Sci U S A 82:4438–4442
23. Van Keuren ML, Gavrilina GB, Filipiak WE, Zeidler MG, Saunders TL (2009) Generating transgenic mice from bacterial artificial chromosomes: transgenesis efficiency, integration and expression outcomes. Transgenic Res 18:769–785
24. Phillips ML, Meagher CC, Johnson DL (2001) What is "powder free"? Characterisation of powder aerosol produced during simulated use of powdered and powder free latex gloves. Occup Environ Med 58:479–481
25. Waller SJ, Ho M-Y, Murphy D (1995) Production of transgenic rodents by microinjection of cloned DNA in fertilized one-cell eggs. In: Glover DM, Hames BD (eds) DNA cloning, vol 4. Oxford University Press, Oxford, pp 184–229
26. Kim S, Koo T, Jee HG, Cho HY, Lee G, Lim DG et al (2018) CRISPR RNAs trigger innate immune responses in human cells. Genome Res 28:367–373
27. Basila M, Kelley ML, Smith AVB (2017) Minimal 2′-O-methyl phosphorothioate linkage modification pattern of synthetic guide RNAs for increased stability and efficient CRISPR-Cas9 gene editing avoiding cellular toxicity. PLoS One 12:e018859328
28. Stickrod G (1979) Ketamine/xylazine anesthesia in the pregnant rat. J Am Vet Med Assoc 175:952–953
29. Cannon CZ, Kissling GE, Goulding DR, King-Herbert AP, Blankenship-Paris T (2011) Analgesic effects of tramadol, carprofen or multimodal analgesia in rats undergoing ventral laparotomy. Lab Anim (NY) 40:85–93
30. De Feo VJ (1966) Vaginal-cervical vibration: a simple and effective method for the induction of pseudopregnancy in the rat. Endocrinology 79:440–442
31. El-Badry HM (1963) Micromanipulators and micromanipulation. Springer, Vienna, pp 48–55
32. Montigny WJ, Phelps SF, Illenye S, Heintz NH (2003) Parameters influencing high-efficiency transfection of bacterial artificial chromosomes into cultured mammalian cells. BioTechniques 35:796–807

Chapter 5

Embryonic Stem Cells and Gene Manipulation in Rat

Masumi Hirabayashi, Akiko Takizawa, and Shinichi Hochi

Abstract

Transgenic technology in rats is increasingly important for the design and implementation of biological and physiological studies in the fields of neuroscience, pharmacology, and toxicology. Pluripotent embryonic stem cells (ESCs) are a useful tool for generation of gene-modified rats. During the last decade, not only foreign DNA introduction but also endogenous DNA modification has been successfully achieved with rat ESCs. Detailed protocols for establishment of bona fide rat ESCs and their use for production of gene-modified rats are described in this chapter.

Key words Blastocyst injection, Bona fide ES cells, Chimeric rats, Electroporation, Germline competency, Homologous recombination, Lipofectamine

1 Introduction

Rats (*Rattus norvegicus*) have been used more extensively than mice in the research fields of neuroscience, pharmacology, and toxicology. There are more than 100 rat strains with various genetic backgrounds, including some useful models for human diseases such as the SHR (spontaneously hypertensive rat) strain [1] and the BB (rats spontaneously developing insulin-dependent diabetes mellitus) strain [2]. Nevertheless, reverse genetic approaches, such as precise and conditional replacements (knocked-in: KI) or loss of gene function (knocked-out: KO) at a specific locus, was considered impossible in rats until the end of 2008 because any protocols to establish stem cell lines conventionally used in mice were not applicable to rats. A breakthrough in establishing rat embryonic stem cells (ESCs) was elicited by the use of a few inhibitors involved in differentiation-related signaling pathways [3, 4]. This protocol was proved reproducible even after slight modifications were added to the culture system [5, 6], and the established ESCs were successfully applied for generation of transgenic (TG) rats via foreign DNA transfection [7, 8] and generation of KI/KO rats via homologous recombination [9–12]. These classical protocols are still

G. Thomas Hayman et al. (eds.), *Rat Genomics*, Methods in Molecular Biology, vol. 2018, https://doi.org/10.1007/978-1-4939-9581-3_5,

important, because recent new genome-editing tools, such as ZFN (zinc-finger nuclease), TALEN (transcription activator-like effector nuclease), and CRISPR/Cas9 (clustered regularly interspaced short palindromic repeats/CRISPR-associated protein 9), often result in off-targeting [13]. Therefore, the new genome-editing tools may lead to disappointment when inducing mutation using pronuclear-stage zygotes directly to produce TG/KI/KO animals.

2 Materials

2.1 Animals

1. Blastocyst donors for ESC establishment: Inbred commercially available rat strains; Fisher 344 (F344/NSlc; Japan SLC, Shizuoka, Japan), Dark Agouti (DA/Slc; Japan SLC), Brown Norway (BN/CrlCrlj; Charles River Japan, Kanagawa, Japan), Lewis (LEW/CrlCrlj; Charles River Japan), SHR (SHR/NCrl; Charles River Laboratories, Wilmington, MA, USA), WKY (WKY/Crl; Charles River Laboratories). The females are mated with stud males of each strain.
2. Unique blastocyst donors for ESC establishment: Rat strains maintained at the Animal Resource Center of the Medical College of Wisconsin, an American Association of Laboratory Animal Care-approved facility; August × Copenhagen Irish (ACI/EurMcwi), Fawn-hooded hypertensive (FHH/EurMcwi), Dahl salt-sensitive (SS/JrHsdMcwi).
3. Immuno-compromised rodents: NOD-SCID mouse (NOD. CB17-Prkdc^{scid}/J; Charles River Japan), Nude rat (F344/ NJcl-rnu/rnu; Clea Japan, Tokyo, Japan).
4. Blastocyst donors for chimera generation: Albino closed-colony strain female rats, such as Wistar (Crlj:WI; Charles River Japan) rats, when non-albino-derived ESCs are injected. Non-albino female rats, such as Crl:LE × ACI/EurMcwi F1 rats, when albino-derived ESCs are injected.
5. Blastocyst recipients, foster mothers: Conventional closed-colony strain female rats, such as Sprague-Dawley (Crl:SD; Charles River Laboratories) or Wistar (Crlj:WI; Charles River Japan).
6. Vasectomized males for pseudopregnancy induction: Conventional closed-colony strain male rats, such as Sprague-Dawley (Crl:SD; Charles River Laboratories) or Wistar (Crlj:WI; Charles River Japan). Although the vasectomized males are commercially available, they can be prepared by aseptic surgery under general anesthesia (https://www.ncbi.nlm.nih.gov/pmc/articles/PMC4141639/).

2.2 Equipment

2.2.1 Microscopes

1. Stereomicroscope for embryo recovery (Nikon SMZ-800 with halogen illumination stand, Model C-DSS115).
2. Inverted microscope with 4×, 10×, 20×, and 40× objectives for ESC injection (TE-2000; Nikon, Tokyo, Japan).
3. Stereomicroscope for embryo transfer (Nikon SMZ-800 with Nikon G-US2 universal stand), with halogen fiber illumination system (MHAB-150W; Moritex Corp., Saitama, Japan).

2.2.2 Polymerase Chain Reaction (PCR) System

1. Thermal cycler for RT-PCR (ABI StepOnePlus™; Applied Biosystems, Thermo Fisher Scientific, Roskilde, Denmark).
2. Thermal cycler for PCR (Applied Biosystems Veriti™; Thermo Fisher Scientific).
3. Electrophoresis apparatus (Mupid®-2plus; M-2P; TaKaRa Bio, Shiga, Japan).

2.2.3 Micromanipulation System

1. A pair of micromanipulators (MO-202S; Narishige Scientific Instrument Laboratory, Tokyo, Japan).
2. Micro injector for embryo holding (IM-5B; Narishige).
3. Micro injector for ESC injection (M1J-1; Prime Tech, Ibaraki, Japan).
4. Piezo-driven micromanipulator (PMM-150FU; Prime Tech) with a pulse controller (PMAS-CT150; Prime Tech).
5. Micropipette for blastocyst holding (PH 10015-30; Prime Tech).
6. Microneedle for ESC injection (PIN 20-20FT; Prime Tech).

2.2.4 DNA Transfection System

1. Electroporater (Gene Pulser Xcell™; 165-2661J1; Bio-Rad Laboratories, Inc., Hercules, CA).
2. Automated cell counter (TC10™; Cat No. 145-0010, Bio-Rad).
3. Cell sorter (SH800; Sony Imaging Products & Solutions Inc., Tokyo, Japan).

2.2.5 Incubators for Cell Culture

1. CO_2 gas incubator (CPD-2702; Hirasawa Works, Tokyo, Japan). Standard setting: 37 °C, 5% CO_2 and 95% air, humidified by a pan containing distilled water.
2. Water bath (Thermal Rabo TR-1A; As One, Osaka, Japan).

2.3 Tools and Supplies

1. Microinjection chamber (P50G-0-30-F, 50 mm glass bottom dishes No. 0; Mat Tek, Ashland, MA, USA).
2. Organ culture dishes (Falcon® 353037; Corning, Corning, NY, USA).
3. 100 and 35 mm Petri dishes (Falcon® 351029 and 351008, respectively; Corning).

4. 100 and 35 mm tissue culture dishes (Falcon® 353003 and 353001, respectively; Corning).
5. 4-well cell culture dishes (Nunc™ 176740; Thermo Fisher Scientific, Roskilde, Denmark).
6. 96-well cell culture microplate (Falcon® 353075; Corning).
7. CellBanker® (Nippon Zenyaku Kogyo, Fukushima, Japan).
8. Cryovials (377224; Thermo Fisher Scientific).
9. RNeasy Mini Kit (74106; Qiagen, Germantown, MD, USA).
10. SuperScript™ III First-Strand Synthesis System (18080051; Invitrogen™, Carlsbad, CA, USA).
11. REDExtract-N-Amp™ Tissue PCR Kit (XNAT-100RXN; Sigma-Aldrich, St. Louis, MO, USA).
12. Electroporation cuvettes (C165-2088; Bio-Rad).
13. Cell counting slides (145-0015; Bio-Rad).
14. Animal Anesthetizer (MK-AT210D; Muromachi Kikai, Tokyo, Japan).
15. Autoclip™ (BD 427630; Fisher Scientific Company LLC, Pittsburgh, PA, USA).

2.4 Media and Chemicals

1. MEF medium: DMEM (Gibco® 11995-065; Life Technologies™, Grand Island, NJ, USA), 10% fetal bovine serum (FBS: Gibco® 10099-141; Life Technologies™), 1× Penicillin-streptomycin-glutamine (Gibco® 10378-016; Life Technologies™).
2. 2iF medium [14]: 1:1 mixture of DMEM/F12 medium (Gibco® 11330-032; Life Technologies™), Neurobasal® medium (Gibco® 21103-049; Life Technologies™), 0.5× B-27® serum-free supplement (50×) (Gibco® 17504-044; Life Technologies™), 0.5× N-2 supplement (100×) (Gibco® 17502-048; Life Technologies™), 0.5× Penicillin-streptomycin-glutamine, 0.1% rat LIF (ESGRO®; Merck Millipore, Darmstadt, Germany), 3 μM CHIR99021 (Axon Medchem, Groningen, The Netherlands), 10 μM PD0325901 (Axon Medchem), 10 μM forskolin (*see* **Note 1**).
3. Modified rat 1-cell embryo culture medium (mR1ECM; ARK Resource, Kumamoto, Japan).
4. Matrigel stock solution: 1:1 mixture of Matrigel (354234; BD Biosciences, Franklin Lakes, NJ, USA) and Knockout DMEM (10829-018; Invitrogen™), stored at −20 °C.
5. Polyvinylpyrrolidone (PVP) solution with mHTF (10% PVP; 9600-002; Nippon Medical & Chemical Instruments, Osaka, Japan).

6. Electroporation medium: 1:1 mixture of DMEM/F12 medium and Neurobasal® medium, 0.5× serum-free supplements (B-27® and N-2).
7. Mineral oil, embryo culture-tested (M8410; Sigma-Aldrich).
8. Mitomycin-C (MMC, 139-18711; Wako).
9. Trypsin (Gibco® 15090-046; Life Technologies™).
10. Taq polymerase (TaKaRa LA Taq®; TaKaRa Bio, Shiga, Japan).
11. Lipofectamine (Lipofectamine®2000; Thermo Fisher Scientific).
12. Epinephrine (245140A1030; Bosumin®; Daiichi Pharmaceutical, Tokyo, Japan).
13. Isoflurane (099-06571; Wako).
14. Kanamycin (6123003M1042; Meiji Seika Pharma, Tokyo, Japan).

2.5 Cells

1. Mouse embryonic feeder cells (MEFs, KBL9284400; Oriental Yeast Co., Ltd. Tokyo, Japan).
2. Neomycin-resistant mouse embryonic feeder cells (Neo-MEFs, KBL9284100; Oriental Yeast).
3. B6-Puro mouse embryonic fibroblasts (Puro-MEFs, A34965; Thermo Fisher Scientific).

3 Methods

3.1 Establishment of Rat ESCs

3.1.1 Colony Isolation, Passage, and Cryopreservation

1. MEFs (2×10^6 cells) are first expanded in 10 mL of the MEF medium in a 100 mm tissue culture dish for 2–3 days. After several passages by 0.25% trypsin/5.3 mM EDTA treatment (5 min, 37 °C), 190 × *g* centrifugation (5 min at 4 °C) and 1:3 dilution with the MEF medium, the MEFs are treated with 10 μg/mL MMC for 2–3 h at 37 °C in 5% CO_2 in air, washed twice with centrifugation, and then cultured in 0.5 mL of the MEF medium in a well of a 4-well cell culture dish at 37 °C in 5% CO_2 in air.
2. Four and a half days after confirmation of mating with fertile males (vaginal plug or sperm detection), female donor rats at 7–13 weeks old are sacrificed by CO_2 gas euthanasia and cervical dislocation. A pair of excised oviducts/uteri is flushed with mR1ECM in a 5-mL syringe with a 26-G sterile needle in a 100 mm Petri dish with 5–10 mm grids. Harvested blastocysts are washed twice with mR1ECM and incubated for 2–3 h at 37 °C in 5% CO_2 in air until reaching expanded blastocyst stage (*see* Fig. 1a, b).

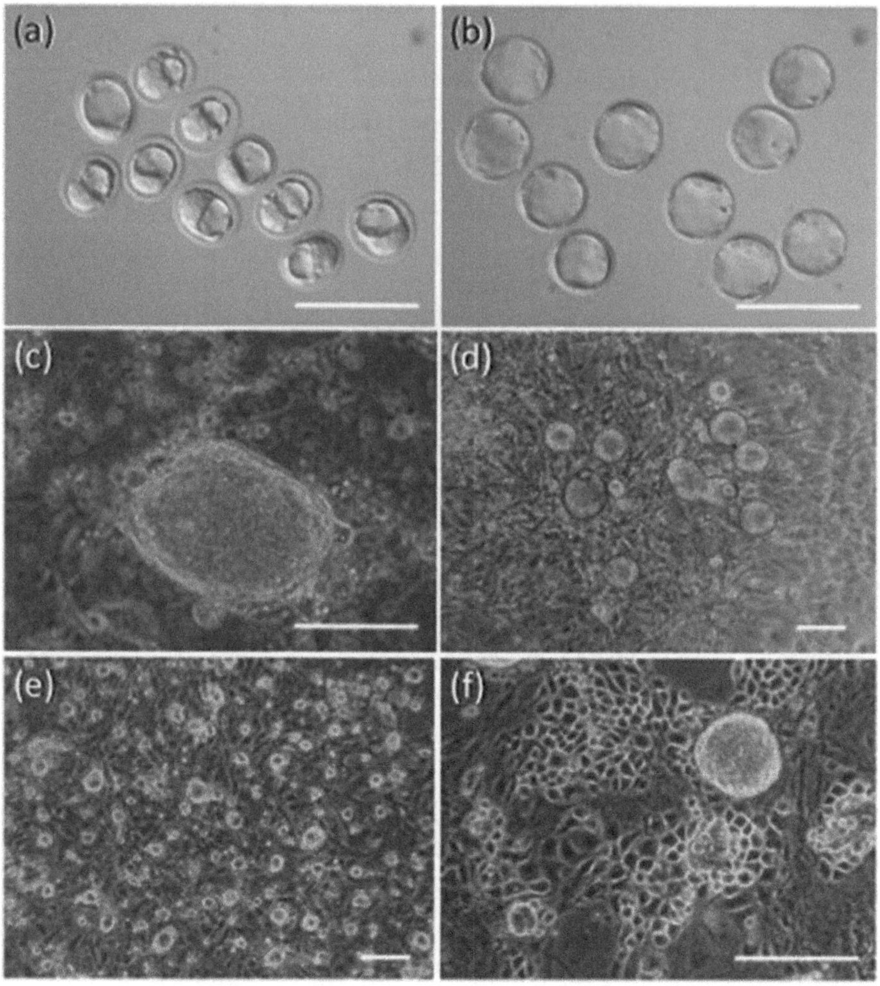

Fig. 1 Establishment of rat ESC lines. (**a**) E4.5 blastocysts immediately after recovery. (**b**) Expanded blastocysts after short-term incubation. (**c**) Outgrowth of a blastocyst on MEFs 7 days after plating. (**d**) ESC colony formation 3 days after the first passage. (**e**) Expanded ESC colonies 2 days after the fourth passage. (**f**) Differentiated extraembryonic cells observed at the fifth passage. Scale bars: 200 μm

3. The E4.5 blastocysts (E0 is defined as the initiation of coupling) are washed with PBS, and freed from their zonae pellucidae in acid Tyrode's solution (5 s at ambient temperature). The zona-free blastocysts are then placed on the MMC-treated MEFs (prepared 1 day before blastocyst seeding) with 0.5 mL of the 2iF medium in a well of a 4-well cell culture dish. After 7 days of culture at 37 °C in 5% CO_2 in air, the outgrowths of the blastocysts (*see* Fig. 1c) are disaggregated by gentle pipetting and transferred to the same MEF/2iF conditions (first passage). The time required until the first passage (5 days to 2 weeks) may depend on rat strains for blastocysts.

4. When ESC-like colonies emerge (*see* Fig. 1d), they are trypsinized in 0.25% trypsin/5.3 mM EDTA solution and then expanded. The tentative ESC lines (*see* Fig. 1e) are maintained in MEF/2iF conditions, with half-volume medium exchange every day and trypsinization/expansion (passage) every 2–3 days (*see* **Note 2**). Medium exchange every other day, especially at advanced passage numbers, often induces differentiation of the ESCs into extraembryonic cell-like cells (*see* Fig. 1f).
5. Alternatively to the repeated passages, the ESCs are cryopreserved to prevent senescence. Liberated ESCs ($2–4 \times 10^5$ cells) are resuspended in 1-mL of CellBanker® and placed in cryovials. The cryovials are transferred into a BICELL® bio-freezing vessel and kept overnight in a −80 °C deep-freezer. The cryovials are then stored in a liquid nitrogen tank until use.

3.1.2 Marker Gene Expression

1. Expression of stem cell marker genes, such as *Oct-4*, *rNanog*, *Fgf-4*, or *Rex-1*, is examined by reverse transcription (RT)-PCR analysis to confirm their undifferentiated characteristics. Expression of the reference gene (*ß-actin*) is examined simultaneously.
2. Total RNA is extracted from 80% confluent ESCs ($1–2 \times 10^6$ cells) using an RNeasy Mini Kit. The cDNA is prepared from total RNA (2 μg) using the SuperScript™III First-Strand Synthesis System, and amplified with TaKaRa LA Taq® using an ABI StepOnePlus™ for 33 cycles at 95 °C for 30 s, 55 °C (or 60 °C in the case of *rNanog*) for 30 s, and 72 °C for 60 s using the primer sets listed (*see* Table 1). The optimal concentration of each primer is 5 pmol/μL.

Table 1
Primer sets used for PCR amplification

Marker gene		Primer sequence	Product size
Oct-4	Forward	5′- GGG ATG GCA TAC TGT GGA C -3′	412 bp
	Reverse	5′- CTT CCT CCA CCC ACT TCT C -3′	
rNanog	Forward	5′- GCC CTG AGA AGA AAG AAG AG -3′	356 bp
	Reverse	5′- CGT ACT GCC CCA TAC TGG AA -3′	
Fgf-4	Forward	5′- CGG GGT GTG GTG AGC ATC TTC -3′	202 bp
	Reverse	5′- CCT TCT TGG TCC GCC CGT TC -3′	
Rex-1	Forward	5′- TTC TTG CCA GGT TCT GGA AGC -3′	297 bp
	Reverse	5′- TTT CCC ACA CTC TGC ACA CAC -3′	
ß-actin	Forward	5′- CAT GGC ATT GTG ATG GAC T -3′	427 bp
	Reverse	5′- ACG GAT GTC AAC GTC ACA CT -3′	

3. The RT-PCR products are separated by 2% agarose gel electrophoresis (100 V/cm, 25 min) with appropriate molecular size markers using a Mupid®-2plus apparatus.

3.1.3 Teratoma Formation

1. Multipotency of the tentative ESCs can be confirmed by their potential to generate teratomas derived from three primary germ layers (endoderm, mesoderm and ectoderm). Immunocompromised, NOD-SCID mice or nude rats (6–8 week-old) are used for injections.
2. Approximately 80% confluent ESCs are harvested after trypsinization/centrifugation, and placed with 2iF medium in a chilled cryovial on ice. The ESCs in 2iF medium are mixed with an equal volume of Matrigel stock solution, and injected into NOD-SCID mice or nude rats (1×10^6 cells; 0.2 mL per mouse, 0.5 mL per rat). Injection sites include beneath the skin on the dorsal rear flank (subcutaneous) or into the abdomen (intraperitoneal).
3. 5–8 weeks after the ESC injection, the NOD-SCID mice or nude rats are examined for the presence of tumors (*see* Fig. 2a). Based on hematoxylin-eosin staining of the histological sections, tumors are identified as teratomas when various tissues including gut-like epithelium or hepatic cells (endoderm origin; *see* Fig. 2b, c), bone, cartilage, or muscle (mesoderm origin; *see* Fig. 2d, e), and neural tissue (ectoderm origin; *see* Fig. 2f) are detected (*see* **Note 3**).

3.1.4 Karyotype-Gender Analysis

1. Karyotyping is performed to determine normal ploidy and gender of the ESC lines. Approximately 80% confluent ESCs are harvested as for passaging, and sent for G-band examination by a karyotyping service (e.g., Nihon Gene Research Laboratories, Miyagi, Japan; Cell Line Genetics, Inc., Madison WI, USA).
2. Alternatively, PCR is performed to determine gender of the ESC lines using a REDExtract-N-Amp™ Tissue PCR Kit. A 104-bp fragment of the rat Y chromosome-specific *Sry* gene is amplified by 1 cycle of 3 min at 94 °C; 38 cycles of 30 s at 94 °C, 30 s at 55 °C and 30 s at 72 °C; and 1 cycle of 10 min at 72 °C with a primer set (forward primer 5′-CAT CGA AGG GTT AAA GTG CCA-3′; reverse primer 5′-ATA GTG TGT AGG TTG TTG TCC-3′) using an Applied Biosystems Veriti™ cycler. The PCR products are separated by 2% agarose gel electrophoresis.

3.1.5 Blastocyst Injection

1. The ESCs are liberated by trypsin treatment and centrifuged. They are then placed in 2 mL of the 2iF medium in a 35 mm petri dish, and kept for 0.5–1 h at 4 °C until the injection into blastocysts.

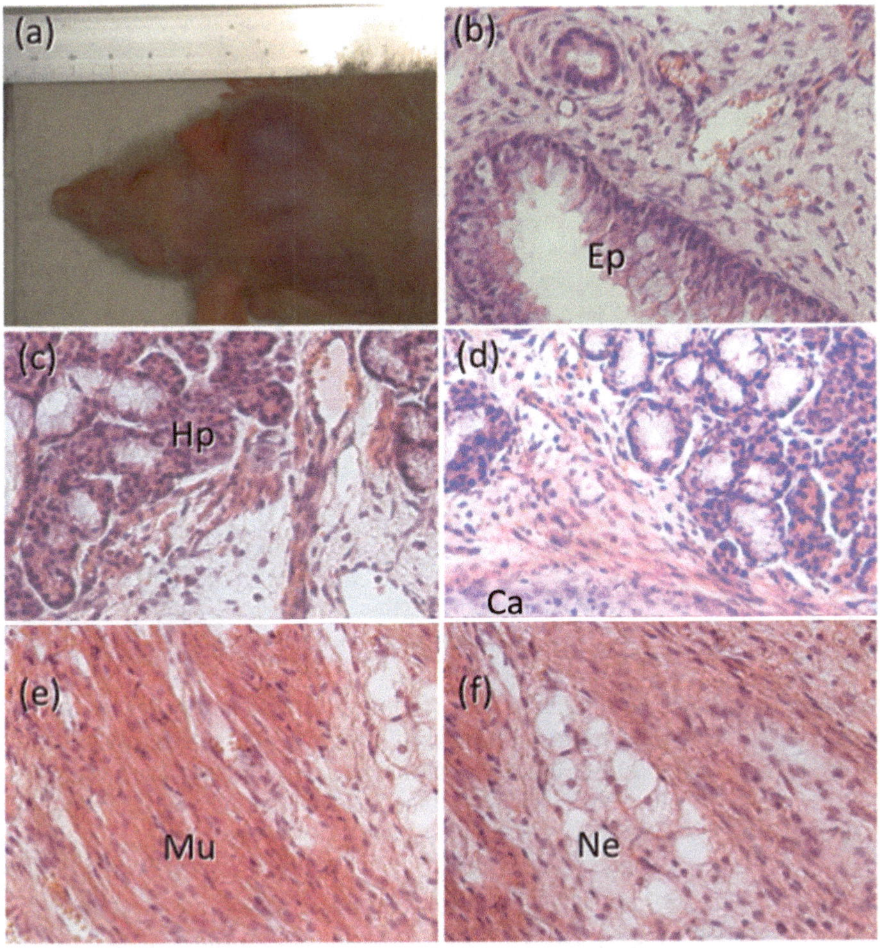

Fig. 2 Teratoma derived from rat ESCs. (**a**) Tumor observed 5 weeks after subcutaneous injection of the ESCs into a nude rat. (**b–f**) Hematoxylin/eosin-stained sections of the tumor. Pluripotency of the ESCs is confirmed by differentiation into three primary germ layers (endoderm, mesoderm and ectoderm). *Ep* Epithelium (gut-like), *Hp* Hepatic cells, *Ca* Cartilage, *Mu* Muscle, *Ne* Neural tissue

2. Two microdrops of 2iF medium (30 μL drop-[I] and 10 μL drop-[II]) and one microdrop of mHTF/10% PVP (6 μL drop-[III]) overlaid with mineral oil are prepared in a microinjection chamber (*see* Fig. 3a). E4.5 blastocysts (10–15 per 1 operation) and ESCs (approximately 10^3 cells) are transferred into drop [I] and drop[II], respectively.
3. A piezo-driven micromanipulator with a pulse controller is used for ESC injection into blastocysts. The inner surface of the injecting pipette is washed by repeated blow-off of mercury and aspiration of mHTF/10% PVP in drop [III]. Then, 10–30 ESCs are aspirated into the injecting pipette in drop [II], standing by for injecting in drop [I]. The holding pipette holds a blastocyst so that the inner cell mass (ICM) is located at the 9:00 o'clock position (*see* Fig. 3b).

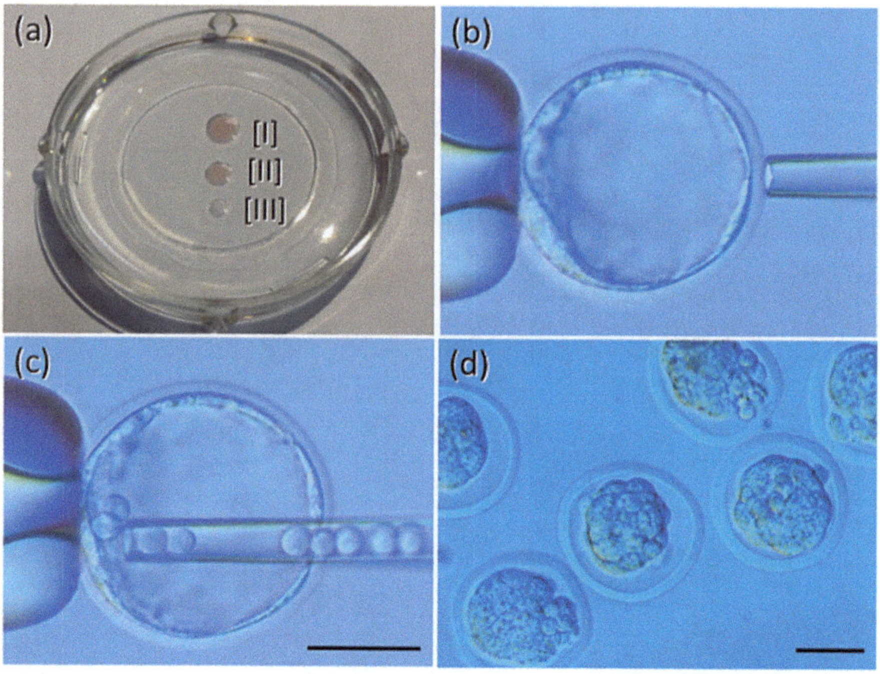

Fig. 3 Microinjection of rat ESCs into blastocysts. (**a**) The injection chamber contains two microdrops of 2iF medium (drop [I] and [II]) and 1 microdrop of mHTF/10% PVP (drop [III]). (**b**) The rat blastocyst is held so that its inner cell mass (ICM) is located at the 9-o'clock position. (**c**) ESCs are deposited near the ICM of the blastocyst. (**d**) The blastocoel cavity often shrinks after the ESC injection, but re-expands 1–3 h later. Scale bars = 50 μm

4. The zona pellucida is drilled by several piezo-pulses (intensity 3, speed 2). Then, the injection pipette is deeply inserted into the blastocyst cavity through the trophectodermal cells. The ESCs are deposited near the ICM, without touching the pipette tip to the ICM (*see* Fig. 3c). The injection pipette is slowly withdrawn.
5. The injected blastocysts are transferred into 800 μL of fresh 2iF medium in an organ culture dish, and kept at 37 °C, 5% CO_2 in air until uterine embryo transfer. The blastocoel cavity often shrinks after the ESC injection (*see* Fig. 3d). The blastocysts will re-expand 1–3 h later, but they are transferred to recipient uteri without confirming their re-expansion.

3.1.6 Uterine Embryo Transfer

1. Mature female rats (>8 week-old) are mated with a vasectomized male rat, and embryo transfer is performed 3 days after the vaginal plug is detected. An isoflurane gas at 2–3% using a Small Animal Anesthetizer anesthetizes the recipient rats. Under full anesthesia, the uterus is exposed on sterile gauze through a lateral incision. One or two drops of 0.1% epinephrine are applied on the upper part of the uterus near the uterotubal junction to constrict blood vessels, and then a 27-G sterile needle is inserted into the uterus.

2. When the opening of the uterus becomes visible, a glass capillary is gently inserted into this opening. The blastocysts (usually 6–8 embryos per uterus) aspirated in the capillary with an outer diameter of 140–160 μm can be loaded into the uterine horn. It is recommended to introduce an air bubble into the capillary to mark the position of the last blastocyst to be introduced.
3. The peritoneal wall (muscle layer) is closed using suture-needle and suture, and then the skin is closed by Autoclip™. After subcutaneous administration of 12.5 mg Kanamycin, the recipients are allowed to recover on a 37 °C warm plate.

3.1.7 Germline Competency of Chimeras

Chimeric rat offspring can be identified by visualization of ESC fluorescent marker (immediately after birth) and/or coat color marker (2 weeks after birth). The potential of each ESC line for germline transmission is examined by the conventional approach using G1 generation offspring via chimeras. It is well known that genetic characteristics of XX-gender donor cells are not transmitted from male chimeras to their offspring [15, 16]. To conclude that the germline competency of a certain ESC line is "negative," at least G1 litters derived from five different chimeric rats should be examined (*see* **Note 4**).

3.2 Rat Transgenesis Using Bona Fide ESCs

3.2.1 Foreign DNA Introduction by Electroporation [7]

Convenient approaches to generate transgenic rats harboring foreign DNA are DNA microinjection into pronuclear zygotes (major) [17] and sperm (ICSI)-mediated DNA transfer (minor) [18]. In the case that bona fide ESCs are used for transgenesis, the foreign DNA can be introduced into genomes by electroporation [19] or liposome-mediated transfection [20], combined with chimera production and germline transmission.

1. One day before ESC electroporation, neomycin-resistant MEFs (Neo-MEFs) are plated in a 60 mm culture dish. Approximately 80% confluent Neo-MEFs are prepared on the day of electroporation.
2. Rat ESCs are harvested after trypsinization and centrifugation (described as in Subheading 3.1.1, **step 1**), and approximately 1×10^6 cells are resuspended in 400 μL of the electroporation medium. The ESC suspension is mixed with foreign DNA solution (25 μg/100 μL, e.g., CAG/huKO-neo plasmid, *see* Fig. 4a) which has been designed to carry at least one selection marker, neomycin-resistance gene and/or fluorescent protein gene.
3. The ESC-DNA mixture (500 μL) is transferred to an electroporation cuvette with a 4 mm gap, and the cuvette is placed in an electroporation holder of the Gene Pulser Xcell™. Following electroporation (800 V, 10 μF), the ESCs are kept for 15–20 min at room temperature.

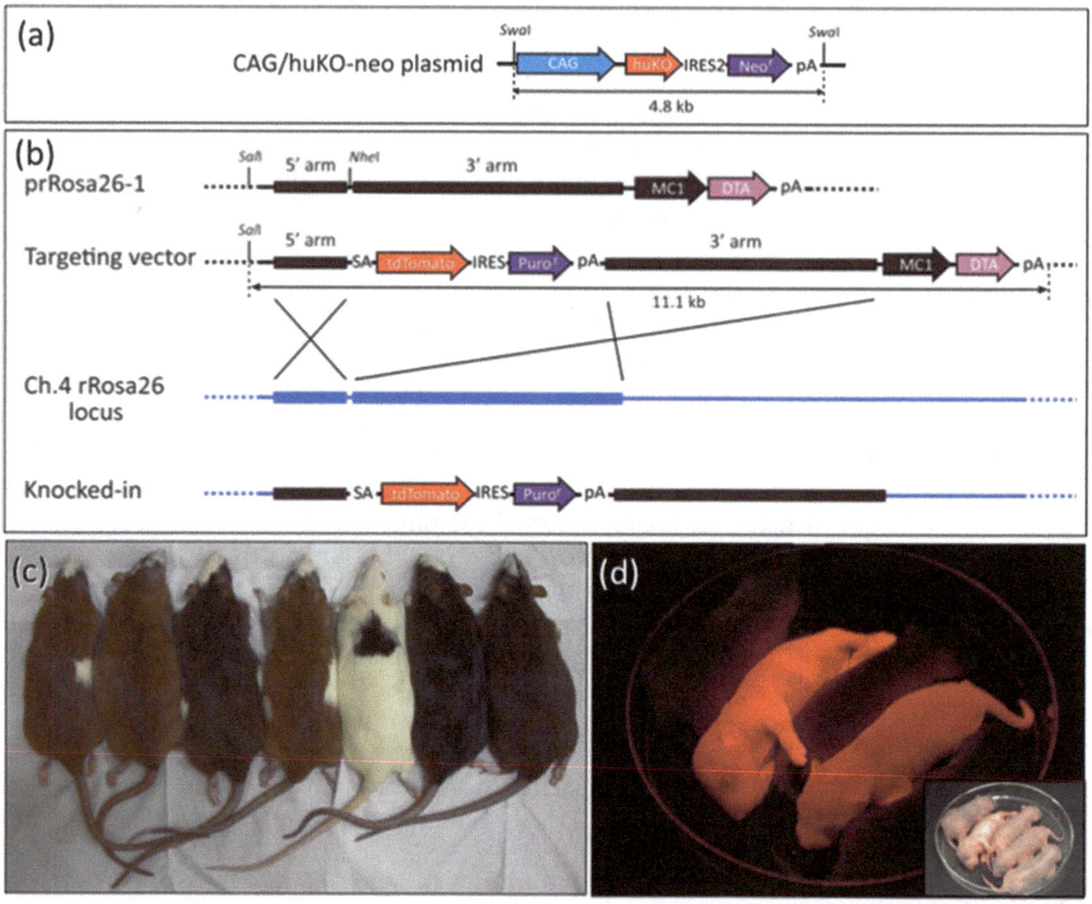

Fig. 4 Examples of rat transgenesis via electroporation of ESCs. (**a**) Construct of CAG/huKO-neo plasmid, used for non-homologous insertion into rat genomes. *CAG* cytomegalovirus/chicken beta-actin promoter, *huKO* humanized kusabira-orange, *Neo^r* neomycin-resistance, *pA* polyadenylation. (**b**) Design of targeting vector to homologously knock-in (KI) rat Rosa26 locus on chromosome 4 with tdTomato. *SA* splicing accepter, *IRES* internal ribosomal entry site, *pA* polyadenylation, *Puro^r* puromycin-resistance MC1: polyoma enhancer/herpes simplex virus thymidine kinase promoter, *DTA* diptheria toxin-A (**c**) Chimeric rats with different contribution of brown-colored coat at G0 generation. (**d**) The tdTomato-positive KI offspring at G1 generation

4. The electroporated ESCs are placed with 3 mL of 2iF medium on Neo-MEFs in a 35 mm tissue culture dish, and the next day the culture medium is exchanged with new 2iF medium. The ESCs are treated with 200 μg/mL neomycin (G418) 2 days after the electroporation (*see* **Note 5**), and cultured for an additional 5 days with daily medium exchanges. Approximately 30–50 neomycin-resistant ESC colonies can be picked up, placed in a 4-well cell culture dish with 2 mL of 2iF medium, and then disaggregated by gentle pipetting (*see* **Note 6**). Thereafter, the ESCs are passaged as described in Subheading 3.1.1, **step 4**.

3.2.2 Foreign DNA Introduction by Lipofectamine Treatment

1. The transfection complex comprises 3 μg foreign DNA in 150 μL of FBS-free medium (1:1 mixture of DMEM/F12 medium and Neurobasal® medium) and 6 μL Lipofectamine®2000 in 150 μL of the FBS-free medium. After complete mixing of the diluted DNA with the diluted lipofectamine, the transfection complex (300 μL) is incubated for >25 min at room temperature.
2. Rat ESCs and Neo-MEFs are prepared as in Subheadings 3.2.1, **step 1** and 3.2.1, **step 2**. An appropriate number of the ESC pellet (5×10^5 cells) is suspended in 300 μL of the transfection complex, mixed well, and incubated for 5 min at room temperature. The transfected ESCs are resuspended in 3 mL of 2iF medium, and cultured on the Neo-MEFs in several 35-mm tissue culture dishes after 1:2 and 2:1 dilution with the 2iF medium. On the next day, half of the culture medium (1 mL) is exchanged with new 2iF medium.
3. Thirty-six hours after the lipofectamine treatment, the ESCs are treated with 200 μg/mL of G418 in 2iF medium. Half the volume of the G418/2iF medium is exchanged 2 days later, and ESC colonies that survived the G418 selection can be recovered with the supernatant after tapping the culture dish 3 days later.
4. The ESC colonies are harvested, dissociated by trypsin/EDTA treatment, centrifuged, and resuspended in 2 mL of MEF medium. Single ESCs are sorted one per well in 96-well cell culture microplates with MEF/2iF medium. One week later, newly formed colonies are trypsinized, centrifuged and expanded in 4-well cell culture dishes. An aliquot of the ESCs is used for genotype analysis and the remaining ESCs are cryopreserved with CellBanker® as described in Subheading 3.1.1, **step 5**.

3.2.3 Endogenous DNA Modification by Homologous Recombination [10, 12]

Homologous recombination (HR), as well as new genome-editing tools (ZFN, TALEN, or CRISPR/Cas9), can induce mutations in endogenous DNA sequence. Double-strand breaks induced at the targeted site result in KOs by non-homologous end-joining (NHEJ) or KIs by HR. In general, KI of a selection marker gene alone at the targeted site is classified as a KO. The new genome-editing tools can be directly used to generate KI/KO animals via pronuclear microinjection, but this often results in off-targeting [13]. Therefore, the traditional HR approach is still convenient to modify the endogenous DNA.

1. One day before electroporation, puromycin-resistant MEFs (Puro-MEFs) are plated in a 60 mm culture dish. Approximately 80% confluent Puro-MEFs are prepared on the day of electroporation.

2. Rat ESCs are harvested after trypsinization and centrifugation (described in Subheading 3.1.1, **step 1**), and approximately 5×10^6 cells are resuspended in 400 μL of electroporation medium. The ESC suspension is mixed with targeting vector solution (20 μg/100 μL; e.g., Rosa26/tdTomato-IRES-Puro^r-pA, *see* Fig. 4b). The ECS-vector mixture (500 μL) is electroporated as described in Subheading 3.2.1, **step 3**.
3. The electroporated ESCs are resuspended with 9 mL of 2iF medium and cultured on Puro-MEFs in three 60 mm culture dishes, with daily exchange of the culture medium. Three days after the electroporation, the ESCs are treated with 1 μg/mL puromycin for 1 day. Approximately 30–50 puromycin-resistant ESC colonies can be harvested. Single colonies are placed into 4-well cell culture dishes with 500 μL MEF/2iF medium and then disaggregated by gentle pipetting.
4. One week later, colonies are trypsinized, centrifuged, and expanded to yield $1–2 \times 10^5$ cells in 4-well cell culture dishes. An aliquot of the ESCs is used for genotype analyses such as PCR and Southern blotting (*see* **Note 7**), and the remaining ESCs are cryopreserved with CellBanker® as described in Subheading 3.1.1, **step 5**.

3.2.4 Identification of TG/KI/KO Rats

1. Foreign DNA- or endogenous DNA-modified rat ESCs (Subheadings 3.2.1, 3.2.2 and 3.2.3) are injected into E4.5 blastocysts, as described in Subheading 3.1.5. The blastocysts are transferred into uteri of E3.5 pseudopregnant females, as described in Subheading 3.1.6.
2. Chimeric rat offspring can be identified by visualization of the ESC fluorescent marker (immediately after birth) and/or coat color marker (2 weeks after birth). Successful transgenesis is confirmed by conventional PCR analysis (TG) or genotyping of the target locus (KI/KO) of F1 heterozygous offspring derived from the chimeras (*see* Fig. 4c, d).

4 Notes

1. The basal media, to which CHIR99021, PD0325901 and ESGRO® have not yet been added, can be stored at 4 °C for 1 month. Once these supplements are added to the basal media, the 2iF medium should not be kept at 4 °C beyond 2 weeks.
2. When rat ESCs are passaged, care should be taken to avoid wasting the ES colonies. Because the attachment of stem cell colonies to the feeder layer is very weak, the supernatant medium must be carefully aspirated preceding the enzymatic digestion.

3. The assistance of a trained pathologist in evaluating the tumor tissues is recommended, because many of these tissues may be immature and definitive identification of the mature tissue can be difficult.
4. High percentages of seeded blastocysts (52–100%) exhibit outgrowth and formation of ESC-like colonies in all the nine inbred rat strains. A majority of the tentative ESC lines can participate in chimeric rat production. The germline competency can be proven in many chimeric rats (>50% of chimeras), but transmission of the ESC characteristics to the G1 progeny varies among the ESC lines (1–21% of G1 progeny). These ESC lines are listed on the Rat Genome Database web site.

   (http://rgd.mcw.edu/rgdweb/search/ge.html?term=ES&speciesType=0).
5. Rat ESCs after electroporation should be cultured for a few days, and then treated with G418. In contrast, mouse ESCs can be treated with the antibiotic immediately after electroporation.
6. Rat ESC single colonies that survive G418 selection must be carefully handled and passaged without trypsin digestive treatment, as if the colonies were originated from the blastocyst ICMs.
7. Antibiotic-resistant rat ESC colonies are likely to have genuine targeted mutations at an approximate proportion of 10%, which is higher than the <3% in mouse ESC colonies.

References

1. Okamoto K (1969) Spontaneous hypertension in rats. Int Rev Exp Pathol 7:227–270
2. Like AA, Kislauskis E, Williams RR, Rossini AA (1982) Neonatal thymectomy prevents spontaneous diabetes mellitus in the BB/W rat. Science 216:644–646
3. Buehr M, Meek S, Blair K, Yang J, Ure J, Silva J et al (2008) Capture of anthentic embryonic stem cells from rat blastocysts. Cell 135:1287–1298
4. Li P, Tong C, Mehrian-Shai R, Jia L, Wu N, Yan Y et al (2008) Germline competent embryonic stem cells derived from rat blastocysts. Cell 135:1299–1310
5. Hirabayashi M, Kato M, Kobayashi T, Sanbo M, Yagi T, Hochi S et al (2010) Establishment of rat embryonic stem cell lines that can participate in germline chimerae at high efficiency. Mol Reprod Dev 77:94
6. Hirabayashi M, Goto T, Tamura C, Sanbo M, Hara H, Hochi S (2014) Effect of leukemia inhibitory factor and forskolin on establishment of rat embryonic stem cell lines. J Reprod Dev 60:78–82
7. Hirabayashi M, Kato M, Sanbo M, Kobayashi T, Hochi S, Nakauchi H (2010) Rat transgenesis via embryonic stem cells electroporated with the Kusabira-Orange gene. Mol Reprod Dev 77:474
8. Kawamata M, Ochiya T (2010) Generation of genetically modified rats from embryonic stem cells. Proc Natl Acad Sci U S A 107:14223–14228
9. Tong C, Li P, Wu NL, Yan Y, Ying QL (2010) Production of p53 knockout rats by homologous recombination in embryonic stem cells. Nature 467:211–213
10. Kobayashi T, Kato-Itoh M, Yamaguchi T, Tamura C, Sanbo M, Hirabayashi M et al (2012) Identification of rat Rosa26 locus enables generation of knock-in rat lines ubiquitously expressing tdTomato. Stem Cells Dev 21:2981–2986

11. Yamamoto S, Nakata M, Sasada R, Ooshima Y, Yano T, Shinozawa T et al (2012) Derivation of rat embryonic stem cells and generation of protease-activated receptor-2 knockout rats. Transgenic Res 21:743–755
12. Goto T, Kobayashi T, Hara H, Sanbo M, Hochi S, Nakauchi H et al (2015) Knock-in of a histone H2B-tdTomato reporter into the Rosa26 locus allows visualization of cell nuclei in rats. Mol Reprod Dev 82:916–917
13. Fu Y, Foden JA, Khayter C, Maeder ML, Reyon D, Joung JK et al (2013) High-frequency off-target mutagenesis induced by CRISPR-Cas nucleases in human cells. Nat Biotechnol 31:822–826
14. Hirabayashi M, Tamura C, Sanbo M, Kato-Itoh M, Kobayashi T, Nakauchi H et al (2013) A retrospective analysis of germline competence in rat embryonic stem cell lines. Transgenic Res 22:411–416
15. Nagashima H, Giannakis C, Ashman RJ, Nottle MB (2004) Sex differentiation of germ cell production in chimeric pigs produced by inner cell mass injection into blastocysts. Biol Reprod 70:702–707
16. Isotani A, Nakanishi T, Kobayashi S, Lee J, Chuma S, Nakatsuji N et al (2005) Genomic imprinting of XX spermatogonia and XX oocytes recovered from XX<->XY chimeric testes. Proc Natl Acad Sci U S A 102:4039–4044
17. Hirabayashi M, Hochi S (2001) Systematic production of transgenic rats by pronuclear DNA microinjection. In: Miyamoto H, Manabe N (eds) Reproductive biotechnology. Hokuto Shobo, Kyoto, pp 243–250
18. Hirabayashi M, Kato M, Hochi S (2006) Transgenesis via intracytoplasmic sperm injection (ICSI) in rodents. J Mamm Ova Res 23:86–90
19. Tompers DM, Labosky PA (2004) Electroporation of murine embryonic stem cells: a step-by-step guide. Stem Cells 22:243–249
20. Bartman CM, Egelston J, Ren X, Das R, Phiel CJ (2015) A simple and efficient method for transfecting mouse embryonic stem cells using polyethylenimine. Exp Cell Res 330:178–185

Chapter 6

Protocols for Cryopreservation and Rederivation of Rat Gametes

Akiko Takizawa and Tomoo Eto

Abstract

New genome-editing tools, such as ZFNs, TALEN, and CRISPR/Cas9, have enabled the generation of gene-modified models effectively in mammals. These technologies are a powerful tool for studying gene function and creating animal models for human diseases. On the other hand, such gene-modified animals are raised in numerous experimental animal facilities, which puts pressure on breeding space and maintenance costs. Embryo and sperm cryopreservation is not only the most simple and cost-effective method available for most gene-modified strains but also the most reliable method to preserve strains to avoid breeding problems and contamination. We have established a reliable, high quality embryo and sperm cryopreservation system for rat strains, ensuring the longevity of these valuable resources for the scientific community. These cryopreserved resources have been successfully used to rederive next generation pups using embryo transfer and intracytoplasmic sperm injection (ICSI). In this chapter, we describe in detail protocols for rat embryo vitrification and sperm cryopreservation followed by pup rederivation using the ICSI procedure and embryo transfer.

Key words Cryopreservation, Embryo transfer, ICSI, Rederivation, Sperm, Two-cell, Vitrification

1 Introduction

Several gene-targeting technologies have been reported in rats, including gene-driven ENU mutagenesis [1], ZFN [2], TALEN [3], and CRISPR/Cas9 systems [4]. Since many mutants such as transgenic (Tg), knockout (KO), and knock-in (KI) rats are generated using these genome-editing technologies, embryos and spermatozoa from the Tg/KO/KI rat strains have been preserved in an effective manner in rat resource projects such as the Rat Resource Research Center, University of Missouri, USA (http://www.rrrc.us/), the National Bioresource Project-Rat, Kyoto University, Japan (http://www.anim.med.kyoto-u.ac.jp/nbr/Default.aspx), and the Gene Editing Rat Resource Center, Medical College of Wisconsin, USA (http://rgd.mcw.edu/wg/gerrc). Cryopreservation of animal gametes (embryos and spermatozoa) provides not

G. Thomas Hayman et al. (eds.), *Rat Genomics*, Methods in Molecular Biology, vol. 2018,
https://doi.org/10.1007/978-1-4939-9581-3_6,

only for efficient laboratory animal maintenance and management, but also for the prevention of genetic drift occurring during animal breeding, a source of early embryos as research materials, insurance against loss of animals due to microbial contamination and unexpected accidents, and straightforward transportation of material all over the world [5]. At various developmental stages, rat embryos can be cryopreserved by slow-freezing [6], 2-step freezing [7], and vitrification procedures [8–11]. More than half of the cryopreserved 2-cell stage embryos are capable of developing to full-term when transferred to pseudopregnant females [7, 11]. On the other hand, very few publications are available describing the cryopreservation of rat spermatozoa. One research group reported the successful production of rat offspring derived from intrauterine insemination of frozen-thawed spermatozoa [12–15], but this method is not practical for most researchers. Improvements in the protocol for cryopreservation of rat sperm are necessary to recover the post-thaw sperm without loss of motility and to apply them to routine in vitro fertilization (IVF). Cryopreserved rat sperm with poor motility can be stably re-derived by intracytoplasmic sperm injection (ICSI) with specialized technical skill. ICSI has been successfully used in mice [16] and rats [17] using a piezo-driven injection pipette, with the offspring rate of >30%. The ICSI technique was applied not only to rescue infertile male strains but also to produce transgenic rodents [18, 19]. In this chapter, successful protocols for rat embryo and sperm cryopreservation and rederivation with ICSI are described.

2 Materials

2.1 Animals

1. Rat embryos for cryopreservation: Donor females are ideally mature (7 weeks of age). The original rat strain that made gene-modified rats can be bought from a commercial breeding company, such as Fisher 344, Dark Agouti, Brown-Norway, Lewis, SHR, WKY, Long-Evans, Sprague-Dawley, and Wistar (*see* **Note 1**).
2. Sperm donor for cryopreservation and stud male: Sperm donors are ideally mature male rats (sperm motility seems to be high in animals aged 15 weeks or older).
3. Recipient for embryo transfer: Conventional outbred strain female rats, such as Sprague-Dawley or Wistar rats (*see* **Note 2**).
4. Vasectomized male for pseudopregnancy induction: Conventional outbred strain male rats, such as Sprague-Dawley or Wistar rats. Vasectomized males can be obtained from a commercial breeding company.

2.2 Equipment

1. Stereomicroscope for embryo handling.
2. Inverted microscope with 4×, 10×, 20×, and 40× objectives (*see* **Note 3**).
3. Stereomicroscope with halogen fiber illumination system.
4. CO_2 gas incubator. Standard setting: 37°C, 5% CO_2 and 95% air, humidified by a pan containing distilled water.
5. Liquid nitrogen storage tank.
6. A pair of micromanipulator and microinjector (for oocyte holding and sperm injection).
7. Piezo-driven micromanipulator with a pulse controller.
8. Mechanical puller and microforge for ICSI microtool preparation (injection and holding pipette).
9. Ultrasonic homogenizer to dissociate sperm heads from midpiece and tails.
10. Straw sealer or sealing powder.
11. Animal anesthetizer.

2.3 Tools and Supplies

2.3.1 Vaginal Smear Sampling and Hormone Injection

1. Glass slide.
2. Glass pencil.
3. Cotton swab.
4. Sterile filter paper.
5. 35 mm Petri dishes.
6. 26G, 27G syringe needles.

2.3.2 Collection of Embryos

1. Scissors.
2. Tweezers.
3. Sterile filter paper.
4. 1 mL syringe.
5. 30G syringe needle.
6. Glass capillary.

2.3.3 Vitrification and Warming of Embryos

1. 35 mm Petri dishes.
2. Embryo manipulation pipette.
3. 1 mL cryotube: Cryogenic vials inner cap type (*see* **Note 4**).
4. Crushed ice.
5. Micropipette, 1000 μL.
6. Micropipette, 100 μL.
7. Micropipette, 10 μL.

2.3.4 Cryopreservation of Epididymal Sperm

1. 1 mL cryotube.
2. 0.25 mL plastic straw (Cassou straw).
3. 18G syringe needle.
4. Sterile 1000 μL tips.
5. Sterile filter paper.
6. Micropipette, 1000 μL.

2.3.5 Intracytoplasmic Sperm Injection (ICSI)

1. Microinjection chamber (lid of 35 mm Petri dish).
2. Glass capillaries (for making sperm injection pipette and oocyte holding pipette).
3. 35 mm Petri dishes.
4. 26G, 27G syringe needle.
5. 1 mL syringe.

2.3.6 Embryo Transfer of Zygote and Embryos

1. Embryo transfer pipette.
2. Scissors.
3. Noyes scissors.
4. Hemostat clamp.
5. Tweezers.
6. Sterile gauze.
7. Synthetic absorbable sterile surgical suture.
8. Autoclip®.

2.4 Media and Chemicals

2.4.1 Embryo Vitrification

1. Embryo handling solution for vitrification: PB1 medium (Table 1) [20] (Commercially available).
2. Vitrification solution (PEPeS): 10% Propylene glycol, 30% Ethylene glycol, 20% Percoll, and 0.3 M Sucrose in PB1.
3. Pretreatment solution (P10): 10% Propylene glycol in PB1.
4. Warming solution (SPB1): 0.3 M Sucrose in PB1.
5. Pregnant mare's serum gonadotropin (PMSG): Dissolved in physiological saline (100 IU/mL) and stored at −20°C until use.
6. Human chorionic gonadotropin (hCG): Dissolved in physiological saline (50 IU/mL) and stored at −20 °C until use.
7. Mineral oil (Mouse embryo culture tested).
8. 0.9% sterile saline.
9. 10% Giemsa solution in ethanol.

2.4.2 Sperm Freezing and ICSI

1. Tris-EDTA (TE) buffer (pH 8.0)
2. Oocyte handling solution for ICSI: Modified Krebs-Ringer Bicarbonate Buffer (mKRB) medium (Table 1) [21].

Table 1
Composition of media

Reagents	PB1	Modified-KRB	Hepes-KRB
NaCl	136.87 mM	94.6 mM	94.6 mM
KCl	2.68 mM	4.78 mM	4.78 mM
KH_2PO_4	1.47 mM	1.19 mM	1.19 mM
$CaCl_2$	0.90 mM	1.71 mM	1.71 mM
$MgCl_2{\cdot}6H_2O$	0.49 mM	–	–
Na_2HPO_4	8.09 mM	–	–
$MgSO_4{\cdot}7H_2O$	–	1.19 mM	1.19 mM
Glucose	5.56 mM	5.56 mM	5.56 mM
$NaHCO_3$	–	25.07 mM	4 mM
Na-Lactate	–	21.58 mM	21.58 mM
Na-Pyruvate	0.33 mM	0.5 mM	0.5 mM
Penicillin G	100 U/mL	75 μg/mL	75 μg/mL
Streptomycin	–	50 μg/mL	50 μg/mL
Hepes (1M)	–	–	22 mM
BSA	3 mg/mL	4 mg/mL	4 mg/mL
Osmolarity		280–290 mOsm	
		pH 7.4	

3. Hepes-KRB medium (Table 1).
4. 10% Polyvinylpyrrolidone (PVP) solution with Hepes-KRB (Hepes-KRB/PVP) (*see* **Note 5**).
5. 0.1 % Hyaluronidase in medium.
6. Mercury (Mercury can be replaced with Fluorinert™: e.g., FC-70).

2.4.3 Embryo Transfer

1. Epinephrine (0.1% Adrenaline): Dilute to 1/3 concentration with saline.
2. Ampicillin (250 mg/1 mL in saline).
3. Isoflurane.
4. Analgesic (e.g., Carprofen and Buprenorphine)
5. 10% Giemsa solution in ethanol.

3 Methods

3.1 Cryopreservation of Rat Embryos

3.1.1 Vaginal Smear Checking and Superovulation

1. Prepare a cotton swab moistened with distilled water and a glass slide with 5–10 mm grids written on it using glass pencil.
2. Vaginal smear samples are collected with the cotton swab and spread on the glass slide.
3. The glass slide is air-dried and stained with Giemsa solution, and then vaginal smears are observed under a microscope (*see* Fig. 1).

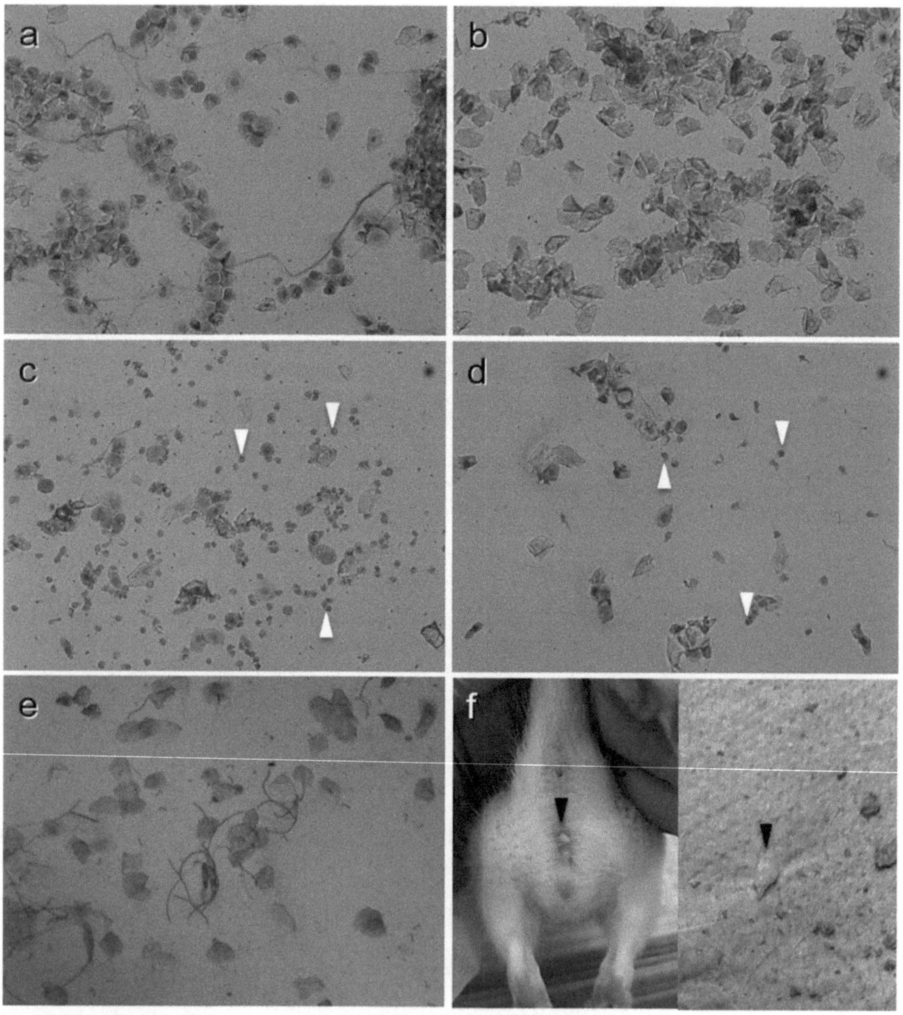

Fig. 1 Vaginal cytology showing each stage of estrus. Stages of estrous include (**a**) proestrus, (**b**) estrus, (**c**) metestrus, and (**d**) diestrus. Three cell types are identified: nucleated epithelial, cornified epithelial, and leukocytes (white arrowhead). (**e**) A lot of sperm in the rat vaginal smear after copulation. (**f**) Vaginal plug (black arrowhead) after copulation

4. The metestrus (*see* Fig. 1c) female rats are superovulated by intraperitoneal injection of 150 iu/kg PMSG and 75 iu/kg of hCG at 48–50 h intervals [22] (*see* **Note 6**). This step can be skipped if superovulation is not performed, such as when using poor breeding performance rats as embryo donors.
5. On the day of the hCG injection, female rats are set up with stud male rats for mating. When using natural ovulation, select females which are in proestrus and directly set up mating with stud males. Next morning, check for sperm or vaginal plug (*see* Fig. 1e, f).

3.1.2 Recovery of Rat 2-cell Embryos

1. Five drops with PB1 are formed in a 35-mm Petri dish and covered with mineral oil (*see* Fig. 2a). PB1 is aspirated into a 1-mL syringe, attached with a 26G syringe needle. These are kept in a CO_2 incubator until medium is warmed up to 37 °C.
2. Hyaluronidase is dissolved with PB1 at 1 mg/mL. The solution is kept in a CO_2 incubator for more than 30 min.
3. Donor females are euthanized by CO_2 inhalation.
4. The oviducts are excised from a female rat on the next day following confirmation of copulation, and placed on a filter paper on a hot plate.
5. After removal of adhered tissue fluid and blood, the oviducts alone are cut out and placed in PB1 drop A.
6. One oviduct is placed in drop B. Gently hold the end of the oviduct and insert the syringe needle into the oviductal infundibulum. Perfuse the oviduct with 0.1–0.2 mL of PB1 (oviduct flushing). The other oviduct is similarly processed.
7. When cumulus cells are adhered to embryos, cells are removed by adding Hyaluronidase solution to drop B.
8. Two-cell stage embryos collected by perfusion are rinsed by transferring embryos from drops C to E. Collected embryos are placed in drop E until cryopreserved.

3.1.3 Vitrification of Rat Embryos

1. Pretreatment solution (P10) is warmed at room temperature. Vitrification solution (PEPeS) is cooled on crushed ice in a styrene foam box (0 °C).
2. Two drops (about 100 μL each) of P10 are formed in a 35 mm dish lid.
3. Two-cell stage embryos are placed in the bottom of a P10 drop with a minimum volume of PB1 (*see* Fig. 2b). Start the timer at this point.
4. The embryos float in the P10-drop and then sink down to the bottom.

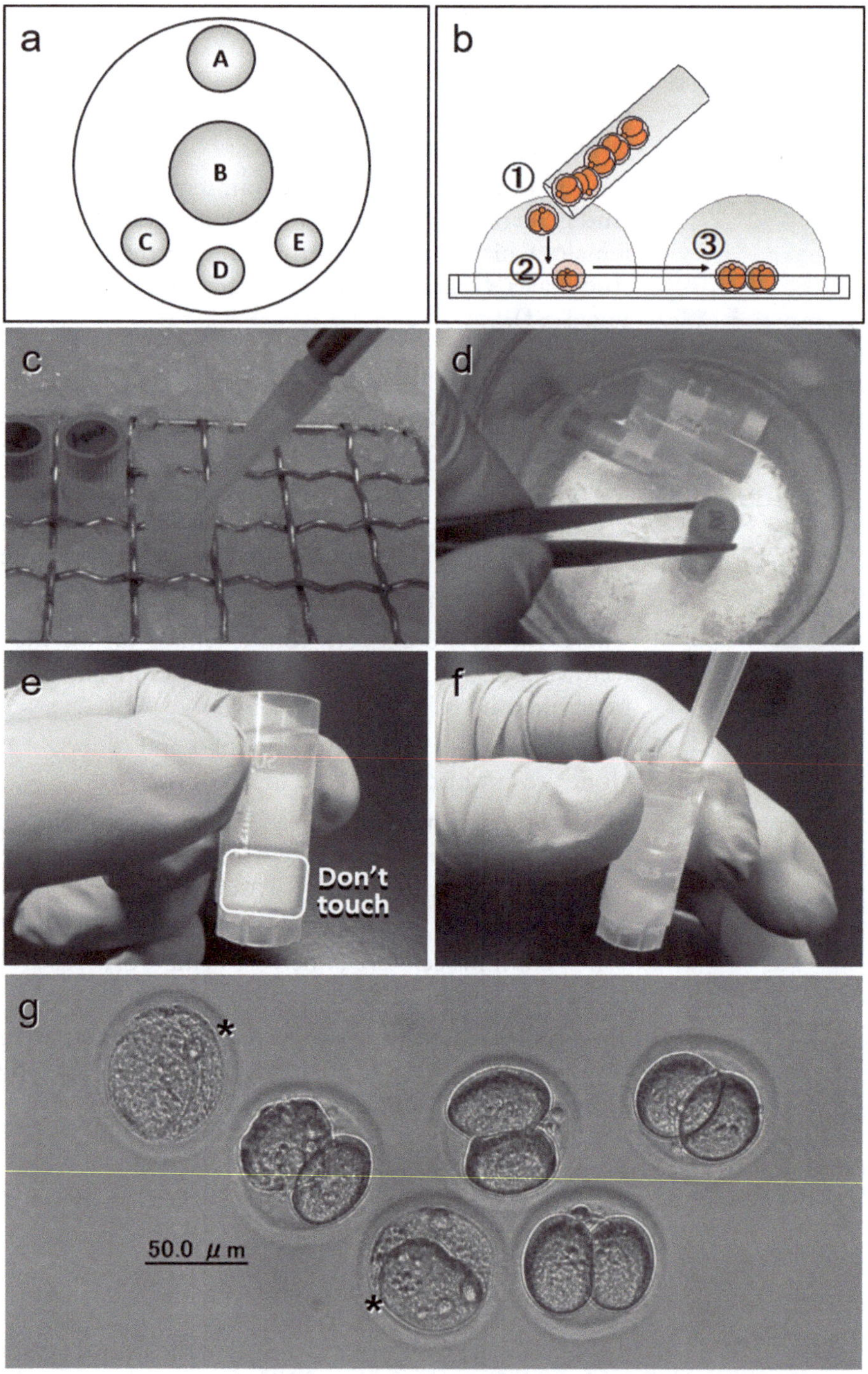

Fig. 2 Two-cell embryo recovery and vitrification. (**a**) For collection dish, make five drops with PB1 of different volumes (A, 50 μL; B, 100 μL; C, D, and E, 30 μL). (**b**) Pretreatment before vitrification. (1) Embryos float to the drop surface. (2) Embryos sink to the dish bottom. (3) Embryos are transferred to the second drop to change the solution. (**c**) Add 95 μL of PEPeS at 0 °C into the cryotube and equilibrate for 1 min. at 0 °C. (**d**) The cryotube is directly plunged into liquid nitrogen. (**e**) Cryotube is taken out of liquid nitrogen and is kept at room temperature. Don't touch the bottom of the cryotube. (**f**) SPB1 solution is added into the tube and immediately stirred 5–10 times. (**g**) Viable embryos and degenerated embryos (∗) post vitrification

5. After transferring the embryos to the other P10 drop, the embryos are transferred with 5 μL of P10 into a cryotube using a P100 Pipetman®.
6. At about 5 min after placing the embryos into the first drop, the cryotube containing the embryos is cooled on crushed ice. Make sure at least half of the tube depth is placed into crushed ice.
7. After cooling for 1 min, 95 μL of PEPeS is added into the cryotube along the wall of the tube (*see* Fig. 2c).
8. After 1 min, the cryotube is placed in liquid nitrogen and vitrified (*see* Fig. 2d).

3.1.4 Warming of Vitrified Rat Embryos

1. Four drops of PB1 (50 μL each) are formed and covered with mineral oil in a 35-mm dish. The dish containing the drops and warming solution (SPB1) are incubated at 37 °C.
2. Cryotube is taken out from liquid nitrogen and is kept at room temperature (*see* Fig. 2e).
3. Previously warmed SPB1 solution (900 μL) is aspirated into a micropipette.
4. About 1–1.5 min after the PEPeS solution surface begins to turn white, add SPB1 solution to the cryotube (*see* Fig. 2f) and warm the sample quickly via pipetting 5–10 times (*see* **Note 7**).
5. The embryo containing liquid in the cryotube is transferred into a 60-mm dish lid.
6. Two-cell stage embryos are collected under a stereoscopic microscope using a glass capillary.
7. The embryos are slowly transferred to and kept at the bottom of a PB1 drop.
8. After 2 min, the embryos are rinsed by transferring them to the second and third drop. A new capillary should be used in each drop.
9. Morphologically normal embryos are collected and placed into mKRB until oviductal transfer (*see* Fig. 2g).

3.2 Cryopreservation of Rat Sperm

3.2.1 Recovery of Rat Sperm from Caudal Epididymides

1. The donor stud male for collection of sperm is euthanized by CO_2 inhalation.
2. The caudal epididymides are excised from a mature male rat (3–6 months of age) and placed on a sterile filter paper.
3. After trimming fat, blood, and tissue, the caudal epididymides are held gently with tweezers (*see* Fig. 3a).
4. The seminiferous tubes in the caudal epididymides are cut with a 18G needle tip (*see* Fig. 3b) and sperm masses are squeezed out (*see* Fig. 3c).

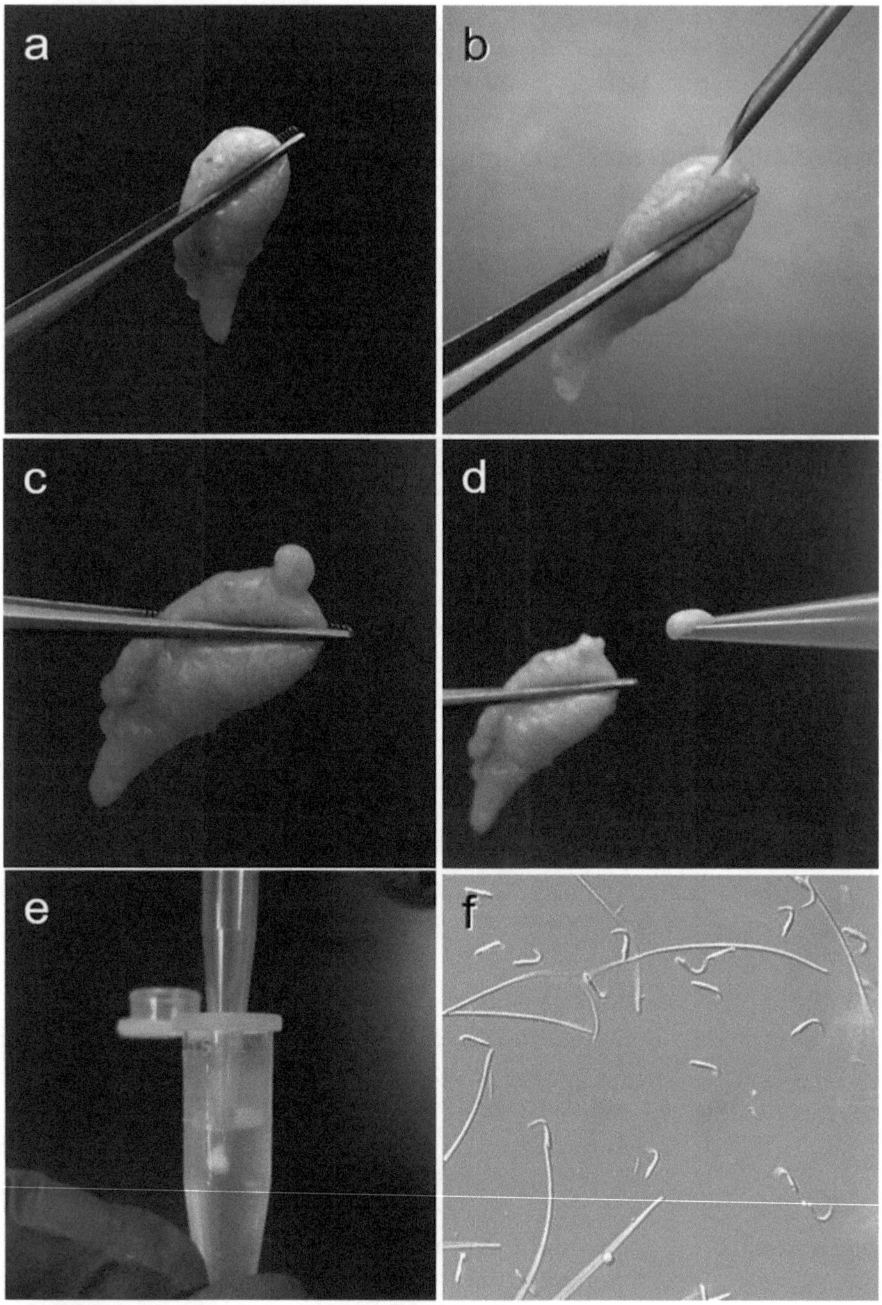

Fig. 3 Sperm preparation from rat epididymis. (**a**) Place the cauda epididymides on a piece of filter paper and completely remove any fat, blood, and connective tissue. (**b**) The seminiferous tubes in the caudal epididymides are cut with a needle tip. (**c**) The sperm masses are squeezed out. (**d**) The sperm masses are squeezed out using pipette tip. (**e**) The sperm mass is transfer into the TE buffer. (**f**) The sperm suspension is sonicated for 10 s using a low power output from an Ultrasonic homogenizer to dissociate sperm head from midpiece and tail

5. Sperm masses are scooped using pipette tip (*see* Fig. 3d) and suspended with 1 mL of TE buffer via pipetting with a micropipette (*see* Fig. 3e).
6. The sperm suspension is aliquoted into cryotubes (200 μL each) or 0.25 mL plastic straw, and cryopreserved in liquid nitrogen until used.

3.2.2 Thawing of Rat Sperm

1. A cryotube is taken out of a liquid nitrogen tank, and inverted with the lid open to remove liquid nitrogen in the tube.
2. The lid is then closed, and the cryotube is immersed in a 20–25 °C water to defrost sperm.
3. After 5–10-fold dilution with Hepes-KRB, the sperm suspension is sonicated for 5–10 s using a minimum power output from an ultrasonic homogenizer to dissociate sperm heads from midpieces and tails (*see* Fig. 3f).

3.3 Intracytoplasmic Sperm Injection (ICSI)

3.3.1 Preparation of Glass Microtools (*see* **Note 8**)

1. Platinum filament of the mechanical puller in the adjustable region of the heater is molded using forceps and precision tools.
2. A glass capillary is set so that the central part of the capillary is enclosed by the filament of the heater.
3. The heater and magnet scales are arbitrarily set. Since the pulling condition varies depending on the puller model, temperature, and humidity, settings are adjusted to prepare an about 1-cm shank region (*see* Fig. 4a).
4. For sperm injecting pipette: The capillary extended by the puller is horizontally attached to the microforge where a glass bead is equipped at the tip of the platinum heater (*see* Fig. 4b).
5. The outer diameter of the capillary contacting with the glass bead should be 3–4 μm, as determined microscopically (*see* Fig. 4c).
6. The tube is appropriately heated, and the power supply of the heater is turned off when fusion of the capillary and glass bead starts, by which the tip is vertically cut off (*see* Fig. 4d).
7. The capillary is positioned vertically to the glass bead, and then bent to approximately 150° using the heated glass bead (*see* Fig. 4e).
8. Before ICSI, use a 27G syringe needle and 1-mL syringe to deliver aliquots of mercury or Fluorinert™ into the barrel of a sperm injection pipette through the open (back) end.
9. For oocyte holding pipette (commercially available): A capillary extended by the puller is scratched with an ampoule-cutter at a site with an outer diameter of 90–120 μm in order to cut the tip vertically (*see* Fig. 4f).

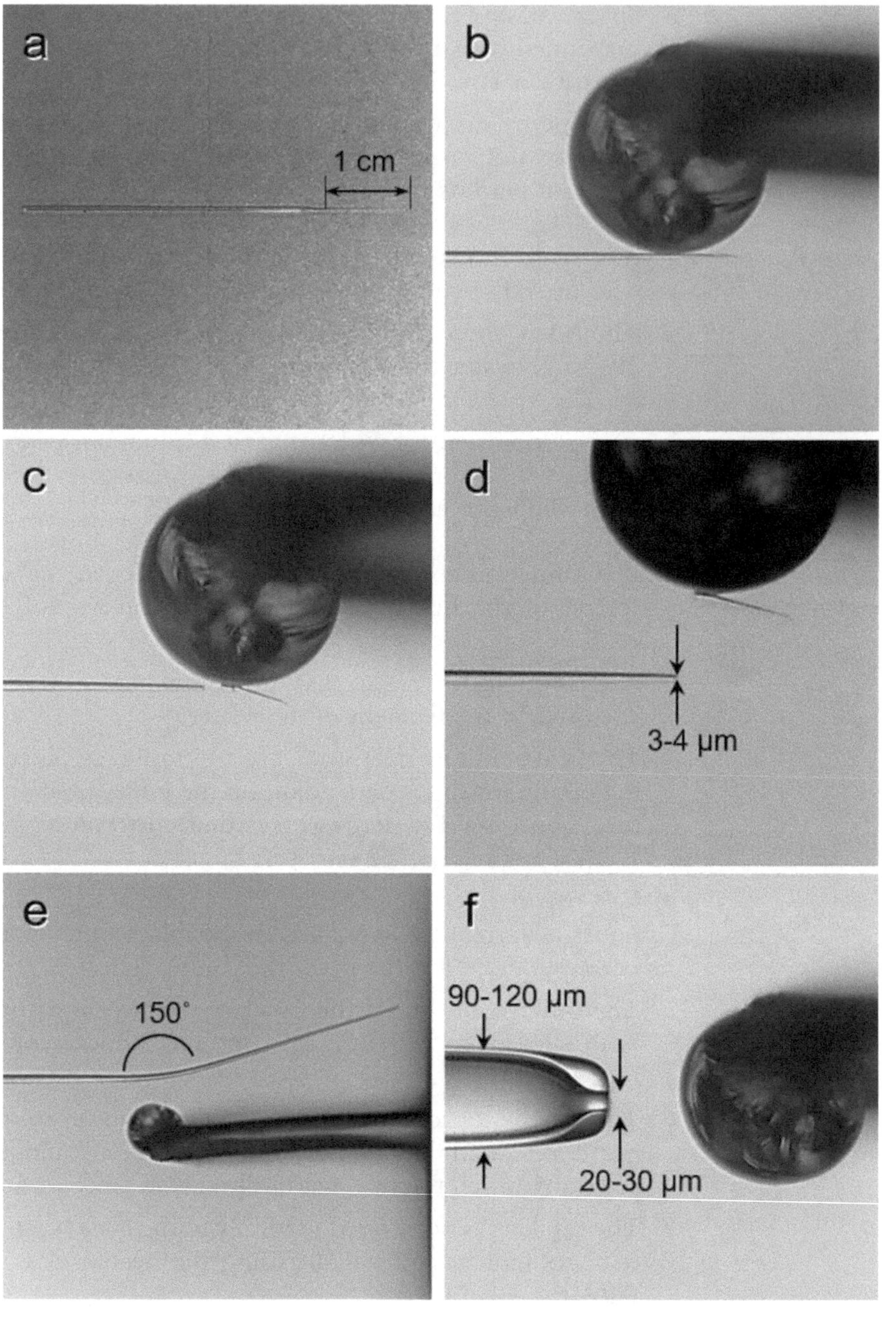

Fig. 4 Microtool processing for rat ICSI. (**a**) A glass capillary is heat-pulled by mechanical puller, to make the shank region approximately 1-cm length. The extent of tapering off is dependent on the puller model. (**b**) The second step is to contact the pulled capillary where the outer diameter is 3–4 μm with the unheated glass bead of the microforge. (**c**, **d**) The heater is turned on/off quickly, so the tip of the capillary is vertically cut off. (**e**) The heated bead is used to bend the injection pipette to give an angle of approximately 150°. (**f**) For an oocyte holding pipette, the blunt end of a pulled and cut capillary (90–120 μm outer diameter) is processed by the heated bead, to give an inner diameter of 20–30 μm

10. The capillary is horizontally attached to a microforge, and the tip is rounded (until the inner diameter of the tip becomes 20–30 μm) by heating with the glass bead equipped at the tip of the platinum heater.
11. Also bend holding pipette to approximately 150°.

3.3.2 Preparation of Oocytes

1. Hyaluronidase is dissolved with mKRB at 1 mg/mL, and sterilized by filtration.
2. Three 50 μL drops of mKRB and one drop of hyaluronidase solution are formed and covered with mineral oil in a 35 mm Petri dish, and equilibrated in an incubator at 37 °C in 5% CO_2 and 95% air together with the remaining culture medium.
3. The donor female rat (*see* **Note 9**) is euthanized by CO_2 inhalation 14 h after hCG administration, and the oviducts are excised and placed on sterile filter paper.
4. After trimming fat, blood, and tissue, the excised oviducts are placed in mineral oil near the hyaluronidase drop, and the ampullae of the oviducts are opened using a 26G syringe needle under a stereoscopic microscope. Released cumulus oocyte complexes in mineral oil are pulled into the hyaluronidase drop using the syringe needle.
5. The collected cumulus oocyte complexes are denuded by mechanically pipetting with a glass capillary, then the oocytes are washed 3 times by transferring them through the mKRB drops. They are then placed in a CO_2 incubator until ICSI (*see* **Note 10**).

3.3.3 Preparation of Spermatozoa

1. One drop (8 μL) each of solutions <1>–<4> is formed in the lid of a 35 mm Petri dish, as shown (*see* Fig. 5a).
2. The inside of an injection pipette is washed by repeated blowing out mercury or Fluorinert™ and aspirating Hepes-KRB/PVP in drop <4>.
3. Sperm are placed in drop <3>, and oocytes are placed in drop <1>.
4. A sperm is aspirated and hooked into the injection pipette and transferred into drop <1> through the drop <2> for sperm injection.
5. An ovum is retained, placing the first polar body at 12:00 or 6:00 position, using the holding pipette.
6. The sperm is released near the zona pellucida (*see* Fig. 5b; arrow head), and only the zona pellucida is penetrated by several piezo-pulses (intensity: 2, speed: 1) (*see* Fig. 5c).
7. The sperm is re-aspirated into the pipette (*see* Fig. 5d). The pipette is then deeply inserted into the egg cytoplasm, and a low positive pressure is loaded on the injector (*see* Fig. 5e).

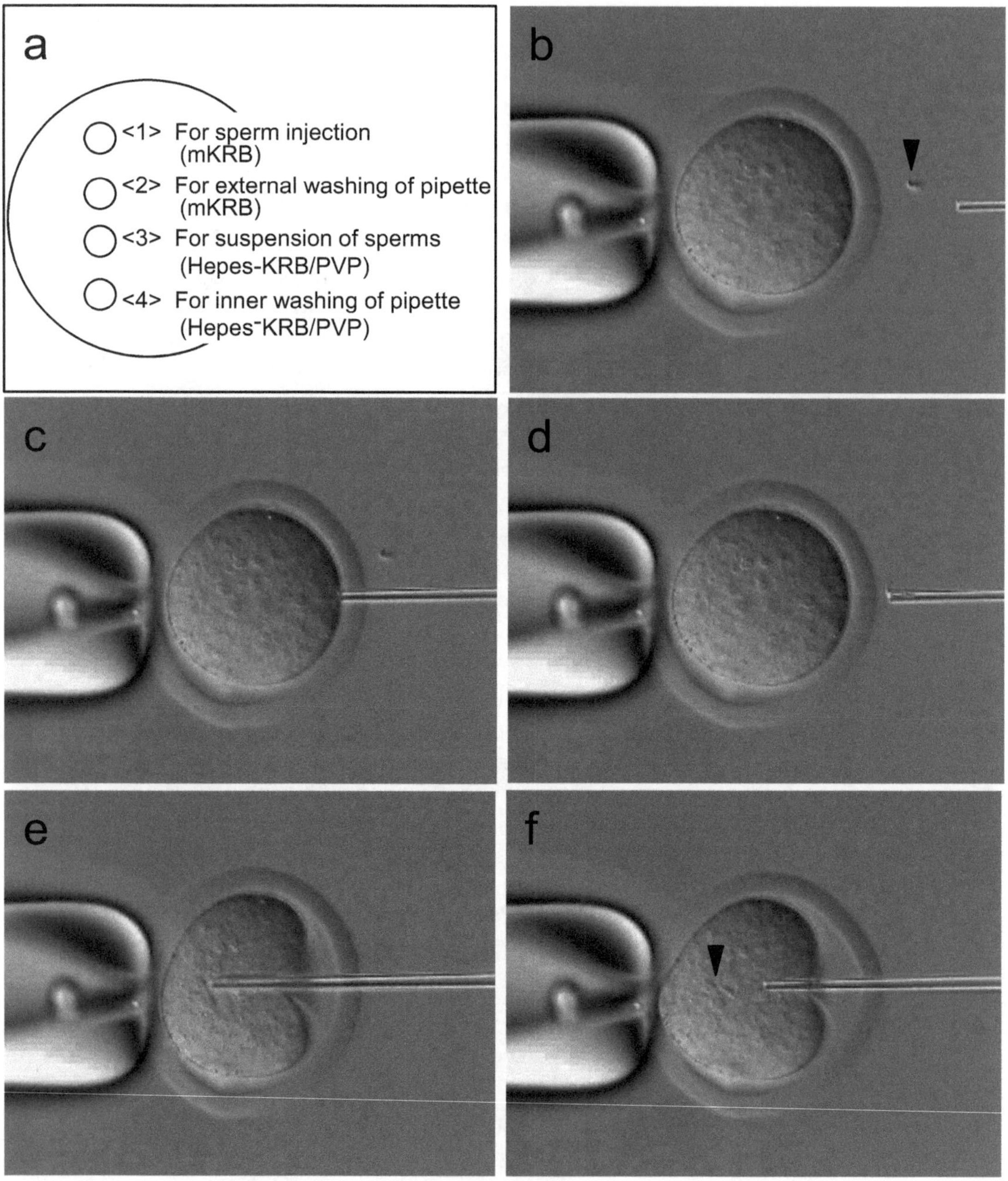

Fig. 5 Protocol for rat ICSI. (**a**) The ICSI chamber contains two 8 μL microdrops of mKRB (<1> and <2>) and two 8 μL microdrops of mKRB/PVP (<3> and <4>). Oocytes and sperm heads are placed in drop-<1> and -<3>, respectively. (**b**) The sperm (black arrowhead) is released near the zona pellucida. (**c**) Zona pellucida is drilled using several piezo-pulses (intensity 2, speed 1). (**d**) Sperm head is hung on the blunt end of the injecting pipette. (**e**) The oolemma is punctured by one piezo-pulse (intensity 1, speed 1) with a light positive pressure. (**f**) The sperm head (black arrowhead) is injected into the ooplasm with a minimum amount of the accompanying medium

8. A single piezo-pulse (intensity: 1, speed: 1) is loaded to make the pipette penetrate the egg cell membrane, and the sperm is simultaneously injected into the egg cytoplasm (*see* Fig. 5f; arrow head), immediately followed by rapidly pulling the pipette out of the oocyte.
9. The ovum containing the injected sperm is placed into Hepes-KRB for recovery, and kept at room temperature for 10 min.
10. After washing three times with mKRB drops, oocytes are cultured in a CO_2 incubator.
11. The viability and fertility are determined 6 h after sperm injection. Oocytes with both the male and female pronuclei and the second polar body are regarded as normal fertilized zygotes.
12. The cleavage rate of fertilized oocytes is determined 24 h after sperm injection, and embryos with normal morphology are subjected to oviduct transfer.

3.4 Oviductal Embryo Transfer

1. The day before embryo transplantation, proestrus (*see* Fig. 1a) female rats are set up with vasectomized male rats for mating.
2. The next morning, the rats are checked for vaginal plugs. The plugged rats can be used as recipient foster mothers for embryo transfer.
3. Recipient rats are anesthetized with 2–3% isoflurane gas using a small animal anesthetizer.
4. Dorsal hair is clipped, and a midline incision is made in the dorsal skin.
5. Pinching fat with forceps, the region from the left ovary to uterus is taken out and placed on sterile gauze.
6. The upper ovarian fat is fixed with a hemostat clamp to position the ovary-uterus.
7. Three air bubbles and 2-cell stage embryos (usually 5–10 per one side of oviduct) are aspirated into an embryo transfer pipette (*see* Fig. 6a). In the case of ICSI embryo transplantation, 20–30 embryos per recipient may be transferred because the successful development of ICSI embryos to pups is not very high.
8. One or two drops of epinephrine are dripped onto the ovarian bursa. The blood vessels on the membrane surface immediately constrict.
9. Pinching the ovarian bursa with forceps, the membrane is widely incised using Noyes scissors.
10. The oviductal infundibulum is exposed using a tweezer, and then an embryo transfer pipette is inserted into the infundibulum.

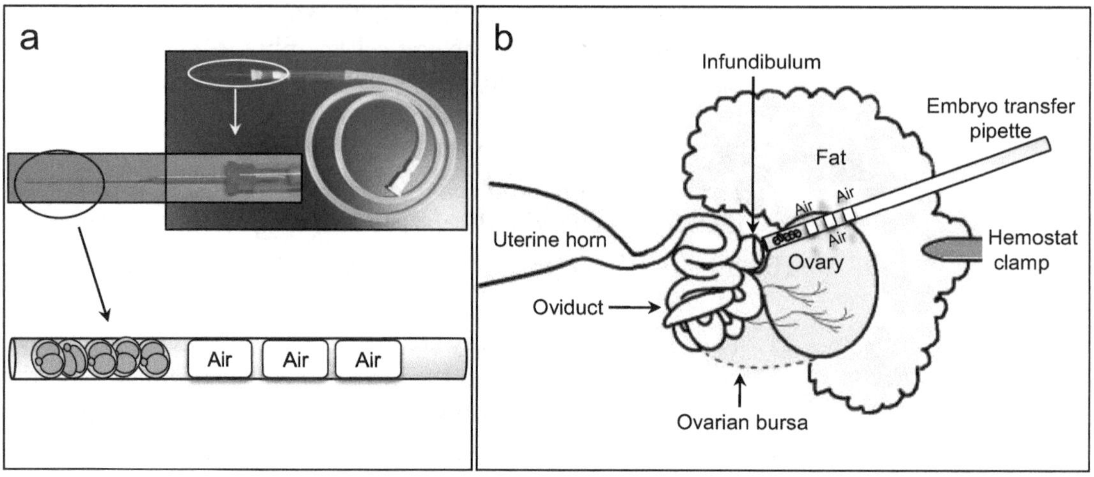

Fig. 6 Oviductal embryo transfer. (**a**) Embryo manipulation pipette is attached to a pipette holder and used. Embryos are aspirated after the 3-air bubble marker in the embryo transfer pipette. (**b**) Cut off the ovarian bursa with Noyes scissors and expose the oviductal infundibulum. Insert the pipette into the infundibulum and blow the embryo with 2–3 of the air bubbles into the oviduct

11. The embryos with air bubble are injected into the oviduct by gentle pressure through the pipette (*see* Fig. 6b).
12. The hemostat clamp is removed, and the ovary-uterus is returned into the abdominal cavity.
13. Embryos are similarly transferred into the oviduct of the other side.
14. The outer skin is closed with Autoclips®. The rat is placed in a cage and warmed under an incandescent lamp, and returned to the animal room when the rat starts moving and is recovered from anesthesia.
15. Analgesic drugs such as Carprofen and Buprenorphine are given during and post-surgery (*see* **Note 11**).

4 Notes

1. Depending on the experimental design or client requests, rat strains for embryo donor females as well as sperm donor and stud males must be carefully chosen (consideration of reproductive performance, fertilization rate, offspring rate, etc.).
2. Generally, it takes approximately 2–3 weeks until post-pubertal rats exhibit stable estrus cycles with a defined 4-day period. It is recommended to select female rats showing a stable estrus cycle.
3. Using the Relief-contrast or Nomarski module is recommended to set up on the microscope for embryo manipulation.

4. The type of cryotube can have a negative impact on embryo cryopreservation. We recommend using Sumitomo tubes for PEPeS vitrification (Sumitomo Bakelite Co., Ltd.).
5. Usually use PVP with an average molecular weight of 360,000. The mKRB/PVP can be stored at −20 °C, but should not be kept beyond 3 months. The Hepes-KRB/PVP is thawed by placing at room temperature, and should be used within one week (ideally within the day of ICSI).
6. Usually adult metestrus female rats (8–13 weeks old) are superovulated by intraperitoneal injection of 150 iu/kg PMSG at 10:00–12:00 AM and 75 iu/kg of hCG at 10:00–12:00 AM. Female rats are set up with stud male rats for mating at 5:00 PM on the day of the hCG injection. When using natural ovulation, proestrus females are selected and directly set up for mating with stud males. The next morning, check the rat for sperm with vaginal smear or vaginal plug. If the vaginal plug cannot be confirmed on the tray, the donor should be checked for sperm with vaginal smear. The 2-cell stage embryos are collected from the oviductal ampullae 44–50 h after the hCG injection using PB1.
7. The timing of adding SPB1 solution varies depending on the lab environment. We recommend measuring the time it takes for the edge of the vitrified solution in the tube to change slightly to white. SPB1 solution must be added before this time.
8. Glass microtools (pipettes for sperm injection and oocyte holding) should be prepared by processing glass capillaries with equipment such as a mechanical puller and a microforge. We use glass capillaries (#B100-75-10, Sutter Instrument Co., Novato, CA), mechanical puller (P-97/IVF, Sutter), and microforge (e.g., MF-900, Narishige Tokyo, Japan). Glass capillaries from other commercial sources (e.g., G-1, Narishige) are also available, as well as mechanical pullers and microforges from different suppliers.
9. For oocyte donors, superovulation treatment of females at immature or younger age is recommended because gonadotropin can be administered regardless of their estrus cycles [23–25].
10. Ovulated rat oocytes spontaneously activate during in vitro culture [26]. Therefore, it is desirable that sperm injection is done within 45 min of oocyte recovery. Reducing the time of oocyte recovery to less than 5 min after sacrificing the donor rats and handling of oocytes at 37 °C reduces the spontaneous activation of the rat oocytes.
11. The use of Carprofen (5 mg/kg) or Buprenorphine (0.01–0.05 mg/kg) is recommended. For example, 0.04 mL

Carprofen (50 mg/mL stock solution) is administered subcutaneously per 200 g body weight of rat. Analgesics are administered to animals undergoing surgery on the day of surgery and the following 2 days. The decision on usage and dosage of an analgesic should be done in consultation with veterinary staff.

Acknowledgments

This work was supported by the National Heart, Lung, and Blood Institute of the National Institutes of Health (R24HL114474).

References

1. Mashimo T, Yanagihara K, Tokuda S, Voigt B, Takizawa A, Nakajima R et al (2008) An ENU-induced mutant archive for gene targeting in rats. Nat Genet 40:514–515
2. Geurts AM, Cost GJ, Freyvert Y, Zeitler B, Miller JC, Choi VM et al (2009) Knockout rats via embryo microinjection of zinc-finger nucleases. Science 325:433
3. Mashimo T, Kaneko T, Sakuma T, Kobayashi J, Kunihiro Y, Voigt B et al (2013) Efficient gene targeting by TAL effector nucleases coinjected with exonucleases in zygotes. Sci Rep 3:1253
4. Li D, Qiu Z, Shao Y, Chen Y, Guan Y, Liu M et al (2013) Heritable gene targeting in the mouse and rat using a CRISPR-Cas system. Nat Biotechnol 31:681–683
5. Festing MFW (1974) Embryo banks in the production of genetically defined laboratory animals: a step towards the concept of type culture collection of defined laboratory animals. In: Zeilmarker GH (ed) Frozen storage of laboratory animals. Gustav Fischer Verlag, Stuttgart, New York, pp 149–155
6. Whittingham DG (1975) Survival of rat embryos after freezing and thawing. J Reprod Fertil 43:575–578
7. Hirabayashi M, Takahashi R, Sekiguchi J, Ueda M (1997) Viability of transgenic rat embryos after freezing and thawing. Exp Anim 46:111–115
8. Kono T, Suzuki O, Tsunoda Y (1988) Cryopreservation of rat blastocysts by vitrification. Cryobiology 25:170–173
9. Takahashi R, Hirabayashi M, Ueda M (1990) Production of transgenic rats using cryopreserved pronuclear-stage zygotes. Transgenic Res 8:397–400
10. Han MS, Niwa K, Kasai M (2003) Vitrification of rat embryos at various developmental stages. Theriogenology 59:1851–1863
11. Eto T, Takahashi R, Kamisako T, Hioki K, Sotomaru Y (2014) A study on cryoprotectant solution suitable for vitrification of rat two-cell stage embryos. Cryobiology 68:147–151
12. Nakatsukasa E, Inomata T, Ikeda T, Shino M, Kashiwazaki N (2001) Generation of live rat offspring by intrauterine insemination with epididymal spermatozoa cryopreserved at −196 degrees C. Reproduction 122:463–467
13. Nakatsukasa E, Kashiwazaki N, Takizawa A, Shino M, Kitada K, Serikawa T et al (2003) Cryopreservation of spermatozoa from closed colonies, and inbred, spontaneous mutant, and transgenic strains of rats. Comp Med 53:639–641
14. Seita Y, Sugio S, Ito J, Kashiwazaki N (2009) Generation of live rats produced by in vitro fertilization using cryopreserved spermatozoa. Biol Reprod 80:503–510
15. Seita Y, Fujiwara K, Takizawa A, Furukawa K, Inomata T, Ito J et al (2011) Full-term development of rats from oocytes fertilized in vitro using cryopreserved ejaculated sperm. Cryobiology 63:7–11
16. Kimura Y, Yanagimachi R (1995) Mouse oocytes injected with testicular spermatozoa or round spermatids can develop into normal offspring. Development 121:2397–2405
17. Hirabayashi M, Kato M, Aoto T, Sekimoto A, Ueda M, Miyoshi I et al (2002) Offspring derived from intracytoplasmic injection of transgenic rat sperm. Transgenic Res 11:221–228
18. Perry ACF, Wakayama T, Kishikawa H, Kasai T, Okabe M, Toyoda Y et al (1999) Mammalian transgenesis by intracytoplasmic sperm injection. Science 284:1180–1183
19. Kato M, Ishikawa A, Kaneko R, Yagi T, Hochi S, Hirabayashi M (2004) Production of transgenic rats by ooplasmic injection of

spermatogenic cells exposed to exogenous DNA: a preliminary study. Mol Reprod Dev 69:153–158

20. Whittingham DG (1974) Embryo banks in the future of developmental genetics. Genetics 78:395–402
21. Toyoda Y, Chang MC (1974) Fertilization of rat eggs in vitro by epididymal spermatozoa and the development of eggs following transfer. J Reprod Fertil 36:9–22
22. Mukumoto S, Mori K, Ishikawa H (1995) Efficient induction of superovulation in adult rats by PMSG and hGH. Exp Anim 44:111–118
23. Hirabayashi M, Ito K, Sekimoto A, Hochi S, Ueda M (2001) Production of transgenic rats using young Sprague-Dawley females treated with PMSG and hCG. Exp Anim 50:365–369
24. Popova E, Krivokharchenko A, Ganten D, Bader M (2002) Comparison between PMSG- and FSH-induced superovulation for the generation of transgenic rats. Mol Reprod Dev 63:177–182
25. Corbin TJ, McCabe JG (2002) Strain variation of immature female rats in response to various superovulatory hormone preparations and routes of administration. Contemp Top Lab Anim Sci 41:18–23
26. Zernicka-Goetz M (1991) Spontaneous and induced activation of rat oocytes. Mol Reprod Dev 28:169–176

Chapter 7

Fluorescent Imaging and Microscopy for Dynamic Processes in Rats

Ruben M. Sandoval, Bruce A. Molitoris, and Oleg Palygin

Abstract

The rat is a favored model organism to study physiological function in vivo. This is largely due to the fact that it has been used for decades and is often more comparable to corresponding human conditions (both normal and pathologic) than mice. Although the development of genetic manipulations in rats has been slower than in mice, recent advances of new genomic editing tools allow for the generation of targeted global and specific cell type mutations in different rat strains. The rat is an ideal model for advancing imaging techniques like intravital multi-photon microscopy or IVMPM. Multi-photon excitation microscopy can be applied to visualize real-time physiologic events in multiple organs including the kidney. This imaging modality can generate four-dimensional high resolution images that are inherently confocal due to the fact that the photon density needed to excite fluorescence only occurs at the objective focal plane, not above or below. Additionally, longer excitation wavelengths allow for deeper penetration into tissue, improved excitation, and are inherently less phototoxic than shorter excitation wavelengths. Applying imaging tools to study physiology in rats has become a valuable scientific technique due to the relatively simple surgical procedures, improved quality of reagents, and reproducibility of established assays. In this chapter, the authors provide an example of the application of fluorescent techniques to study cardio-renal functions in rat models. Use of experimental procedures described here, together with multiple available genetically modified animal models, provide new prospective for the further application of multi-photon microscopy in basic and translational research.

Key words Intravital imaging, Fluorescence microscopy, Multi-photon, Kidney, Rat, Cardio-renal, Rat models, Vascular, Albumin

1 Introduction

Breakthroughs in biomedical fields often follow discoveries in technological advances that afford researchers opportunities to probe physiologic processes where mechanisms were previously deduced using more indirect methods. Intravital multi-photon microscopy (IVMPM), as applied to the study of renal anatomy and physiology, represents such a disruptive technology. The technique allows for the direct visualization of dynamic pathways and labeled structures in vivo at a level of spatial resolution to discriminate small,

G. Thomas Hayman et al. (eds.), *Rat Genomics*, Methods in Molecular Biology, vol. 2018,
https://doi.org/10.1007/978-1-4939-9581-3_7,

individual structures such as endosomes [1] with temporal resolution approaching real time through the use of resonant scanners and smaller viewing regions. IVMPM was first used to study neuronal calcium dynamics in a living rat by Svoboda and colleagues in 1997 [2]. The brain represents an ideal organ to study, having a lower opacity than the kidney thus minimizing photon scatter, allowing for deep tissue imaging. Also, the stereotactic skull apparatus allows for greater stability and minimal effect from breathing to induce motion artifacts. In 2002, Peti-Peterdi et al. brought IVMPM technology to the renal field and utilized two-photon excitation to study the isolated, perfused juxtaglomerular apparatus [3]. Our group was the first who applied multi-photon excitation to study the intact, attached rat kidney showing three-dimensional data at subcellular resolution while also capturing dynamic physiological processes in vivo [1].

The rat as a research model provides an excellent platform in which to study complex processes simultaneously in conjunction with IVMPM. Rat organ size allows for ease of use on most standard inverted microscope stages. Surgical procedures and disease models, some complex such the 5/6 nephrectomy model [4], although initially challenging, can be mastered with practice. Similar procedures in mice would be more challenging and require highly specialized tools for manipulation. Furthermore, blood volumes in rats allow for infusion of more feasible quantities of fluorescent compounds than in mice, which may require the undesirable need to mix compounds to minimize the total volume delivered to avoid hypervolemia.

In this chapter, we will provide a basis for the experienced surgeon and research microscopist with which to expand their methods "toolbox" and incorporate IVMPM to augment their research. We will discuss microscope settings and the importance of keeping detector settings constant between experiments to produce consistent data that allow the best comparison even though acquisition occurs on different days. We will briefly touch upon the advantages of using multi-photon excitation over single-photon excitation for intravital studies. One important consideration relating to microscope settings is the type of data that will be collected and the subsequent analysis that will be conducted. Simplified data analysis such as morphological changes or regions a structure occupies allows for more flexibility in acquisition parameters. In contrast, intensity-based comparisons demand more exacting attention be placed on acquisition settings since the microscope's ability to consistently detect fluorescence emanating from a region of interest will vary depending on these settings. Detector gain and offset, and laser power transmissivity used for fluorophore excitation must be kept consistent. We will also cover the necessary materials needed to anesthetize the rat and perform the surgery to expose the kidney for imaging with an inverted microscope. The final sections of the chapter will expand on the list of fluorescent dyes and compounds needed to study a host of dynamic processes occurring within the renal cortex, and methods by which to quantify them. Scoring renal injury arising from disease models should be done imposing stringent parameters and careful selection of the fluorophores which is imperative to provide meaningful data. Finally utilizing a strain of rat with surface glomeruli, glomerular permeability of large and small molecular weight compounds can be determined using ratiometric intensities in the capillary loops and Bowman's space.

The described techniques represent a small sample of those available to study and quantify kidney processes. Great strides have been made in advancing this technique since this group first used it in the descriptive publication in 2002 [1]. They have occurred through the use of novel fluorescent compounds and advances in the development of transgenic models expressing fluorescent protein conjugates allowing visualization of morphological structures and dynamic processes. There are, however, still challenges ahead such as the limited ability of current technology to penetrate opaque tissues like the kidney. The ability to image through an entire glomerular volume using higher magnification objectives represents an elusive but tangible goal that may be realized through the use of adaptive optics which employs deformable mirrors capable of compensating for scatter associated with more opaque tissues. Newer lasers generating longer excitation wavelengths with better tissue penetration may enhance the ability to efficiently excite fluorescence from deeper regions in the kidney to minimize the scattering of excitation photons.

To summarize, IVMPM can serve as an invaluable research tool to enhance the goals of laboratories studying dynamic physiologic processes. One of the key benefits of technological advances is the ability to produce more robust hardware components, while lowering cost, making systems such as multi-photon confocal microscopes turnkey and affordable. When utilized correctly this disruptive technology can create novel insights into complex physiologic processes via their direct visualization in vivo.

2 Materials

2.1 Rat Strains

A variety of rat strains are available to study specific disease processes. The Sprague Dawley strain is the most widely used outbred rat used in research, with study applications ranging from aging to behavior and reproductive/developmental biology. The one key aspect this strain lacks with regard to the kidney is superficial glomeruli. The Frömter and Simonsen strains of Munich Wistar rats possess superficial glomeruli immediately adjacent to the surface which allow for direct visualization of renal filtration and the study of glomerular structures through the use of fluorescent dyes or transgenic animals expressing fluorescent proteins. The structural density and opacity of the kidney do not allow for the deep imaging attainable in other more transparent organs such as the brain [2, 5]. A publication by Castrop et al. [6] surveyed the distance of glomeruli to the renal surface in mice, and none were found to be within range for IVMPM as with the two Munich Wistar strains of rats. Consequently, although some structural information may be visualized in mice with near surface glomeruli, photon scatter of light emanating from those focal planes equates to less robust quantitative information and images with reduced resolution.

Despite improved therapy, the increased prevalence of metabolic and physiologic disorders, such as hypertension and diabetes in the general population, makes rat models with similar diseases highly desirable to study mechanisms and test therapeutics [7]. A widely used, since the derivation in 1962, research model of salt-induced hypertension—the Dahl SS rat provides numerous phenotyping studies demonstrating the importance of the kidney in the regulation of blood pressure. The Dahl SS rat is a naturally occurring model of salt-sensitive hypertension that recapitulates many aspects of progressive human hypertension and has provided key insights into the mechanisms underlying salt sensitivity [8, 9]. Moreover, with the injection of streptozotocin, Dahl SS rats can be used as the model of diabetes-induced renal disease and display renal histological lesions characteristic of those seen in patients with diabetic nephropathy [10–12]. Our recent studies confirmed that the Dahl SS rat kidneys contain superficial glomeruli suitable for IVMPM application [13] and corresponding direct measurements of albumin filtration at the glomerulus and reuptake at the proximal tubule during the development of salt-sensitive hypertension. Importantly, this is an inbred rat strain commonly used to manipulate specific genes to test their function and recently utilized for the creation of a tissue-specific Podocin-Cre (glomeruli podocyte-specific) line which we believe can be used for future IVMPM studies.

2.2 Microscope Settings/Renal Morphology

Fluorescence microscopy of thick tissue presents a unique problem in that unlike cultured cells which are relatively flat and have fluorescence emanating from a confined focal area, images from tissue also contain out of focus fluorescence from above and below the focal plane generating images with drastically reduced resolution and contrast. Confocal microscopy addresses this issue through the use of a pinhole aperture which rejects out-of-focus light, and unfortunately scattered light from the focal plane. This leads to a reduction in the ability to capture information from deeper regions of the tissue since scattering increases at deeper depths. Multiphoton excitation of fluorophores is performed through the use of two or three longer wavelengths, with lower energy photons,

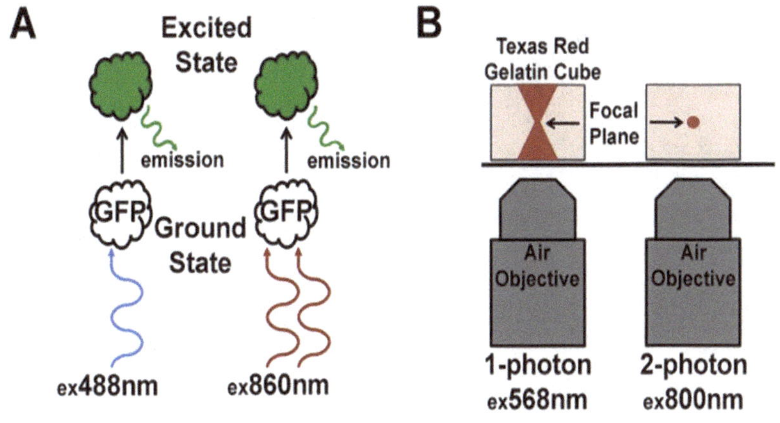

Fig. 1 Multi-photon excitation of a fluorophore. The schematic in panel **a** shows two GFP molecules (clear, no color) in the ground state. These two fluorophores can be excited by either single photon excitation (left) or multi-photon excitation (right) at their respective peak excitation wavelengths. After absorption of the photon or photons, the fluorophore jumps to the excited state where it remains there for a duration lasting in the single digit to tens of nanoseconds. Upon decay from the excited state, a small amount of heat is given off along with a photon of lower energy than the peak single photon excitation wavelength; in this case, emission is typically around 520 nm. The schematic in panel **b** illustrates the extremely low probability event of multi-photon excitation. The objective and gelatin cube on the left demonstrate how single photon excitation of a fluorophore occurs at any photon density since absorption occurs at that specific wavelength, in this case, 568 nm for Texas Red. Note how the column of illumination widens above and below the focal plane, where even though the photon density drops off, excitation nonetheless occurs via single photon excitation. In contrast, the objective and cube on the right illustrate multi-photon excitation, where only the focal plane condenses the photon density to such a high degree that this extremely low probability event becomes a certainty. Above and below the focal plane where the requisite photon density is not established, excitation does not occur making the image inherently confocal

absorbed simultaneously (*see* Fig. 1). For example, Fluorescein, which is typically stimulated by a 488 nm photon, can also be excited by two 800 nm photons. This phenomenon of multi-photon excitation is an incredibly low probability event, occurring only at the focal plane where photon density is sufficient to assure this occurs. Extended wavelength photons passing above and below the focal plane, lacking the requisite energy to excite the fluorophore, go undetected; hence fluorescence is generated only from the focal plane. This allows for the collection of all emitted photons, which in turn produces enhanced quality and higher contrast images; *see* Fig. 1 for details. Three-dimensional imaging of live tissue is an application ideally suited for multi-photon imaging. Over a given volume, photo-excitation and possible phototoxicity occur only at the focal plane with multi-photon excitation. This aids in the

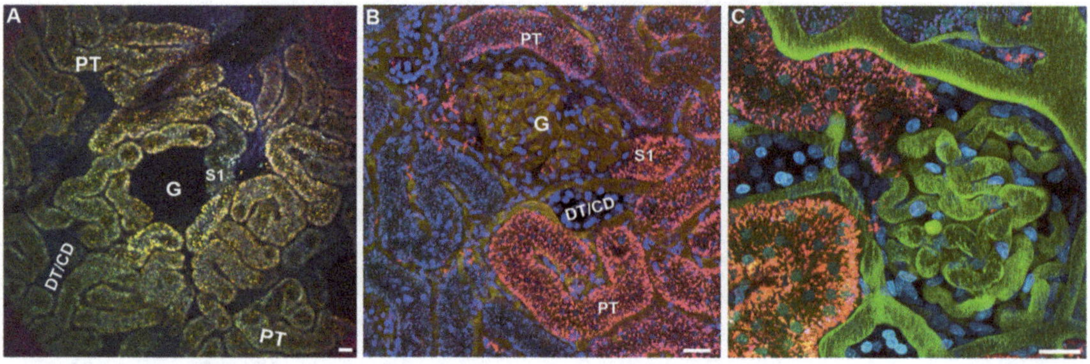

Fig. 2 Autofluorescence of the renal outer cortex. Autofluorescence associated within the lysosomes of the proximal tubule epithelia help demarcate landmarks and localize other nonfluorescent structures. Panel **a** is a low power 20× image showing the prevalence of proximal tubule cells (PT). A glomerulus is seen in the center (G) with the corresponding S1 segment (S1). Distal tubules and collecting ducts (DT/CD) lack autofluorescence and their presence is noted by a tubular-shaped empty space. Surrounding the PTs, the peritubular microvasculature can be seen meandering between the tubules. Panels **b** and **c** represent the outer cortex having been pre-labeled 24 h prior with a small Cascade Blue dextran (blue) and given a Texas Red-labeled and Fluorescein-labeled albumin, successively, filling up the plasma. Panel **b** is a 40× image showing a glomerulus (G) in the center with an S1 segment adjacent, as well as a distal tubule/collecting duct (DT/CD) below. Early proximal tubule segments (PT) can be seen having taken up the filtered albumins. Due to fluorescence quenching of Fluorescein at the lower pH of the endosomes/lysosomes, the localized albumin appears mainly red/magenta. Panel **c** shows a 60× high magnification image with the glomerulus and open S1 segment at the center. Note the clearly delineated peritubular vasculature with flowing red blood cells. Distal tubules/collecting ducts can be identified by their brighter staining with Hoechst (cyan), and lack of albumin accumulation. A dual pass FITC/Rhodamine cube is recommended to see epifluorescence images similar to those seen in panel **a**. The technical specs for the cube are FITC exBP490/15, emBP560/25; Rhodamine exPB560/25, emBP605/30. (Bar = 20 μm)

prevention of photo-bleaching of fluorophores above and below focal planes, allowing for the repeated collection of the same volume with reduced phototoxicity.

When viewing the unlabeled renal cortex through the microscope eyepieces, using standard epifluorescence, distinct tubule types and other structures can be readily distinguished (*see* Fig. 2a). It is recommended to equip the multi-photon microscope with a fluorescein/rhodamine dual pass cube to simultaneously view these commonly used fluorophores as well as the autofluorescence associated with the kidney. The orange/yellow autofluorescence associated in lysosomes of proximal tubules quickly distinguishes them from the surrounding distal tubules, collecting ducts and surface glomeruli (in Munich Wistar rats) which lack any lysosomal fluorescence. The peritubular vasculature appears as dark tracts surrounding the proximal tubules, distal tubules, and collecting ducts which have a faint blue tint. The glomeruli in both strains of Munich Wistar rats typically have the appearance of a circular void, and are distinguishable from distal tubules and collecting ducts whose space generally is more tube-like and linear; *see* Fig. 2 for a detailed example and technical specifications on the dual pass cube.

The majority of data acquired in our studies is accomplished using an excitation wavelength of 800 nm. This wavelength efficiently excited many common fluorophores in the blue, green, and red spectra such as Cascade Blue, Fluorescein, and Rhodamine, respectively. To produce images that closely match what is seen through the eyepieces, the detector gain settings for our systems are kept at a ratio of 750/630 for the green and red emission detectors, respectively. These values may not be applicable across all systems and the operator should determine what proportions work best on their individual system. It is then important to keep those numbers consistent for day-to-day reproducibility, understanding that certain situations will require some deviation. In our system laser power transmissivity is kept low, between 12% and 15% at 800 nm. The use of longer wavelengths causes a reduction in power output from the laser and transmissivity can be increased accordingly (*for more details and examples see* Fig. 2).

3 Methods

The experimental procedures described below were approved by the Institutional Animal Care and Use Committee at the Indiana University School of Medicine and were in accordance with the National Institutes of Health Guide for the Care and Use of Laboratory Animals.

3.1 Anesthesia and Exposure of the Kidney

The type of anesthesia utilized for any study should correspond to the desired length of anesthesia and whether or not recovery will be the eventual outcome after imaging. The two forms of anesthesia used in our laboratory are the long-acting thiobutabarbital derivative Inactin, used in terminal studies, and isoflurane used in serial imaging studies. For detailed information see the chapter by Rhodes in the journal *Methods* [14].

3.2 Isoflurane

1. Initial induction of anesthesia occurs inside an induction chamber with manifold set at 5% with an oxygen flow rate of 0.5 L/min.
2. Assure the rat is adequately sedated and carefully shave the left flank between the left arm and leg as well as the neck or inner thigh where a venous or arterial access line will be placed. It may be necessary to re-sedate the rat to assure complete shaving of the areas mentioned above.
3. After shaving is complete switch the anesthetic feed line from the induction chamber to the closed rodent circuit, which will

provide a constant administration of isoflurane to the closed circuit (*see* **Note 1**).

4. Once the rat is sedated on the closed circuit, reduce the manifold to 1.5–2% at 0.5 L/min oxygen flow rate and proceed with the necessary surgical procedures.

3.3 Inactin

This anesthetic is made up at a concentration of 130 mg/mL in 0.9% saline and administered at a dose of ~130 mg/Kg intraperitoneally, and the surgeon must access the efficacy after about 10 min by toe pinch to check for reflexes, assuring complete anesthesia (*see* **Note 2**). In our recent experiences, it is advantageous to utilize isoflurane as described above in the preparation of the rat. Care must be taken to assure the rat is taken off the isoflurane circuit once the inactin takes effect. Typically it takes less than 1 min, and reintroduction of isoflurane can be administered as needed.

3.4 Surgery to Expose the Kidney (see Note 3)

1. Place the rat on the right side. The shaved left flank faces up with the legs pointed to the left. Assure the rat is perfectly flat and straight with the right and left front paws touching and the rear paws touching each other. Any shift in the rat's body plane from being perfectly flat will lead to a less than ideal initial incision.
2. With your thumb, fore and middle finger gently palpate to feel the kidney in the peritoneal cavity and determine the natural position therein (*see* **Note 4**).
3. At the center where the kidney naturally lays, carefully pick up the skin with a pair of toothed forceps and pinch the skin with a pair of hemostats along the line and hold for approximately 5–10 s; this will crush the tissue and almost wholly prevent bleeding.
4. Cut along the incision using a pair of surgical scissors. Crushing the outer skin and the muscle layers before cutting will dramatically reduce and typically eliminate bleeding.
5. Carefully repeat previous step for the thin outer muscle layer (*see* **Note 5**). Using the two handling forceps, widen the incision to prepare for cutting into the inner muscle layer.
6. Cutting into the inner, final muscle layer will expose the peritoneum and care must be taken to assure the incision will be made in the optimal location, so re-palpate to feel for the kidney and assess the size. Pinch a smaller section of the kidney with the hemostats directly over the center of the organ and make the incision; you should now see the peritoneal space. If you are using isoflurane and wish to switch over to inactin, now is the time to administer the correct dose into the peritoneal space. This initial incision was kept small since it is easier to

widen the incision than suturing it if the size is too large (*see* **Note 6**).

7. Peering into the peritoneal cavity should reveal a white/tan collection of fat surrounding the kidney, which should have a light reddish/purple appearance; other organs in proximity such as the spleen or liver will be much darker.
8. Using a pair of handling forceps, grab the fat and lift it close to the incision so that you can grip the region below the first pair of forceps with the second pair of forceps. The goal here is to carefully work your way down the fat in a hand-over-hand fashion until you have reached the lower pole of the kidney while nicely, but firmly, gripping the fat so that it does not tear.
9. Once you have reached the lower pole, put down the forceps in your free hand and gently place the thumb and forefinger around the incision. In a fluid motion very gently push down on the abdominal wall while delicately squeezing below the medial plane of the kidney. Gently pull the kidney out of the peritoneal cavity with the forceps (*see* **Note 7**).
10. Moisten a 2 × 2 gauze with warmed 0.9% saline and place over the exposed kidney to keep it moist during transport and preparation of the microscope stage (*for schematic details see* Fig. 3). For a video on this procedure see our publication in JoVE [15].

3.5 Placing the Rat on the Inverted Microscope Stage

1. Carefully transport the rat with the kidney draped in the moistened gauze to the microscope and place the organ in the 50 mm coverslip bottom dish (*see* **Note 8**). The rat will be placed with the exposed kidney inside the 50 mm dish; the head should be positioned toward the right as you face the microscope stage (*see* **Note 9**).
2. To assure the thorax lies far enough away from the kidney and minimize motion artifact from breathing, place the kidney up against the edge closest to you with the ventral side touching the coverslip and the dorsal rotated side up against the body. Here, the goal is to place the most considerable distance between the thorax and the kidney.
3. If a motion is detected while looking through the eyepiece, stretch the rat along the sagittal axis to assure it is not hunched up (*see* **Note 10**).

3.6 Synthesizing Fluorescent Albumin

The application of IVMPM to an initial study to follow renal handling of fluorescent albumin leads to the observation that a far greater amount is filtered across the glomerular capillary loops [10, 16]. IVMPM can also be used to quantify the amount of albumin that is efficiently reclaimed from the urinary space by the proximal tubules [17]. Previous methods used to study renal

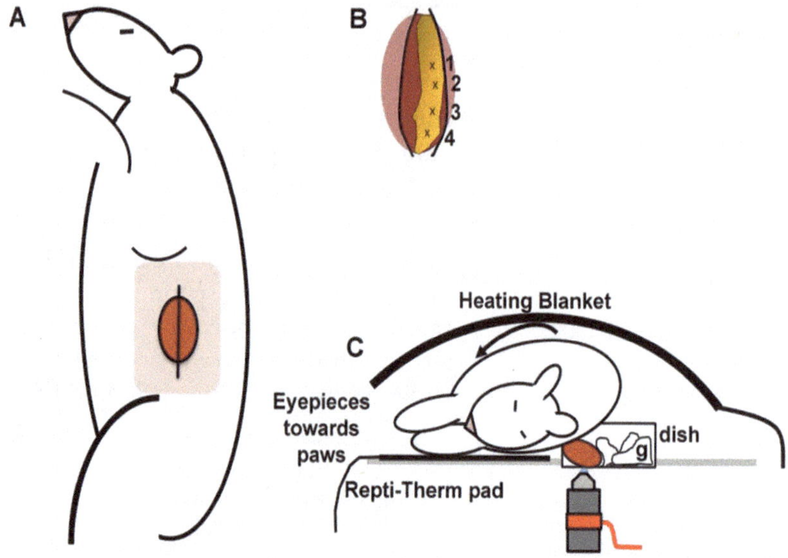

Fig. 3 Exteriorization of kidney and placement on the microscope stage. Panel **a** shows an anesthetized rat, placed with its right side down. The pink square indicates the shaved area, with the position of the kidney underneath the skin and muscle layers. Once sequential incisions are made to penetrate successive muscle layers, the kidney, and surrounding fat should appear as the schematic in panel **b**. Carefully grip the fat (in yellow) in positions 1–4, successively moving down in a hand-over-hand fashion until the lower pole is reached (position 4). Gently grip the side of the incision and pop the kidney out. Placement of the kidney inside the coverslip bottom dish is shown in panel **c**, with the kidney as far forward to the edge of the dish as possible to minimize motion artifact. Three to four pieces of 2″ × 2″ gauze generously moistened with warm sterile 0.9% saline should be placed behind the kidney, to further stabilize movement and keep the organ wet

albumin handling such as fractional clearance [18] and micropuncture have reported very low values [19]. In the case of micropuncture, sampling of the filtered fluid is done further downstream of the glomerulus along the nephron, while fractional clearance samples concentrations of albumin in the plasma and urine. Both of these techniques, however, do not account for the avid uptake of albumin that occurs in the S1 segment immediately adjacent to the glomerulus, which we have shown in numerous studies [10, 15–17, 20]. This would deplete the detectable level of albumin collected from tubule segments more distal to the S1 or the urine, producing a permeability value that is artificially low. Here, fluorescent albumin will be used to demonstrate quantitative analysis of glomerular permeability and uptake by proximal tubules within the same set of images.

1. Dissolve 96 mg of Rat Serum Albumin (RSA) in 6.4 mL of 100 mM Sodium Bicarbonate pH 8.3; final concentration 15 mg/mL in a 15 mL conical tube.
2. The reaction will be carried out at room temperature.
3. Add 100 μL of high quality anhydrous Dimethyl Formamide (DMF) to a 2 mg vial of Texas Red-X- Succinimidyl Ester, single isomer (TR-X); vortex on medium for 15 s.
4. While vortexing the uncapped 15 mL conical tube with the RSA solution on medium (vortex at a speed that keeps the solution from spilling out), add the 100 μL solution of DMF with the dissolved TR-X.
5. Cap the 15 mL conical tube, cover with foil (Parafilm can be wrapped around the cap to assure a tight seal), and place on a rocker at low speed to gently agitate the reaction for 1 h at room temperature.
6. Add 45 g of NaCl to 5 L of ddH_2O to make a significant volume of normal saline to remove unconjugated TR-X fluorophore via dialysis. Wet/reconstitute per instructions a 50 kDa molecular weight cut off membrane of choice; sufficient to hold the ~7 mL of reaction solution.
7. Carefully inject/pipette the reaction into the dialysis membrane/chamber and place in the 5 L container of normal saline, dialyze overnight at 4 °C with gentle agitation using a stir bar.
8. Change the dialysis solution with five more liters of normal saline in the morning and again in the afternoon with five more liters of normal saline and allow to dialyze again overnight; all dialysis should occur at 4 °C.
9. In the morning, the final 5 L volume should be clear of any free TR-X. Carefully remove the solution from the dialysis membrane/chamber and measure the volume. Divide the initial 96 mg of RSA by the volume to give you concentration, which will be approximately 12 mg/mL due to swelling (*see* **Note 11**).

3.7 Glomerular Permeability

1. Using a lower powered objective, such as a 20× water immersion (NA 0.7), prior to infusion of TR-RSA, scan the kidney surface in a raster pattern and mark the positions where glomeruli are suspected to be present in a Munich Wistar rat of either strain (*see* **Note 12**).
2. Acquire a 3D volume of the glomeruli before delivery of TR-RSA to be used as a background subtraction image during quantitation (*see* Fig. 4). Carefully adjust the detector offset or black level (*see* **Note 13**). The Bowman's space in the background images should have a few random pixels displaying

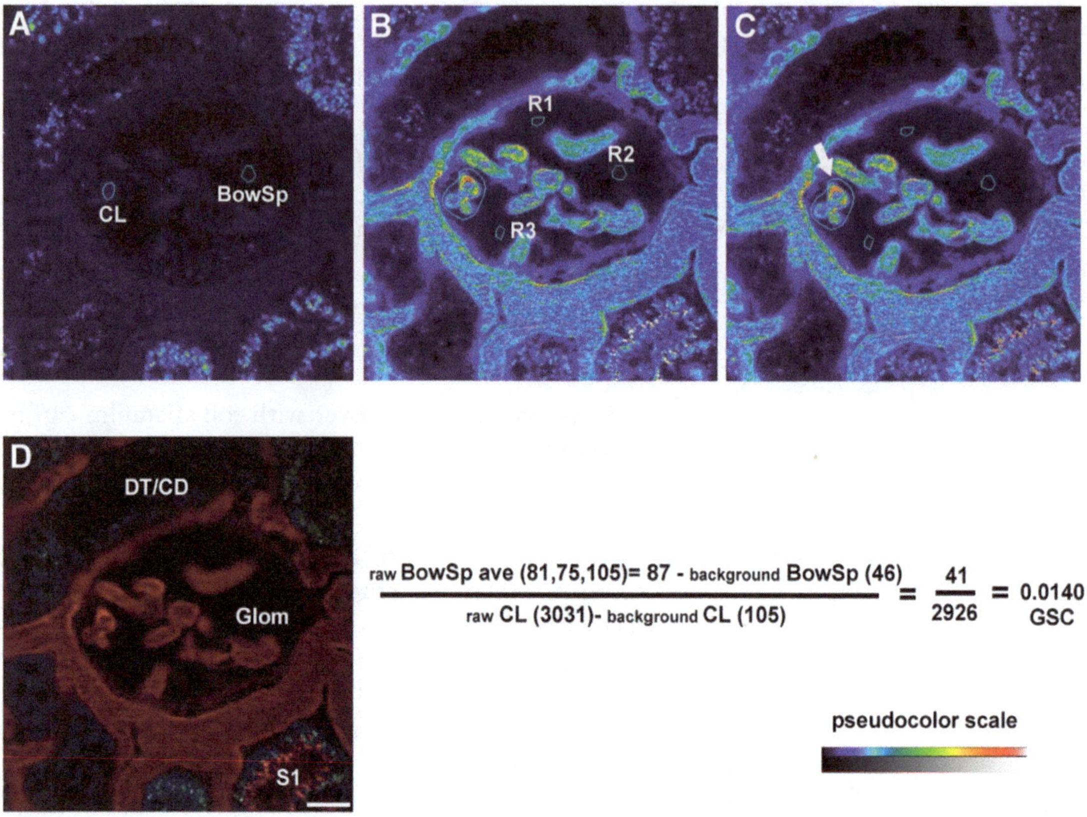

Fig. 4 Glomerular permeability in Munich Wistar Frömter rats. Calculating the degree to which a fluorescence compound can filter across the glomerular capillary loops, glomerular sieving coefficient (GSC) is done by rationing the intensity in the Bowman's space to fluorescence within the capillary loops. Panel **a** shows a background image with a glomerulus in the center and two outlined regions, the capillary loops (CL) and Bowman's space (BowSp) selected to determine background fluorescence. In panel **b**, an image taken approximately 15 min after infusion of Texas Red rat serum albumin (TR-RSA) shows the albumin filling the plasma along a large vessel surrounding the glomerulus and capillary loops. Three regions (R1–R3) have been selected within the Bowman's space, and the fluorescence values noted. Panel **c** shows the same image, highlighting the brightest capillary loops (arrow), with thresholding applied to select regions within the plasma to note the fluorescence intensity. The formula below panels **b** and **c** shows the calculation to determine the GSC for TR-RSA (for this particular glomerulus it is 0.014). Panel **d** shows a color image of the same region, note the S1 segment (lower right) with an accumulation of filtered TR-RSA at the apical surface. (Bar = 20 μm)

values of zero (typically less than 10 or so total pixels) when the offset/black level is correctly set (*see* **Note 14**).

3. For the infusion of TR-RSA, a glomerulus showing capillary loops or a larger superficial vessel can be selected to monitor plasma concentration; be sure to use a LUT or palette that will inform you if saturation occurs. Set the desired number of frames to collect for the infusion time series, 50–200 frames is typical. In one hand take the syringe containing the TR-RSA, connected to an indwelling IV line; it is best to anchor it by

resting it near the eyepieces or some other stable feature of the microscope. Take the free hand and start a timer or stopwatch to mark the start of the albumin infusion, click the microscope software to start acquiring the time series and finally start the slow infusion of the TR-RSA. You will notice there will be fluctuations in plasma intensities characterized by peaks as material is pushed in, followed by ebbs as systemic vascular distribution occurs. In total, approximately 3.0–5.0 mg of TR-RSA will be infused (*see* **Note 15**).

4. Wait approximately 5–10 min before starting the collection of images from the list of glomeruli marked in **step 1**. This will assure complete distribution of the TR-RSA. To collect images, focus on the area approximately 5–15 μm below the Bowman's capsule where a few capillary loops are visible, surrounded by an abundant amount of empty Bowman's space, taking either a single image or a 3D volume of the glomerulus (*see* **Note 16**).
5. Save and transfer the images for analysis using image-processing software such as Metamorph (*see* Figs. 4, 5, and 6) or Image J (*see* Fig. 7); the analysis below will be conducted using Metamorph.
6. To analyze the images, match the focal plane of the background image taken before TR-RSA infusion to the corresponding image after infusion (*see* **Note 17**). In your background image, draw a region of interest closely around one of the capillary loops (which will be very faint) and a small region in the Bowman's space away from the Bowman's capsule and capillary loop and note the average intensity value for each. For the image with circulating TR-RSA, select three regions in the Bowman's space and mark the average of the three average intensity values. For the capillary loop select the one with the brightest intensity and draw a region to loosely encompass all of the brightest region. Use the threshold tool to select the very bright plasma intensities along the capillary loop edge, avoiding the red blood cells appearing as streaks, or the regions between the streaks (*see* **Note 18**).
7. The glomerular sieving coefficient (GSC) is calculated below using the average intensity values obtained above (*see* **Note 19**):

$$\frac{\textbf{Raw}\text{ Bowman's Space (ave 3 regions)} - \textbf{Background}\text{ Bowman's Space}}{\textbf{Raw}\text{ Capillary Loop} - \textbf{Background}\text{ Capillary Loop}} = \textbf{GSC}$$

Using GSC values obtained by IVMPM and known GFR values for the experimental rat model total filtered albumin can be determined. By determining PT uptake total 24 h urinary

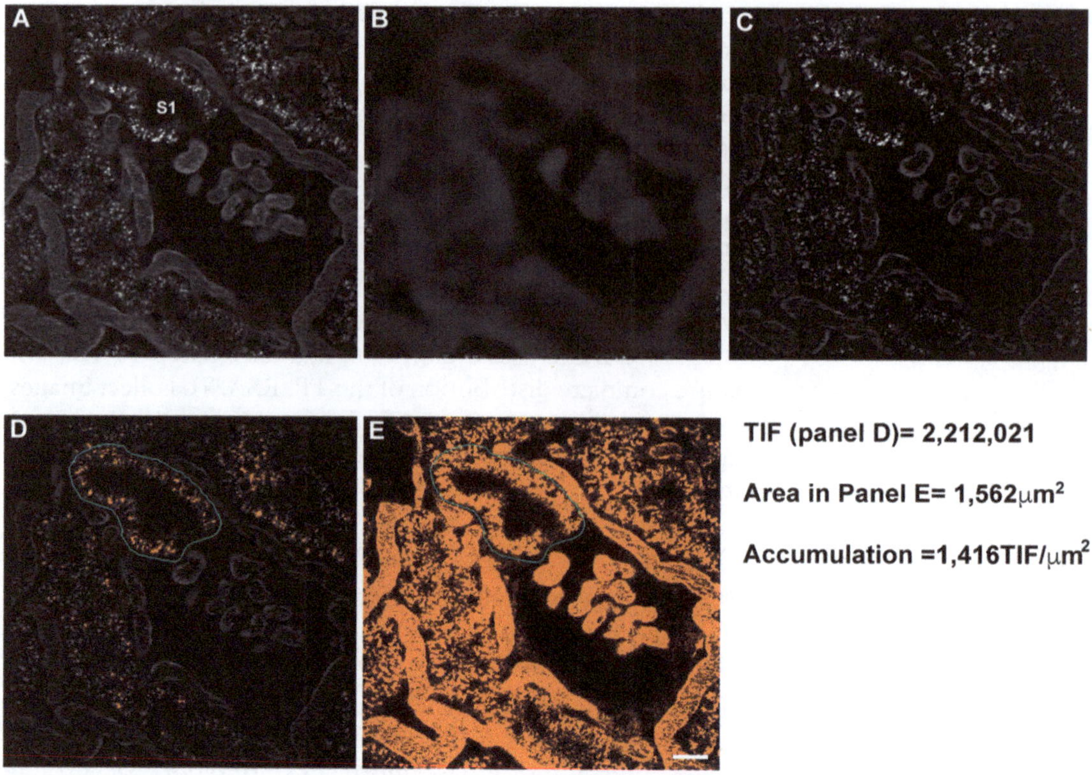

Fig. 5 Quantifying uptake of albumin in proximal tubules. Measuring uptake of albumin or any fluorescent compound can be used to gauge temporal accumulation, derive a rate or stratify differences between untreated or disease models. This process starts by applying a 32×32, subset $= 1$ median filter to generate a background image (panel **b**) from the original (panel **a**). The background image is subtracted from the original image (**a–b**) to generate a clean background corrected image (**c**). For this study, a region of interest (cyan) was around the S1 segment of the proximal tubule, and the remaining albumin signal was thresholded to encompass the associated pixels (in orange). The total integrated fluorescence (TIF) for this S1 segment (the product of average fluorescence intensity x number of highlighted pixels) $= 2{,}212{,}021$. Since this value can vary widely based on the area highlighted, it is essential to normalize this value to the area such as μm^2. The region of interest from panel **d** is transferred to the original images (panel **a**), and the region is thresholded to encompass all of the proximal tubule area (shown in panel **e**, orange). The total number of pixels in panel **e** $= 9114$. The area of an individual pixel is determined by multiplying the X/Y pixel size ($0.414\ \mu m \times 0.414\ \mu m$) to arrive at $0.171\ \mu m^2$. Multiplying the number of pixels (9114) by the area (0.171) $= 1562\ \mu m^2$. The value for this image region $= 1416\ TIF/\mu m^2$. (Bar $= 20\ \mu m$)

albumin can be calculated by the following equation (*for more details see* [17]):

$$\left(\mathbf{GFR} \times 1440\ \text{min} \times \mathbf{serum\ albumin} \left(\frac{\text{mg}}{\text{mL}}\right) \times \mathbf{GSC} \times (1 - \%\mathbf{PT\ uptake}\right) = 24\ \text{h}\ \mathbf{urinary\ albumin}$$

3.8 Proximal Tubule Endocytosis

1. Accumulation of fluorescent albumin or other compounds via endocytosis, when studied over a protracted period, can lead to

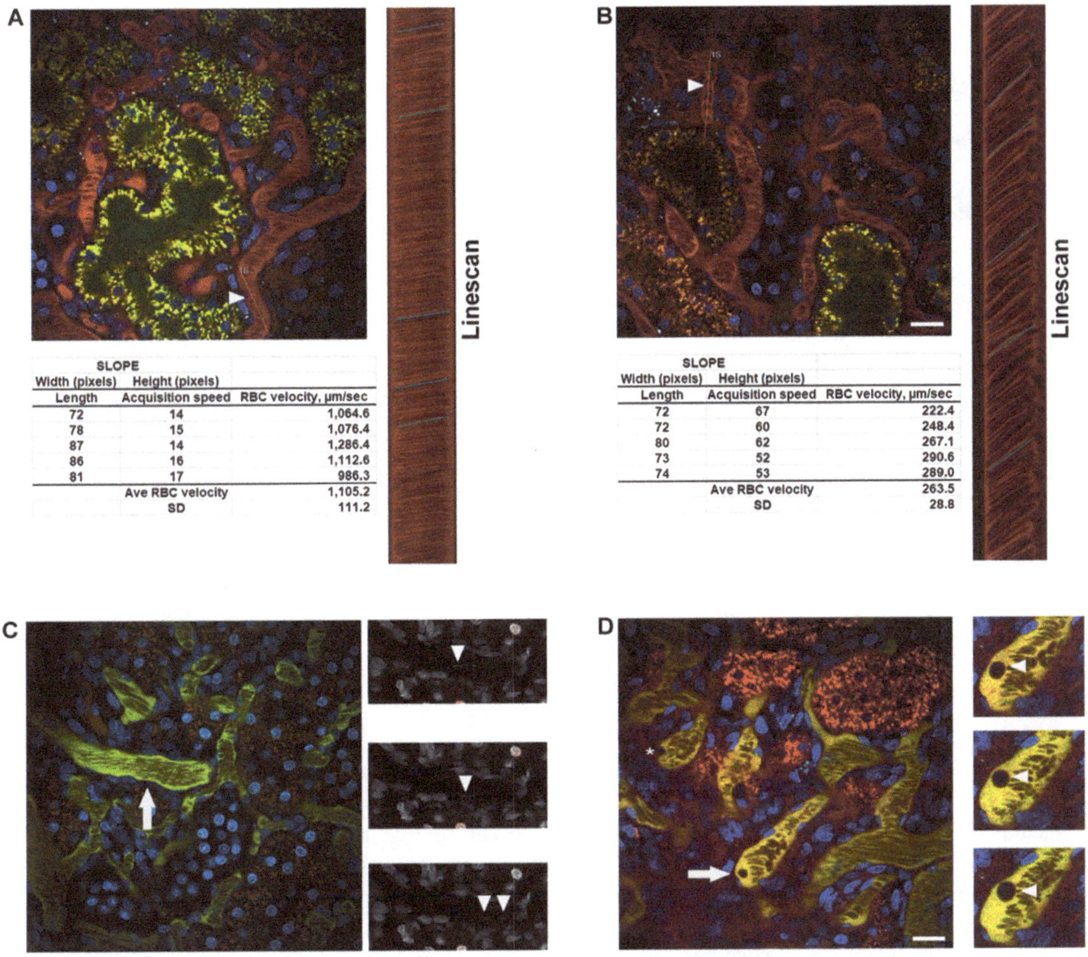

Fig. 6 Determining red blood cell (RBCs) velocity utilizing the linescan function and quantifying white blood cell (WBC) dynamics. The velocity of RBCs can be determined by utilizing the motion artifact generated where the RBCs appear as streaks within peritubular and glomerular capillary loops. The slope of the streak is dependent on speed; RBCs with higher velocities will have shallower slopes while cells with slower velocities will have steeper slopes. Panel **a** shows the reference image from an untreated rat with a highlighted vessel (arrowhead, "1S") and the corresponding linescan to the right. The linescan corresponds to 1000 scans placed on top of each other to form a column generating the slopes associated with the flowing RBCs. In the linescan, the Y-axis relates to time while the X-axis relates to distance. Five slopes are drawn along the column to generate an average RBC velocity for the vessel. In panel **b**, RBC velocity is appreciably slower following ischemia and reperfusion. The highlighted vessel (arrowhead, "1S") and associated linescan show steeper slopes and slower RBC velocity. Panel **c** shows a photomicrograph of an untreated rat with a series of images of the large vessel (arrow). Hoechst 33342 stains the nuclei of WBCs, which produce streaks within the vessels seen here (arrowheads) in the blue channel lacking any vascular marker to highlight the streaks better. Counting the number of streaks and factoring in the length of time the temporal acquisition occurred gives a rate of occurrence per minute. In panel **d**, activated WBCs can adhere to the inner vessel wall and show no movement during a time series (asterisk). Activated WBCs can also roll along the vessel wall. The highlighted WBC in panel **d** (arrow and cropped regions, arrowheads) show the motion of a WBC as it rolls along the vessel wall. (Bar = 20 µm)

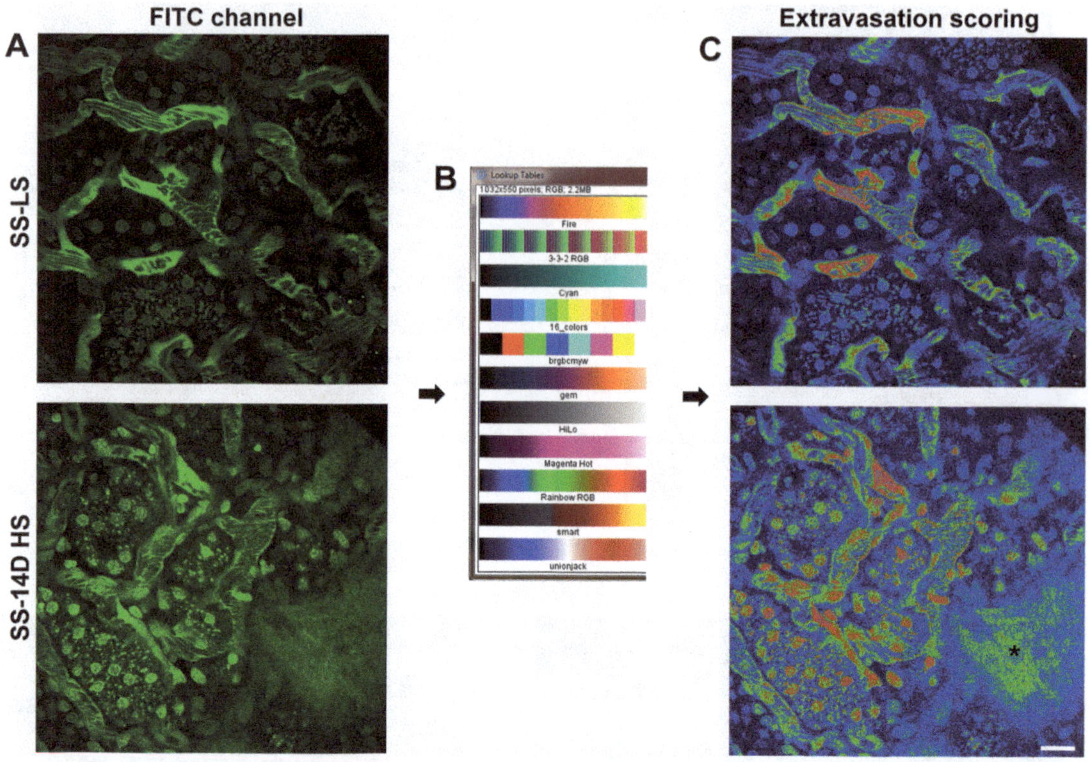

Fig. 7 Quantifying vascular permeability in Dahl salt-sensitive rats. Extravasation of large compounds normally retained within the vasculature into the interstitial space can be quantified using semiquantitative scoring. Photomicrographs were given a score from 0 to 3, where increasing values indicated more severe damage. A score of zero indicates an image was having no leakage of the large molecular weight compound into the interstitial space. A score of one showed some leakage into the interstitial space, while a score of two indicated leakage into the majority of the interstitial space within an image. A score of three was reserved for the most severe damage, where the intensity of the compound within the interstitial space was very near or equal the intensity of the plasma. In panel **a**, a photomicrograph of the kidney cortex from the salt-sensitive Dahl S rats on a low salt (SS LS, upper image) and after 14 days on high salt (8% NaCl; SS 14D HS, lower image). Note the lack of fluorescence in areas adjacent to the peritubular microvasculature as compared to the 14D high salt image below showing a high degree of extravasation at the lower left. Panel **b** shows the "Rainbow RGB look up table (LUT)" available using ImageJ software (National Institute of Health; 1.47v) and the corresponding images in panel **a** now displayed using the LUT in panel **c**. This LUT display allows for better visualization of subtle differences in intensities that are not readily discernible in images using single color or B/W display LUTs. (Bar $= 20\ \mu$m)

fluorescence saturation within the endosomal pool, in particular, the lysosomes. To correct for this, collect a series of identical background reference images at different laser power transmissivity (*see* **Note 20**). It is also essential to design the experiment beforehand to plan for the collection of images at the appropriate time points after delivery of the fluorescent probes.

2. Select an image to analyze and apply a 32 × 32 (pixel width × pixel height) Median filter with the subsample ratio set to 1; this will be used as a background image (*see* Fig. 5).
3. Take the original image and subtract out the background image (*see* Fig. 5a–c).
4. Draw a region around the proximal tubule you wish to analyze (*see* Fig. 5d); it can be done by tracing around the basolateral side of the proximal tubule (*see* **Note 21**).
5. Threshold the image to select the bright endosomes/lysosomes and mark the total integrated fluorescence (TIF, expressed as fluorescence units FU) (*see* Fig. 5e), which is the product of the total number of pixels x of the average intensity (*see* **Note 22**).
6. Metamorph software allows for the transfer of the outline of a drawn region of interest (ROI) from one image to another in the corresponding position; transfer the ROI from the background subtracted image to the original.
7. In the original image, threshold the proximal tubule inside the ROI until all of the area occupied by the proximal tubule is highlighted, not just the endosomes/lysosomes (*see* **Note 23**). Make a note of the total number of pixels in the thresholded region, not the entire region.
8. Multiply the number of pixels by size dimensions associated with the microscope (*see* **Note 24**).
9. Proximal tubule accumulation is therefore calculated by taking the TIF/area with the value expressed as TIF/μm^2. This method is preferred since it normalizes uptake regardless of the size of the region analyzed (*for more details see* Fig. 5).

3.9 Renal Blood Flow Dynamics

Injury, either related to ischemia, exposure to a toxin, or arising from progression associated with a disease model, is often characterized by alteration in blood flow dynamics within the peritubular microvasculature or glomerular capillary loops, and on occasion the degradation of vascular integrity. Through the use of a large molecular weight dextran (more than 150 kDa) which is retained in the bloodstream, the velocity of red blood cells (RBCs) can be calculated by exploiting a motion artifact that occurs during image acquisition [21]. The typical multi-photon microscope uses a galvo (galvanometer-based) scanner to direct the laser across the sample in a raster pattern to generate an image. The velocity of RBCs traveling across a peritubular vessel is far higher than the acquisition speed of the microscope; this causes the RBCs to appear as thin, slanted streaks within the blood vessels. The slope of the streaks is dependent on the speed at which the RBCs travel across the blood vessel; a shallow slope indicates a rapid RBC velocity while a steeper slope is associated with a slower speed. Changes in

RBC velocity are not the only changes in vascular dynamics that can occur in injury and disease models; often white blood cells (WBCs) become activated, leading to localized obstruction of vascular flow [22]. This phenomenon can also be quantified when analyzing a time series of the blood flow and looking for WBCs whose nuclei are labeled with Hoechst 33342. The results from these data stratify the WBCs as freely flowing, rolling, or adhered based on motion. Of note, the type of WBCs cannot be determined through the simple use of a nuclear dye; this would require the use of specific markers or transgenic rats expressing fluorescent proteins in certain WBC populations.

1. Infuse in approximately 300 μg of Hoechst 33342 to label the systemic nuclei including WBCs (*see* **Note 25**).
2. After placing the rat on the microscope, slowly infuse in approximately 2.0–5.0 mg of a large 150 kDa dextran (*see* **Note 26**).
3. Red blood cell velocity will be measured on the microscope using the “linescan” setting (*see* Fig. 6). Briefly, this allows the operator to draw a line down the middle of the blood vessel parallel to the length and adjust the settings to acquire 1000 lines (*see* **Note 27**).
4. In Metamorph, open the dialog box “Show Region Statistics” along with any linescan image acquired. Using the line tool, draw a line across the linescan across the slope of either the RBC or the plasma in between to best approximate that slope and note the Width and Height values which are displayed in pixels.
5. An excel spreadsheet can be set up to calculate RBC velocity using the following formula:

$$\frac{\textbf{Distance} = (\text{Width in pixels}) \times (\text{Pixel dimension})}{\textbf{Time} = (\text{Height in pixels}) \times (\text{line scan period})} = \textbf{RBC velocity in } \mu m/\text{s}$$

Example:

$$\frac{50 \times 0.414\mu m}{11 \times 0.002\ \text{s}} = 940.9\ \mu m/\text{s}$$

6. As previously mentioned the nuclei of WBCs will stain with Hoechst 33342; to study WBC dynamics, disengage the “line-scan” function and return to full frame acquisition mode on the microscope, typically 512 × 512 pixels. Set up a time series to capture at least 30 frames (*see* **Note 28**), and select a focal plane where the majority of blood vessels are shown in cross section and acquire the time series.
7. Flowing WBCs are the most difficult to quantify since they appear as faint streaks inside the microvascular tracts

surrounding the tubular epithelia; to aid in counting, break up the image into four 256 × 256 quadrants and count the nuclear streaks in all 30 frames for each quadrant and total the tally (*see* **Note 29**).

8. This data can be presented as occurrences per minute after multiplying by the appropriate time factor (*for a detailed practical application of Protocol 3.9 see* Fig. 6).

3.10 Vascular Permeability

Assessing vascular dynamics such as RBC velocity and WBC adhesion produces values based on more direct physical parameters captured by the multi-photon microscope; there is little room for error in the determination of the slope of the slant in an image caused by a flowing RBC or the visualization of adhered WBCs obstructing renal blood flow. Scoring, as a quantitative method, may initially seem a more arbitrary way of assessing an observed abnormality in physiologic function or morphologic perturbation; however as long as stringent parameters are defined and followed it can produce a tangible method of quantifying an observable alteration. Changes in vascular permeability present a unique challenge in the way in which they can be quantified. These alterations are rarely global in nature, and to compound the problem a suitable marker or markers to define these alterations must be carefully selected. In this section, we present a method for scoring vascular permeability that relies on not a single marker but two markers to address the heterogeneity in severity [13, 23]. We utilize Texas Red labeled albumin to localize moderate alterations in vascular permeability. The red-emitting fluorophore is used because it is downstream of any possible bleed through emissions and will likely be the marker with the highest amount of extravasation into the interstitial space. A large narrowly dispersed 150 kDa Fluorescein dextran is used to localize severe perturbations in vascular integrity (*see* **Note 30**).

1. Using the motorized stage function on your microscope, select at least 10 random fields to study. Infuse in 2.0–5.0 mg of a 150 kDa fluorescein dextran and approximately 3.0–5.0 mg of TR-RSA (*see* **Note 31**).
2. Acquire images at 5, 15, and 30 min post infusion of the selected fields. Since we are quantifying the interstitial space between the vasculature and tubular epithelia, a glancing focal plane just below the surface will provide the largest interstitial area to study.
3. Permeability is assessed based on a score between 0 and 3 with increasing numbers corresponding to greater vascular leak. Zero indicates no leakage of material into the interstitial space. A score of one indicates a small amount of the material in the interstitial space in a small region of the image. Two

indicates material has leaked into the interstitial space throughout a majority of the region. A score of three means enough material has leaked into the interstitial space where the fluorescence intensity closely matches that in the circulating plasma. This indicates severe damage, particularly if this occurs with the large molecular weight dextran (*for a detailed practical application of Protocol 3.10 see* Fig. 7).

4 Notes

1. Anesthesia must be safely maintained as a closed circuit to provide scavenging of excess isoflurane and to prevent inhalation by laboratory personnel.
2. On occasion, an unintended delivery of the anesthetic to the guts can significantly increase the time it takes for the drug to take effect, or in some cases re-dosing may be required. In some instances, this can lead to a slow, unintentional overdose on the microscope stage characterized by a reduction in renal physiologic parameters, such as reduced peritubular blood flow and collapsing of the tubular epithelial lumen which will be evident in the acquired images.
3. It is advantageous to use male rats at approximately 8–12 weeks of age. Older males will have more substantial fat deposits that will make **step 9** (Protocol 3.4) difficult; females regardless of age will have much more fat surrounding the smaller kidneys with the ovaries attached to the lower pole.
4. Until you become familiar and comfortable with the procedure, use a sharpie to draw a line down the body from the ribcage to the upper thigh.
5. The rat may reflexively respond to pinching of the inner muscle layers (this is normal); a toe pinch with toothed forceps will confirm anesthesia is still useful.
6. Making the incision too close to the ventral side will induce motion artifact from breathing while making it too close to the dorsal side will cause a reduction or cessation of renal blood flow.
7. Do not force it! If the incision seems too small and the kidney will not "pop out," widen the incision by no more than a few millimeters at a time and try again. With practice, you will learn the incision should be approximately 75% the length of the kidney to prevent it from receding into the peritoneal space once externalized.
8. For these studies a 50 mm cell culture dish with a 40 mm coverslip bottom allows the largest area for visualization and

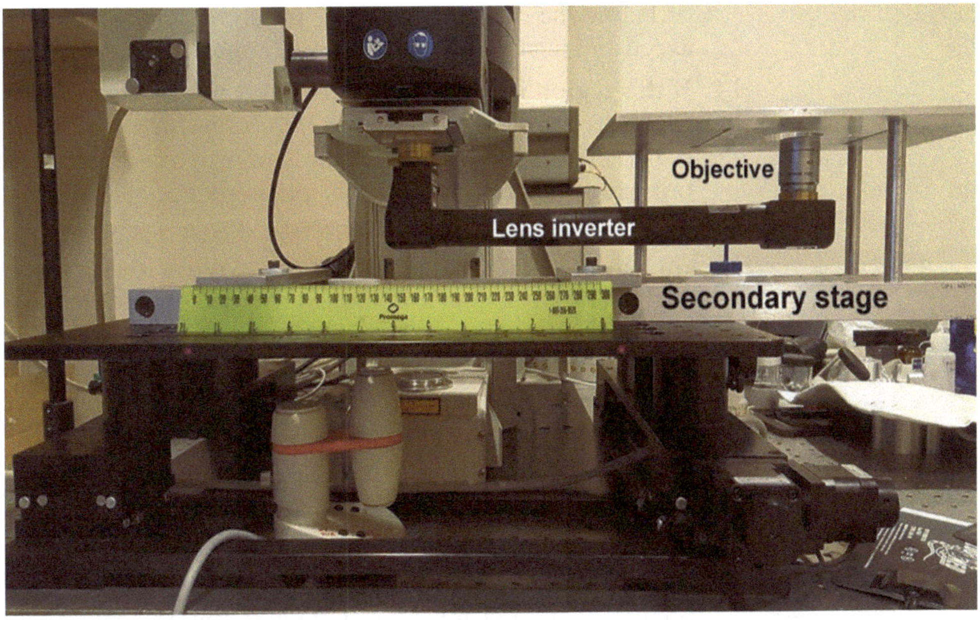

Fig. 8 The example of reconfiguration of an upright microscope to the inverted platform by using a commercially available lens inverter (LSM tech). A secondary stage was custom made to anchor the new inverted stage to the automated table equipped initially with the upright multi-photon Leica SP-8 system

should be used. Inverted microscope stages that accommodate 50 mm dish inserts are commercially available.

9. *See* Fig. 8 for a description of an objective inverter to change the objective orientation on an upright platform.
10. Try aligning the paws so that the two front and two rear paws touch. If motion persists, roll the rat very slightly, so the ventral/abdominal side is closer to the microscope stage insert.
11. The reaction with succinimidyl Ester should give a 1:1 dye: protein ratio. In a typical experiment, approximately 4 mg of TR-RSA is used per rat. A single conjugation can provide enough material for over 20 individual tests. For long-term storage it is best to lyophilize into individual tubes containing 5–10 mg each.
12. Most current microscope systems have a motorized stage that allows the operator to mark locations for computerized relocation. If unavailable, a rudimentary map can be drawn marking positions in relation to landmarks on the surface such as large blood vessels or fat pockets.
13. Setting the detector sensitivity value incorrectly can cause the detector to collect lower intensity values, especially in regions with lower fluorescence intensities such as the Bowman's space. Use a look up table (LUT, *see* Fig. 7b) or palette on the microscope images that informs you if your settings fall above

or below the dynamic range. Typically saturated values appear as red and pixels with values of zero appear either blue or green.

14. If this region is displaying a large percentage of pixels at zero, the detector will fail to acquire photons efficiently from these regions resulting in an artificially lower permeability value; see our study characterizing this phenomenon [24]. This will result in a low and inaccurate sensitivity.
15. Introduce the TR-RSA slowly to avoid saturation, should slight saturation occur, it will likely clear once the serial image collection starts approximately 10 min later. If saturation persists, the experiment is over; lowering the laser power transmissivity to get plasma levels below saturation will invalidate the background images taken at the higher laser transmissivity.
16. Analysis is much easier at more shallow depths because images taken further into the glomerulus suffer from a greater amount of photon scatter and have decreased resolution and contrast.
17. For the visualization of TR-RSA it will be easier to use a pseudocolor LUT image, especially in the background.
18. Thresholded pixels should appear in a different highlight color than the surrounding dimmer pixels. When recording the average intensity value for the capillary loops plasma, make sure you are selecting the average intensity for the thresholded pixels and not the entire region; *see* Fig. 4 for details.
19. The approach can be used to measure the GSC of fluorescent compounds having a wide range of molecular weights. However, compounds that are freely, or nearly freely, filtered should be studied under a continuous IV infusion to maintain constant intensity values in the plasma. The widely fluctuating intensities within the capillary loops produced via a single bolus infusion would produce inaccurate values.
20. For example, if the initial microscope transmissivity is 15%, once a field is selected showing proximal tubule autofluorescence, take identical images at 15%, 13%, 11%, and 9% transmissivity. Generate a correction factor by making the average fluorescence intensity for a small region of the image; keep the region identical for all the images. Next, divide the average intensity of the higher transmissivity image by the average intensity of the lower one; this will give a positive integer greater than one. The obtained correction factor will be applied when comparing the lower transmissivity images to the higher one to correctly scale up the values.
21. For the analyses of proximal tubule endocytosis it is not necessary to circle inside the lumen to locate the apical surface since you will threshold the endosomes to determine the integrated fluorescence therein (*see* Fig. 5d).

22. During the threshold procedure be sure you are selecting the data from the thresholded values and only from the region of interest, not the entire image.
23. The goal is to determine the total area associated with the proximal tubule the TIF file was derived from. During the selection try to highlight the tubule and not the tubular lumen if the proximal tubule is in cross-section.
24. For example, Olympus FLUOVIEW FV1000, when using a 60× water immersion objective, has pixel dimensions of $0.414\ \mu m \times 0.414\ \mu m = 0.171\ \mu m^2$.
25. Hoechst infusion can be done on the bench after surgery to expose the kidney and prior to placement on the microscope stage.
26. Similarly to fluorescent albumin injection (described in Subheading 3.7), use caution to avoid detector saturation in the plasma. Although, in this case laser transmissivity can be reduced, since current measurement procedures are not intensity-based.
27. Linescan acquisition mode requires complete stability of the rat on the stage to produce a suitable image for analysis. The microscope will scan the length of the line drawn across the vessel 1000 times, building a tall column as the image builds line by line with distance represented on the X-axis and time on the Y-axis (*see* Fig. 6a, b).
28. In the current configuration of the Olympus imaging system the acquisition rate is approximately 1 frame per second.
29. Rolling WBCs will move along the wall of the peritubular vasculature for at least 4 frames if subsequent dislodging occurs. Cells are counted as being adherent if they stay in place without moving for at least 4 frames if they become dislodged during the time series.
30. Since 150 kDa Fluorescein dextran produces fluorescent emissions "upstream" from Texas Red, it will not confound determination of moderate alterations of permeability. Moreover, it is essential to understand the nature of the compounds used. Large molecular weight dextrans, despite having a large listed size average, can be a product of large and small sized polymers around the average mean. This means unless a narrowly dispersed or purified compound is used, leakage of the low molecular weight components into the interstitial space will produce erroneous data suggesting severe damage where none exists.
31. Both compounds (TR-RSA and 150 kDa fluorescein dextran) can be introduced in rapid succession to avoid saturation of one of the compounds. Systemic distribution should occur within 5 min.

Acknowledgments

The authors acknowledge grant support from the National Institutes of Health (NIH) (DK091623 and DK079312), the Veterans Administration through a Merit Review award (to B.A.M.), and the American Heart Association 17SDG33660149 (to OP).

References

1. Dunn KW, Sandoval RM, Kelly KJ, Dagher PC, Tanner GA, Atkinson SJ et al (2002) Functional studies of the kidney of living animals using multicolor two-photon microscopy. Am J Phys Cell Phys 283(3):C905–C916. https://doi.org/10.1152/ajpcell.00159.2002
2. Svoboda K, Denk W, Kleinfeld D, Tank DW (1997) In vivo dendritic calcium dynamics in neocortical pyramidal neurons. Nature 385 (6612):161–165. https://doi.org/10.1038/385161a0
3. Peti-Peterdi J, Morishima S, Bell PD, Okada Y (2002) Two-photon excitation fluorescence imaging of the living juxtaglomerular apparatus. Am J Physiol Ren Physiol 283(1): F197–F201. https://doi.org/10.1152/ajprenal.00356.2001
4. Ferrell N, Sandoval RM, Bian A, Campos-Bilderback SB, Molitoris BA, Fissell WH (2015) Shear stress is normalized in glomerular capillaries following (5/6) nephrectomy. Am J Physiol Ren Physiol 308(6):F588–F593. https://doi.org/10.1152/ajprenal.00290.2014
5. Kleinfeld D, Mitra PP, Helmchen F, Denk W (1998) Fluctuations and stimulus-induced changes in blood flow observed in individual capillaries in layers 2 through 4 of rat neocortex. Proc Natl Acad Sci U S A 95 (26):15741–15746
6. Schiessl IM, Bardehle S, Castrop H (2013) Superficial nephrons in BALB/c and C57BL/6 mice facilitate in vivo multiphoton microscopy of the kidney. PLoS One 8(1):e52499. https://doi.org/10.1371/journal.pone.0052499
7. Chobanian AV (2009) Shattuck Lecture. The hypertension paradox--more uncontrolled disease despite improved therapy. N Engl J Med 361(9):878–887. https://doi.org/10.1056/NEJMsa0903829
8. Cowley AW Jr (1997) Genetic and nongenetic determinants of salt sensitivity and blood pressure. Am J Clin Nutr 65(2 Suppl):587S–593S
9. Mattson DL, Dwinell MR, Greene AS, Kwitek AE, Roman RJ, Jacob HJ et al (2008) Chromosome substitution reveals the genetic basis of Dahl salt-sensitive hypertension and renal disease. Am J Physiol Ren Physiol 295 (3):F837–F842. https://doi.org/10.1152/ajprenal.90341.2008
10. Russo LM, Sandoval RM, Campos SB, Molitoris BA, Comper WD, Brown D (2009) Impaired tubular uptake explains albuminuria in early diabetic nephropathy. J Am Soc Nephrol 20(3):489–494. https://doi.org/10.1681/ASN.2008050503
11. Ilatovskaya DV, Levchenko V, Lowing A, Shuyskiy LS, Palygin O, Staruschenko A (2015) Podocyte injury in diabetic nephropathy: implications of angiotensin II-dependent activation of TRPC channels. Sci Rep 5:17637. https://doi.org/10.1038/srep17637
12. Slaughter TN, Paige A, Spires D, Kojima N, Kyle PB, Garrett MR et al (2013) Characterization of the development of renal injury in Type-1 diabetic Dahl salt-sensitive rats. Am J Phys Regul Integr Comp Phys 305(7): R727–R734. https://doi.org/10.1152/ajpregu.00382.2012
13. Endres BT, Sandoval RM, Rhodes GJ, Campos-Bilderback SB, Kamocka MM, McDermott-Roe C et al (2017) Intravital imaging of the kidney in a rat model of salt-sensitive hypertension. Am J Physiol Ren Physiol 313(2):F163–F173. https://doi.org/10.1152/ajprenal.00466.2016
14. Rhodes GJ (2017) Surgical preparation of rats and mice for intravital microscopic imaging of abdominal organs. Methods 128:129–138. https://doi.org/10.1016/j.ymeth.2017.07.003
15. Sandoval RM, Molitoris BA (2013) Quantifying glomerular permeability of fluorescent macromolecules using 2-photon microscopy in Munich Wistar rats. J Vis Exp (74). https://doi.org/10.3791/50052
16. Russo LM, Sandoval RM, McKee M, Osicka TM, Collins AB, Brown D et al (2007) The normal kidney filters nephrotic levels of albumin retrieved by proximal tubule cells: retrieval is disrupted in nephrotic states. Kidney Int 71

(6):504–513. https://doi.org/10.1038/sj.ki.5002041

17. Wagner MC, Campos-Bilderback SB, Chowdhury M, Flores B, Lai X, Myslinski J et al (2016) Proximal tubules have the capacity to regulate uptake of albumin. J Am Soc Nephrol 27(2):482–494. https://doi.org/10.1681/ASN.2014111107
18. Asgeirsson D, Venturoli D, Rippe B, Rippe C (2006) Increased glomerular permeability to negatively charged Ficoll relative to neutral Ficoll in rats. Am J Physiol Ren Physiol 291(5):F1083–F1089. https://doi.org/10.1152/ajprenal.00488.2005
19. Tojo A, Endou H (1992) Intrarenal handling of proteins in rats using fractional micropuncture technique. Am J Phys 263(4 Pt 2): F601–F606. https://doi.org/10.1152/ajprenal.1992.263.4.F601
20. Sandoval RM, Wagner MC, Patel M, Campos-Bilderback SB, Rhodes GJ, Wang E et al (2012) Multiple factors influence glomerular albumin permeability in rats. J Am Soc Nephrol 23(3):447–457. https://doi.org/10.1681/ASN.2011070666
21. Yamamoto T, Tada T, Brodsky SV, Tanaka H, Noiri E, Kajiya F et al (2002) Intravital videomicroscopy of peritubular capillaries in renal ischemia. Am J Physiol Ren Physiol 282(6): F1150–F1155. https://doi.org/10.1152/ajprenal.00310.2001
22. Sharfuddin AA, Sandoval RM, Berg DT, McDougal GE, Campos SB, Phillips CL et al (2009) Soluble thrombomodulin protects ischemic kidneys. J Am Soc Nephrol 20(3):524–534. https://doi.org/10.1681/ASN.2008060593
23. McCurley A, Alimperti S, Campos-Bilderback SB, Sandoval RM, Calvino JE, Reynolds TL et al (2017) Inhibition of alphavbeta5 integrin attenuates vascular permeability and protects against renal ischemia-reperfusion injury. J Am Soc Nephrol 28(6):1741–1752. https://doi.org/10.1681/ASN.2016020200
24. Sandoval RM, Wang E, Molitoris BA (2014) Finding the bottom and using it: offsets and sensitivity in the detection of low intensity values in vivo with 2-photon microscopy. Dermatol Int 2(1):e23674. https://doi.org/10.4161/intv.23674

Chapter 8

Library Preparation for Multiplexed Reduced Representation Bisulfite Sequencing with a Universal Adapter

Yong Liu, Alison J. Kriegel, and Mingyu Liang

Abstract

Reduced representation bisulfite sequencing (RRBS) is an efficient approach for estimating cytosine methylation and hydroxymethylation at single nucleotide resolution with a focus on CpG sites located in CpG islands. Commonly used methods for multiplexing RRBS libraries involve a different indexed adapter, which is expensive to generate, for each library. Here, we describe a library preparation method that utilizes a universal adapter and labels samples with unique indexed PCR primers, significantly reducing the cost of multiplexed RRBS.

Key words RRBS, DNA methylation, Methylated cytosine, Illumina Next-Generation Sequencing, MspI digestion, Rat, Human

1 Introduction

Sodium bisulfite is a compound that deaminates unmethylated cytosine (C) residues of denatured DNA, converting them to uracil (U) which will be recognized as thymine (T) in the subsequent PCR amplification step. The 5-methyl cytosine (5-mC) and 5-hydroxymethyl cytosine (5-hmC) nucleotides, however, are not altered by sodium bisulfite. When compared to other methods for determining DNA methylation, such as endonuclease digestion or affinity enrichment, the bisulfite conversion provides a simplified way to estimate the 5-mC + 5-hmC ratio at single nucleotide resolution. Many DNA methylation analysis techniques, such as methylation-specific PCR and Sanger bisulfite sequencing, rely on bisulfite conversion to distinguish 5-mC + 5-hmC from unmethylated cytosine.

The bisulfite conversion can also be coupled with next-generation sequencing platforms to characterize genome-wide DNA methylation patterns with single-base pair resolution. In

G. Thomas Hayman et al. (eds.), *Rat Genomics*, Methods in Molecular Biology, vol. 2018,
https://doi.org/10.1007/978-1-4939-9581-3_8,

whole genome bisulfite sequencing (WGBS), the methylation status of each C can be evaluated by comparing the DNA sequence following bisulfide conversion to that of the reference genome (non-converted). In order to estimate the methylation ratio for every C, and statistically identify differentially methylated regions (DMRs), a total coverage of 30× is recommended by the NIH Roadmap Epigenomics Project [1]. The large size of mammalian genomes and a need for high sequencing coverage has limited large-scale application of WGBS due to its high cost.

Reduced Representation Bisulfite Sequencing (RRBS) utilizes restriction enzymes such as *MspI* to digest genomic DNA. *MspI* is a methylation insensitive restriction endonuclease that recognizes the sequence of CCGG sites. The CpG sites, especially the CpG islands (CGIs), are disproportionally enriched in the short fragments of *MspI* digestion products, making it possible to preferentially sequence those sites. RRBS significantly reduced the cost per sample by sequencing only ~10% of the total human and mouse genomes, while effectively cover more than 80% of the CGIs [2, 3]. In rat, about 47.7% of the CpG sites detected by RRBS are located in CGIs [4].

The first RRBS protocol utilized *BglII* to digest the genome [5]. In 2011, the first RRBS protocol based on *MspI* digestion and gel size-selection was reported [3]. The following year, a simplified *MspI*-based RRBS protocol was developed which eliminated the gel size selection and utilized indexed adaptors to multiplex samples for sequencing [6]. Currently, commercially available RRBS kits are largely based on this simplified protocol, relying on a set of indexed adapters. The need for these adaptors to be synthesized with many modifications on the nucleotides contributes to the high cost of this approach.

Here, we introduced a new RRBS protocol that utilizes a universal adapter. In this protocol, the unique samples are labeled with index sequences during the very last step of final PCR amplification using a set of indexed PCR primers. As the PCR primers are much less expensive than the modified indexed adapters, our protocol significantly reduces the initial cost for setting up the platform, as well as the final cost per sample. Furthermore, additional indexes can be introduced by simply synthesizing new PCR primers without the need to generate new, expensive modified adapters. We have used this protocol, together with a version of the protocol that incorporates an oxidation step, to generate one of the first single-base resolution maps of 5hmC in any in vivo disease model [4].

2 Materials

2.1 Generation of Methylated Fork Structure Adapter

1. Synthesized oligonucleotides (oligos): adapter-top: 5-ACACT CTTTCCCTACACGAC GCTCTTCCGATC*T-3 adapter-bottom: 5-pGATCGGAAGAGC ACACGTCTGAACTCCAG TCA-3.

   All cytosine (Cs in sequence above) are 5-methylcytosine. * on 3′ end of top oligo indicates

   (phosphorothioated) S-oligo linkage between the two bases. p on the 5′ end of the bottom oligo donates phosphate (*see* **Note 1**).
2. Low TE buffer: 10 mM Tris–HCl (pH 8.0) and 0.1 mM EDTA. Stored at room temperature (RT) for up to 6 months.
3. 5 μM NaCl.
4. Bio-Rad Criterion™ Cell system.
5. Bio-Rad 4–20% Criterion™ TBE Polyacrylamide Gel, 18 well, 30 μL, Bio-Rad, 3450060 or 5% Criterion™ TBE Polyacrylamide Gel, 18 well, 30 μL, Bio-Rad 3450048.
6. O'RangeRuler 20 bp DNA Ladder, ready-to-use, ThermoFisher Scientific SM1323.
7. SYBR™ Gold Nucleic Acid Gel Stain (10,000× Concentrate in DMSO) ThermoFisher Scientific S11494.

2.2 Genomic DNA Isolation and Quantification

1. PureLink™ Genomic DNA Mini Kit, ThermoFisher Scientific, K182001.
2. NanoDrop™ 2000/2000c Spectrophotometers, ThermoFisher Scientific.
3. Qubit™ dsDNA BR Assay Kit, ThermoFisher Scientific, Q32850.

2.3 MspI Digestion

1. *MspI* (20 U/μL), New England BioLabs, R0116S.
2. 10× NEB buffer 2, New England BioLabs, B7002S (*see* **Note 4**).
3. 37 °C water bath.
4. Low Molecular Weight DNA Ladder, New England BioLabs, N3233S.
5. Bio-Rad Criterion™ Cell system.
6. Bio-Rad 4–20% Criterion™ TBE Polyacrylamide Gel, 18 well, 30 μL #3450060 or 5% Criterion™ TBE Polyacrylamide Gel, 18 well, 30 μL, Bio-Rad 3450048.
7. SYBR™ Gold Nucleic Acid Gel Stain (10,000× Concentrate in DMSO) ThermoFisher Scientific #S11494.

2.4 Nick End Gap Filling and A-Tailing of MspI-Digested DNA

1. dNTP mixture: add 10, 1, and 1 μL of 100 mM dATP, dGTP, and dCTP, respectively, into a DNase-free Eppendorf tube, respectively; add 88 μL nuclease-free water into the same Eppendorf tube, mix, and quick spin by centrifugation.
2. Klenow fragment (3′ 5′exo−) (5 U/μL), New England BioLabs, M0212L.
3. Phenol/chloroform/isoamyl alcohol (25:24:1 pH = 7.9), ThermoFisher Scientific, AM9730. Adjust the pH value to 7.9 with included buffer before use.
4. Glycogen (5 mg/mL), ThermoFisher Scientific, AM9510.
5. 3 M sodium acetate (pH 5.2), ThermoFisher Scientific, R1181.
6. Pure ethanol and 70% ethanol made with pure ethanol and nuclease-free water.
7. Buffer EB, Qiagen, 19086. Can also get from many Qiagen DNA/plasmid extraction kits.

2.5 Ligation of the A-Tailed DNA with Methylated Fork Structure Adapter

1. 1.5 μM Methylated Fork Structure Adapter: diluted from 15 μM stock solution with nuclease-free water. Save at −80 °C.
2. T4 DNA ligase (2000 U/μL), New England BioLabs, M0202M.
3. Thermocycling plate, VWR, 89049-178.
4. AMPure XP beads, Beckman Coulter, A63881.
5. 70% ethanol: diluted from 200 proof pure ethanol with nuclease-free water.
6. DynaMag™-96 Side Magnet, ThermoFisher Scientific, 12331D.
7. Buffer EB, Qiagen, 19086. Can also get from many Qiagen DNA/plasmid extraction kits.

2.6 Testing DNA Ligation Efficiency by PCR

1. PfuTurbo Cx Hotstart DNA Polymerase, Agilent, 600410.
2. 25 mM each dNTPs: add equal volume 100 mM dATP, dTTP, dGTP, and dCTP, respectively, into a DNase-free Eppendorf tube, mix, and centrifuge. Save at −20 °C.
3. PE 1.0 and PCR primer index (*see* Table 1).

   Resuspend the primers with nuclease-free water to create a 100 μM stock solution. Further dilute the stock to a 2.5 μM working solution with nuclease-free water. Save stock and working solution at −80 °C.
4. Low Molecular Weight DNA Ladder, New England BioLabs, N3233S.
5. Bio-Rad Criterion™ Cell system.

Table 1
PE 1.0 and PCR primer index

Oligo name	Sequence 5′ to 3′	Purification	Modification
PE 1.0	5-AATGATACGGCGACCACCGAGA TCTACACTCTTTCCCTACACGACGCTCTTCCGATC*T-3	DESALT	*Indicate (phosphorothioated) S-oligo linkage between the two bases where the link should occur; DESALT: Deprotected and Desalted
PCR primer index 1	5-CAAGCAGAAGACGGCATACGAGAT T CGTGAT G TGACTGGAGTTCAGACGTGT GCTCTTCCGATC*T-3	DESALT	
PCR primer index 2	5-CAAGCAGAAGACGGCATACGAGAT T ACATCG G TGACTGGAGTTCAGACGTGT GCTCTTCC GATC*T-3	DESALT	
PCR primer index 3	5-CAAGCAGAAGACGGCATACGAGAT T GCCTAA G TGACTGGAGTTCAGACGTGT GCTCTTCCGATC*T-3	DESALT	
PCR primer index 4	5-CAAGCAGAAGACGGCATACGAGAT T TGGTCA G TGACTGGAGTTCAGACGTGT GCTCTTCCGATC*T-3	DESALT	
PCR primer index 5	5-CAAGCAGAAGACGGCATACGAGAT T CACTGT G TGACTGGAGTTCAGACGTGT GCTCTTCCGATC*T-3	DESALT	
PCR primer index 6	5-CAAGCAGAAGACGGCATACGAGAT T ATTGGC G TGACTGGAGTTCAGACGTGT GCTCTTCCGATC*T-3	DESALT	
PCR primer index 7	5-CAAGCAGAAGACGGCATACGAGAT T GATCTG G TGACTGGAGTTCAGACGTGT GCTCTTCCGATC*T-3	DESALT	
PCR primer index 8	5-CAAGCAGAAGACGGCATACGAGAT T TCAAGT G TGACTGGAGTTCAGACGTGT GCTCTTCCGATC*T-3	DESALT	
PCR primer index 9	5-CAAGCAGAAGACGGCATACGAGAT T CTGATC G TGACTGGAGTTCAGACGTGT GCTCTTCCGATC*T-3	DESALT	
PCR primer index 10	5-CAAGCAGAAGACGGCATACGAGAT T AAGCTA G TGACTGGAGTTCAGACGTGT GCTCTTCCGATC*T-3	DESALT	
PCR primer index 11	5-CAAGCAGAAGACGGCATACGAGAT T GTAGCC G TGACTGGAGTTCAGACGTGT GCTCTTCCGATC*T-3	DESALT	
PCR primer index 12	5-CAAGCAGAAGACGGCATACGAGAT T TACAAG G TGACTGGAGTTCAGACGTGT GCTCTTCCGATC*T-3	DESALT	
PCR primer index 13	5-CAAGCAGAAGACGGCATACGAGAT G TTGACT G TGACTGGAGTTCAGACGTGT GCTCTTCCGATC*T-3	DESALT	
PCR primer index 14	5-CAAGCAGAAGACGGCATACGAGAT C GGAACT G TGACTGGAGTTCAGACGTGT GCTCTTCCGATC*T-3	DESALT	
PCR primer index 15	5-CAAGCAGAAGACGGCATACGAGAT C TGACAT G TGACTGGAGTTCAGACGTGT GCTCTTCCGATC*T-3	DESALT	

(continued)

Table 1
(continued)

Oligo name	Sequence 5′ to 3′	Purification	Modification
PCR primer index 16	5-CAAGCAGAAGACGGCATACGAGAT G GGACGG G TGACTGGAGTTCAGACGTGT GCTCTTCCGATC*T-3	DESALT	
PCR primer index 18	5-CAAGCAGAAGACGGCATACGAGAT T GCGGAC G TGACTGGAGTTCAGACGTGT GCTCTTCCGATC*T-3	DESALT	
PCR primer index 19	5-CAAGCAGAAGACGGCATACGAGAT G TTTCAC G TGACTGGAGTTCAGACGTGT GCTCTTCCGATC*T-3	DESALT	
PCR primer index 20	5-CAAGCAGAAGACGGCATACGAGAT A GGCCAC G TGACTGGAGTTCAGACGTGT GCTCTTCCGATC*T-3	DESALT	
PCR primer index 21	5-CAAGCAGAAGACGGCATACGAGAT C CGAAAC G TGACTGGAGTTCAGACGTGT GCTCTTCCGATC*T-3	DESALT	
PCR primer index 22	5-CAAGCAGAAGACGGCATACGAGAT A CGTACG G TGACTGGAGTTCAGACGTGT GCTCTTCCGATC*T-3	DESALT	
PCR primer index 23	5-CAAGCAGAAGACGGCATACGAGAT T CCACTC G TGACTGGAGTTCAGACGTGT GCTCTTCCGATC*T-3	DESALT	
PCR primer index 25	5-CAAGCAGAAGACGGCATACGAGAT T ATCAGT G TGACTGGAGTTCAGACGTGT GCTCTTCCGATC*T-3	DESALT	
PCR primer index 27	5-CAAGCAGAAGACGGCATACGAGAT A AGGAAT G TGACTGGAGTTCAGACGTGT GCTCTTCCGATC*T-3	DESALT	

6. Bio-Rad 4–20% Criterion™ TBE Polyacrylamide Gel, 18 well, 30 μL #3450060 or 5% Criterion™ TBE Polyacrylamide Gel, 18 well, 30 μL, Bio-Rad 3450048.
7. SYBR™ Gold Nucleic Acid Gel Stain (10,000× Concentrate in DMSO) ThermoFisher Scientific #S11494.

2.7 Bisulfite Conversion of the Adapter-Ligated DNA

1. EpiTect Bisulfite Kits, Qiagen, 59104.

2.8 Determination of PCR Cycles for Final RRBS Library by Semi-Quantitative PCR

1. PfuTurbo Cx Hotstart DNA Polymerase, Agilent, 600410.
2. 25 mM each dNTPs.
3. PE 1.0 and PCR primer index.
4. Low Molecular Weight DNA Ladder, New England BioLabs, N3233S.
5. Bio-Rad Criterion™ Cell system.
6. Bio-Rad 4–20% Criterion™ TBE Polyacrylamide Gel, 18 well, 30 μL #3450060 or 5% Criterion™ TBE Polyacrylamide Gel, 18 well, 30 μL, Bio-Rad 3450048.
7. SYBR™ Gold Nucleic Acid Gel Stain (10,000× Concentrate in DMSO) ThermoFisher Scientific #S11494.

2.9 Final RRBS Library Preparation

1. PfuTurbo Cx Hotstart DNA Polymerase, Agilent, 600410.
2. 25 mM dNTP mixture.
3. PE 1.0 and PCR primer index.
4. Thermocycling plate, VWR, 89049-178.
5. AMPure XP beads, Beckman Coulter, A63881.
6. 70% ethanol.
7. DynaMag™-96 Side Magnet, ThermoFisher Scientific, 12331D.

2.10 Library Validation and Quantification

1. Agilent 2100 Bioanalyzer, Agilent, G2939BA.
2. Agilent DNA 1000 Kit, Agilent, 5067-1505.
3. Quant-iT™ PicoGreen™ dsDNA Assay Kit, ThermoFisher Scientific, P11496.
4. Qubit™ dsDNA HS Assay Kit, ThermoFisher Scientific, Q32851.

3 Methods

3.1 Generation of Methylated Fork Structure Adapter

1. Dissolve the top and bottom adapter oligos with low TE buffer to generate 100 μM stock solution.
2. In a 1.5 mL DNase-free Eppendorf tube, add 30 μL of the 100 μM top and bottom oligos each. Next, add 2 μL of 5 M NaCl and 138 μL of low TE buffer. Incubate the reaction in a 98 °C water bath for 5 min, then turn off the water bath and let the water cool down to room temperature (RT). Split the annealed adapter solution into 20 μL aliquots. Annealed 15 μM adapters can be stored at −80 °C for up to 2 years.
3. The size of the fork structure adapter (*see* Fig. 1) should be evaluated by running it on a 4–20% or 5% Criterion precast PAGE-TBE gel on a Criterion Cell per manufacturer's instructions. To run the gel, set up Criterion precast PAGE-TBA gel and fill the buffer chamber with 0.5% TBE buffer. Load 1 ul of the 15 μM annealed adapter on the gel, load 20 bp DNA ladder to flanking wells. Run the gel at 120 V for 90 min if using 4–20% Criterion gel. Stain the gel with 1:10,000 dilutions of SYBR Gold nucleic acid gel stain for 15 min on an orbital shaker at low speed. Cover the gel container with an opaque lid or aluminum foil to protect SYBR Gold stain from light. Wash the gel with tap water 2–3 times and visualize and calculate the fork structure adapter size under a UV light. The expected size of the fork structure adapter will be about 65 bp (*see* Fig. 2) (*see* **Note 2**).

3.2 Genomic DNA Isolation and Quantification

1. We use the PureLink™ Genomic DNA Mini Kit for genomic DNA isolation and elute the DNA from the column with 30–100 μL of DNA elution buffer provided by the kit (*see* **Note 3**). The purified genomic DNA can be stored at 4 °C

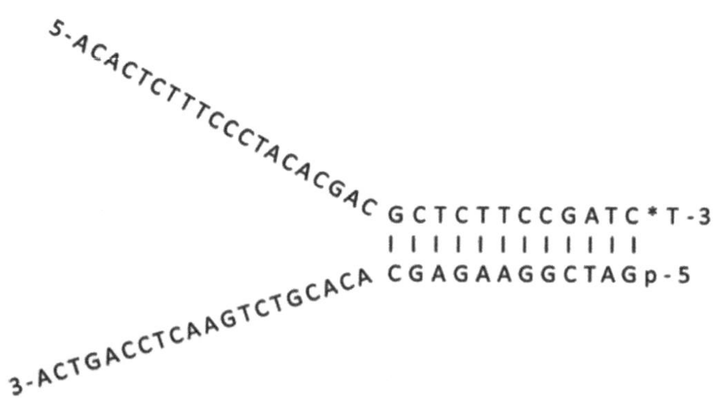

Fig. 1 The fork-like structure of adapter generated from annealing the top and bottom adapter oligos

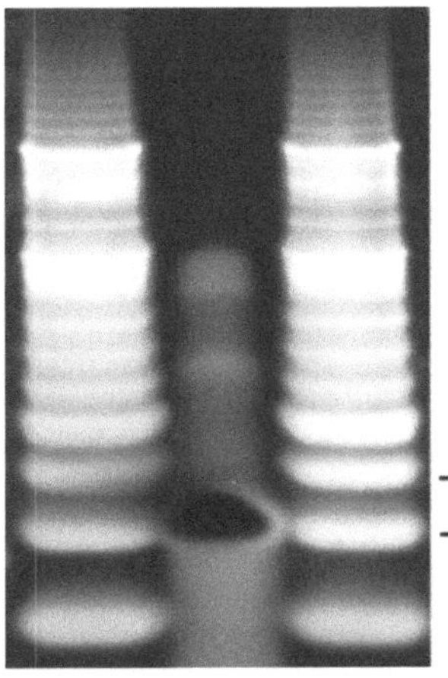

Fig. 2 The annealed fork structure appears as a ~65 bp band (shown as a dark band in the middle lane due to overloading) on a 4–20% Criterion precast PAGE-TBA gel. Left and right lanes are 20 bp DNA ladders

for instant use. For long-term storage, store the DNA at −80 °C.

2. Determine the genomic DNA concentration with Nanodrop or Qubit™ dsDNA BR Assay Kit per the manufacturer's instructions. Adjust the DNA concentration to 100 ng/μL with nuclease-free water.

3.3 Msp I Digestion

1. In a 0.5 mL DNase-free Eppendorf tube, set up 40 μL *MspI* digestion reaction mixture in the following order:
   33 μL nuclease-free water.
   4.0 μL 10× NEB buffer 2.
   1.5 μL 100 ng/μL gDNA (*see* **Note 5**).
   1.5 μL *MspI* (20 U/μL).
2. Mix gently by pipette for 6–8 times and quick spin contents.
3. Incubate at 37 °C water bath overnight for full digestion of genomic DNA.
4. Take out 10 μL mixture to run a diagnostic gel to check the *MspI* digestion efficiency (*see* **Note 6**):
   During this period, the remaining 30 μL reaction mixture can be kept on ice for a couple hours or at −20 °C for longer time storage.
   Following Subheading 3.1, **step 3** to run a 4–20% or 5% Criterion precast polyacrylamide TBE gel. Loading 0.5 μL/lane low molecular weight marker on both side of the gel. *MspI* fully digested genomic DNA will show species-specific microsatellite band(s) with a large molecular size smear at the top of the gel (*see* Fig. 3).

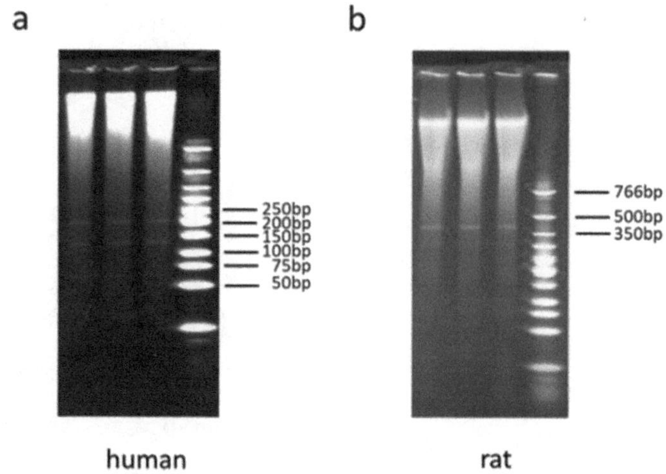

Fig. 3 Typical results of *MspI*-digested human (**a**) and rat (**b**) genomic DNA. For fully digested human gDNA, three satellite bands at about 60, 125, and 200 bp, respectively, are shown. For rat, only one satellite band at about 400 bp is shown

3.4 Nick End Gap Filling and A-Tailing of MspI-Digested DNA

1. In a 0.5 mL DNase-free Eppendorf tube, set up reaction mixture in the following order (*see* **Note** 7):

   30 μL *MspI*-digested DNA mixture.

   1.0 μL dNTP mixture (10 mM dATP; 1 mM dGTP; 1 mM dCTP).

   0.2 μL 10× NEB buffer 2.

   1.0 μL 5 U/μL Klenow fragment (3′ 5′exo−).
2. Mix gently by pipette for 6–8 times and spin down contents for a few seconds.
3. Perform the reaction in a PCR thermal cycler without the heat lid with the program of: 30 °C for 20 min followed by 20 min at 37 °C, then 4 °C infinite hold.
4. A-tailed DNA purification with ethanol precipitation: after A-tailing, take the tube out of the thermal cycler, add 167.8 μL TE buffer to bring the total volume to 200 μL. Vortex and centrifuge briefly. Add 200 μL of phenol/chloroform/isoamyl alcohol (25:24:1 pH = 7.9) and vortex vigorously for at least 30 s. Spin at 14,000 × *g* for 5 min at room temperature. Transfer about 200 μL upper aqueous phase carefully without carryover of any phenol to a 1.5 mL EP tube and add the following reagents sequentially: 4 μL 5 mg/mL glycogen, 20 μL 3 M sodium acetate (pH 5.2), and 500 μL cold (−20 °C) pure ethanol. Mix vigorously by vortex and precipitate the DNA at −80 °C for 30 min or −20 °C overnight. Centrifuge at 16,000 × *g* for 30 min at 4 °C and decant

the supernatant. Add 500 μL of 70% (V/V) ethanol to wash the pellet and make sure the pellet has floated by tapping the bottom of the EP tube. Centrifuge at 16,000 × *g* for 10 min at 4 °C. Decant the supernatant. Spin down at 16,000 × *g* for 2 min at room temperature to collect the residual liquid on the wall of the tube. Carefully remove all of the liquid using a 20 μL pipette without disturbing the pellet. Air dry the pellet for 3–5 min until approximately half of the pellet changes its appearance from milky white to clear and film-like. Do not overdry the pellet as this will reduce the DNA solubility. Resuspend the pellet in 20 μL EB buffer. At this point, the extracted DNA can either be stored at −20 °C or used immediately in the next step.

3.5 Ligation of the A-Tailed DNA with Methylated Fork Structure Adapter

1. In a 0.5 mL DNase-free Eppendorf tube, set up reaction mixture in the following order:

   5.4 μL nuclease-free water.

   20 μL Purified end-repaired and A-tailed DNA.

   3 μL T4 DNA ligase buffer (10×).

   0.6 μL Methylated Fork Structure Adapter (1.5 μM).

   1 μL T4 DNA ligase (2000 U/μL).

   Mix well by pipetting 6–8 times and centrifuge for a few seconds (*see* **Note 8**).

2. Incubate the ligation reaction at 16 °C overnight on a PCR machine without using the heat lid.
3. Deactivate the enzyme at 65 °C for 20 min.
4. Bead purification of adapter-ligated DNA: Transfer the reaction to a 96-well thermocycling plate. Fully mix the SPRI Ampure XP beads and then a 2× concentration of AMPure XP beads (60 μL beads for 30 μL sample) is added to wells with sample. Beads and samples should be mixed by pipetting at least five times. The mixtures are incubated at room temperature for 30 min for DNA binding with the beads. The 96-well plate is placed onto a DynaMag™-96 Side magnet for 5 min. Carefully remove the supernatant by pipetting from the side opposite from the accumulated beads. Wash the beads with 200 μL of 70% ethanol and incubate for 30 s. Aspirate out the ethanol with a 200 μL pipette with setting at 190 μL and discard. Add 200 μL of 70% ethanol to repeat wash. Use a 200 μL pipette at setting of 190 μL and a 20 μL pipette at setting of 20 μL sequentially to remove all of the liquid without loss of the beads. Dry the beads for <5 min at room temperature. Remove plate from the magnet apparatus and add 23 μL of EB buffer to each well and resuspend the beads by pipetting 10 times. Place the reaction plate back onto the magnet plate

for 1 min to separate beads from solution. Without disturbing the beads, carefully transfer 21 μL liquid into a new 0.5 mL DNase-free Eppendorf tube.

5. Save the purified adapter-ligated DNA at −20 °C or process to next step immediately.

3.6 Testing DNA Ligation Efficiency by PCR

1. In a 0.5 mL DNase-free Eppendorf tube, set up PCR reaction mixture on ice in the following order:

   30.5 μL nuclease-free water.

   5 μL PfuTurbo Cx hotstart DNA polymerase buffer (10×).

   0.5 μL dNTPs (25 mM each).

   6 μL PE 1.0 (2.5 μM).

   6 μL PCR primer index (2.5 μM) (*see* **Note 9**).

   1 μL purified adapter-ligated DNA.

   1 μL Pfu Turbo Cx hotstart DNA polymerase (2.5 U/μL).

   Mix gently by pipette 6–8 times and centrifuge shortly (*see* **Note 10**).
2. Dispense 10 μL of the 50 μL PCR mixture into 3–4 0.5 mL DNase-free Eppendorf tubes.
3. Set up 3–4 PCR programs on different PCR thermal cyclers, or individual heads of same thermal cycler, as follows: 95 °C, 2 min followed by 10, 13, 16, or 19 cycles, respectively, of 95 °C for 30 s, 65 °C for 30 s, 72 °C for 45 s, with final extension of 72 °C for 7 min (*see* **Note 11**).
4. Follow steps in Subheading 3.1, **step 3,** to run a diagnostic PAGE-TBE gel to check the ligation efficiency. Load the PCR products from the same sample with different cycles close to each other for easy comparison. Make sure to include a size ladder.
5. Check the results of the gel by imaging. DNA with efficient adapter ligation will show species-specific microsatellite band (s) usually at 10 cycles (*see* Fig. 4).

3.7 Bisulfite Conversion of the Adapter-Ligated DNA

1. For the 20 μL remaining adapter-ligated DNA from Subheading 3.5, **step 4**, follow the Qiagen EpiTect Bisulfite kit protocol designated for DNA extraction from FFPE tissue for bisulfite conversion. Elute the DNA with 22 μL of 65 °C preheated EB buffer included in the kit.
2. Perform a second round of bisulfite conversion with the 20 μL of eluted DNA from **step 1** following the same process with the **step 1**. Elute the DNA with 40 μL of 65 °C preheated EB buffer included in the kit (*see* **Note 12**).
3. Store the bisulfite-converted DNA at −20 °C or process to next step.

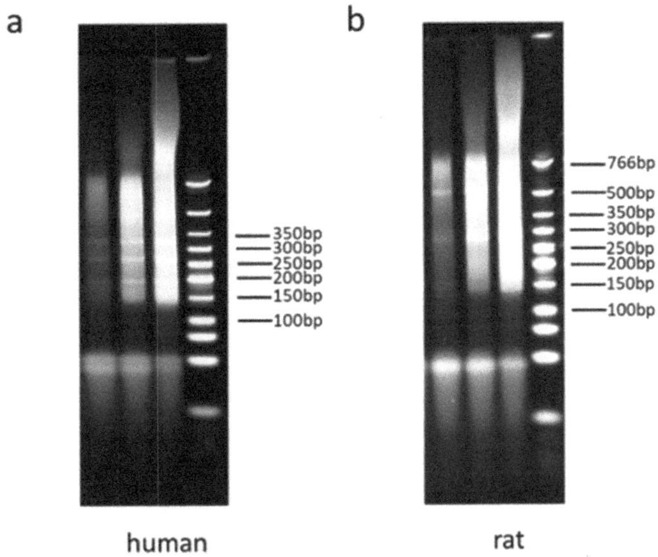

Fig. 4 Characteristic results of a PCR amplification test for adapter ligation efficiency. For typical efficiently adapter-ligated human gDNA (**a**), three satellite bands at about 200, 275, and 325 bp, respectively, are shown (10,13, and 16 cycles, from left to right lane). For rat (**b**), two satellite bands at about 280 and 550 bp are shown (10, 13, and 16 cycles, from left to right lanes). Less than 100 bp primer dimers are also shown, which are subject to change with the PCR primer index used

3.8 Determine PCR Cycles for Final RRBS Library by Semi-Quantitative PCR

1. Repeat the steps in Subheading 3.6 with the following modification: use 1 μL DNA from Subheading 3.7, **step 2**, as the template. Adjust the cycle numbers to 13, 16, and 19 (*see* **Note 13**).
2. Check the results by running and imaging the gel. Identify the fewest number of PCR cycles that still contain enough amplified material to be detected (*see* Fig. 5). This cycle number will be selected as the optimal number of PCR cycles for final RRBS library amplification.

3.9 Final RRBS Library Preparation

1. In a 0.5 mL DNase-free Eppendorf tube, set up PCR reaction mixture on ice in the following order:

   25.5 μL nuclease-free water.

   5 μL PfuTurbo Cx hotstart DNA polymerase buffer (10×).

   0.5 μL dNTPs (25 mM each).

   6 μL PE 1.0 (2.5 μM).

   6 μL PCR primer index (2.5 μM) (*see* **Note 15**).

   6 μL bisulfite-converted DNA.

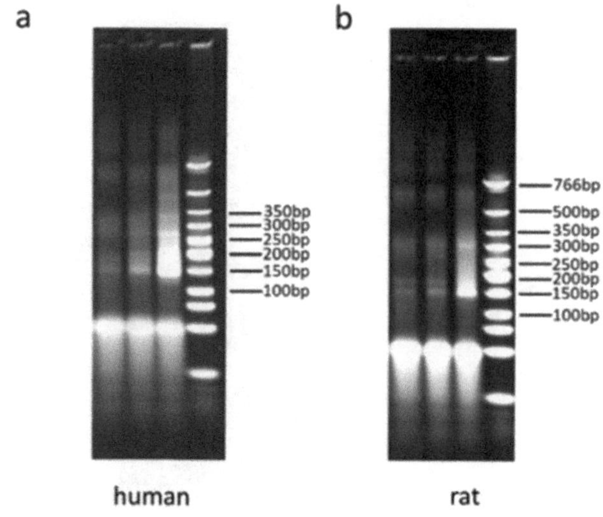

Fig. 5 Characteristic results of a semi-quantitative PCR test for cycle number determination for the final RRBS library. The PCR cycle for each sample is determined by the lowest number of cycles with clear amplification (shown with a smear between the 150 and 350 bp region) and species-specific satellite band(s). For the human sample (**a**), at least two satellite bands at about 275 and 325 bp, respectively, are shown. For rat (**b**), only one band at about 280 bp is shown (*see* **Note 14**). Here, human and rat sample amplification products from 13, 16, and 19 cycles (from left to right) were imaged. In these examples the 16 and 17 cycles are selected for the human (**a**) and rat (**b**) sample, respectively

1 μL Pfu Turbo Cx hotstart DNA polymerase (2.5 U/μL).

Mix gently by pipette 6–8 times and briefly spin.

2. Set up PCR program on PCR thermal cycler: 95 °C for 2 min followed by determined cycles for each sample of 95 °C for 30 s, 65 °C for 30 s, 72 °C for 45 s, with final extension of 72 °C for 7 min.
3. Following the process of Subheading 3.5, **step 4**, the PCR reaction mixture will be cleaned up by adding a 1.2× concentration of SPRI AMPure XP beads (60 μL beads for 50 μL PCR reaction mixture) first and eluting in 42 μL EB buffer for 40 μL of recovery. Put the samples through a second round of 1.5× beads, clean up, and elute the final library with 40 μL EB buffer (*see* Fig. 6).

3.10 Library Validation and Quantification

1. Run the samples on a DNA 1000 chip on an Agilent Bioanalyzer to validate and semi-quantify the RRBS library. A successfully made library will show a "dome" between 150 and 300 bp, with peak at about 190 and 160 bp for human and rat samples, respectively (*see* Fig. 7) (*see* **Note 16**).

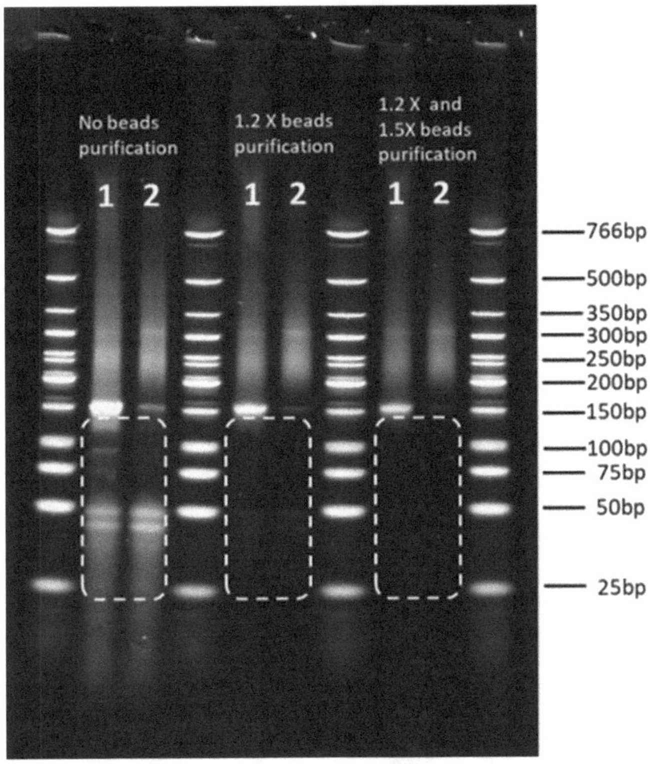

Fig. 6 Cleanup of the PCR reaction product with 1.2 and 1.5× beads to sequentially remove all of the primer dimers (<100 bp) while effectively preserving the 150–350 bp target region fragments. Samples 1 and 2 (labeled with 1 and 2) are rat DNA with high (0.25 μM) and low (0.03 μM) final adapter concentrations used in the adapter ligation step. We use 0.03 μM final concentration of adapter in Subheading 3.5, **step 1**. The short fragment regions of primer dimers are circled with white dashed lines. The initial round of 1.2× bead purification removed most of the primer dimers, while an additional round of 1.5× bead purification removed them all

2. Optional: a Quant-iT™ PicoGreen™ dsDNA Assay Kit (for 50 pg to 2 μg range) or a Qubit™ dsDNA HS Assay Kit (for 0.2–100 ng range) can be used for more accurate quantification. However, this step is not always necessary as more accurate quantification by qPCR will always be performed by the next-generation sequencing core before sequencing.

4 Notes

1. Top and bottom oligos are synthesized with high-performance liquid chromatography (HPLC) purification. All cytosine (Cs) are methylated to form 5-methylcytosine to avoid being converted to Us (Ts by PCR) in the bisulfite conversion step.

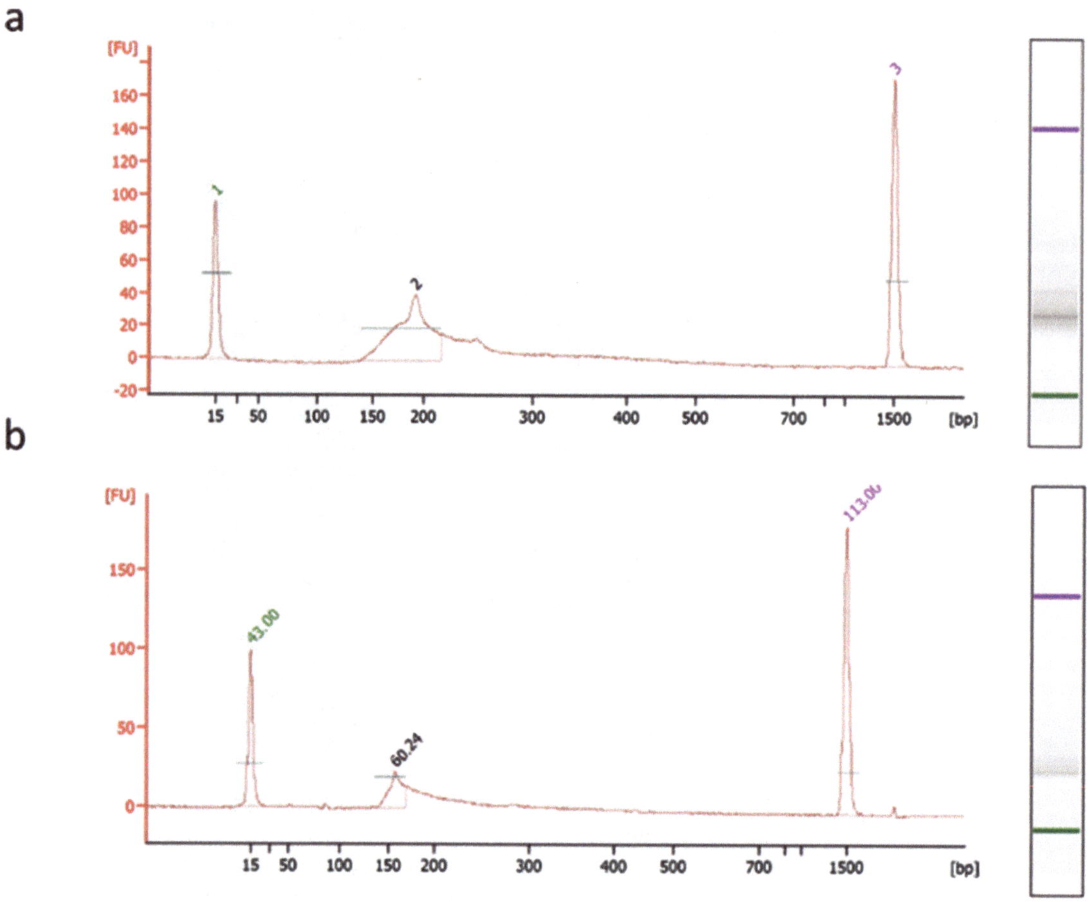

Fig. 7 Characteristic DNA 1000 chip results of the RRBS final library from human (**a**) and rat (**b**) samples

The top oligo has (phosphorothioated) S-oligo linkage (∗) at the 3′ end, which will reduce nucleolytic degradation of the 3′ end T overhang. The T overhang is critical for high efficiency adapter ligation. The 5′ end p (phosphate) of the bottom oligo facilitates the ligation of the annealed fork structure adapter with the double strand DNA fragments.

2. A high quality fork structure adapter is critical for the effective ligation with the double strand DNA fragment. A strong band at about 65 bp with less smear or bands at other sizes (*see* Fig. 2) indicate that the adapter is effectively annealed.

3. Other commercial kits for genomic DNA isolation can also be used. We prefer kits that use Protease K to lyse the cells and digest the protein, which usually yield high quality DNA that is free of proteins.

4. 10× NEB buffer 2 can also be obtained from the NEB Klenow Fragment (3′ → 5′ exo-) package.

5. Use as much genomic DNA as possible if the total DNA available is less than 150 ng. Adjust the nuclease-free water volume if needed.
6. Avoid this step if the total DNA input is very low. At this time, a control sample with normal DNA input can be used to check the digestion efficiency.
7. A master mixture without the *MspI*-digested DNA can be prepared if processing multiple samples. Prepare a master mix with an extra 5–10% of the desired volume to make up for loss with pipetting. Mix gently by pipetting and quickly spin before distributing to each reaction tube.
8. A master mixture without the purified end-repaired and A-tailed DNA can be prepared if processing multiple samples. Prepare a master mix with an extra 5–10% of desired volume to make up for loss with pipetting. Mix gently by pipette and quickly spin shortly before distributing the master mix into each reaction tube.
9. Any one of the 24 PCR primer indexes can be used. However, the polyacrylamide TBE gel results might be slightly different, especially for the pattern of the primer dimer at the lower part of the gel.
10. A master mixture without the purified adapter-ligated DNA can be prepared if processing multiple samples. Prepare a master mix with an extra 5–10% of the desired volume to make up for loss during pipetting. Mix gently by pipetting and quickly spin before distributing to each reaction tube.
11. If the *MspI* digestion begins with the optimal quantity of 150 ng total DNA per sample, then there is no need to test 19 cycles of PCR reaction. However, with less than optimal total DNA input, testing with higher cycles (19 cycles) is always preferred for an accurate estimation of ligation efficiency.
12. We usually reach a 99.9% or higher bisulfite conversion rate with two rounds of bisulfite conversion using the Qiagen EpiTect Bisulfite kit. However, other commercial kits for bisulfite conversion can also be used. For example, with one round of bisulfite conversion with EpiTect Fast Bisulfite Conversion Kits, we can get around a 99.8% bisulfite conversion rate.
13. If the *MspI* digestion started with 150 ng total DNA, then testing with 13, 16, and 19 will be sufficient to estimate an appropriate cycle number for final library amplification. However, with less than optional total DNA input, testing with higher cycles, such as 22 cycles, is preferred.
14. Due to the massive DNA damage from the bisulfite conversion, the satellite band(s), especially large-sized band(s), might not

be visible with gel imaging, especially in lanes with a lower number of PCR cycles.

15. Individual samples that will be pooled for sequencing in the same lane of a flow cell will need to be labeled using different PCR primer indexes.
16. The "peak" from the "dome" might be too small to be quantified by DNA 1000 chip. However, in our experience, many of those RRBS libraries that only show a small "dome" actually have far more DNA than what is required by the NGS platform and can be successfully sequenced.

References

1. http://www.roadmapepigenomics.org/files/protocols/data/dna-methylation/MethylC-SeqStandards_FINAL.pdf
2. Smith ZD, Gu H, Bock C, Gnirke A, Meissner A (2009) High-throughput bisulfite sequencing in mammalian genomes. Methods 48:226–232
3. Gu H, Smith ZD, Bock C, Boyle P, Gnirke A, Meissner A (2011) Preparation of reduced representation bisulfite sequencing libraries for genome-scale DNA methylation profiling. Nat Protoc 6:468–481
4. Liu Y, Liu P, Yang C, Cowley AW Jr, Liang M (2014) Base-resolution maps of 5-methylcytosine and 5-hydroxymethylcytosine in Dahl S rats: effect of salt and genomic sequence. Hypertension 63:827–838
5. Meissner A, Gnirke A, Bell GW, Ramsahoye B, Lander ES, Jaenisch R (2005) Reduced representation bisulfite sequencing for comparative high-resolution DNA methylation analysis. Nucleic Acids Res 33:5868–5877
6. Boyle P, Clement K, Gu H, Smith ZD, Ziller M, Fostel JL et al (2012) Gel-free multiplexed reduced representation bisulfite sequencing for large-scale DNA methylation profiling. Genome Biol 13:R92. https://doi.org/10.1186/gb-2012-13-10-r92

Chapter 9

Characterization of the Rat Gut Microbiota via 16S rRNA Amplicon Library Sequencing

Aaron C. Ericsson, Susheel B. Busi, and James M. Amos-Landgraf

Abstract

It is becoming increasingly apparent that microbiota have measurable effects on numerous phenotypes in laboratory animals. This "second genome" has often been disregarded or ignored due to its commensal nonpathogenic nature, and the difficulty, expense, and analysis of sequence. Recent advances in sequencing methods and analyses of large datasets have made characterization of microbiota populations routine and have uncovered previously unknown relationships of microbial communities and host biological systems. The largest and most diverse microbial community in the laboratory rat is in the gut, and has been shown to affect the physiology of the whole animal, and genetic disease penetrance. We present here a cost-effective method for the characterization of the rat fecal microbiota through multiplexed 16S ribosomal sequencing and freely available software.

Key words Microbiota, 16S rRNA gene, Operational taxonomic unit (OTU), Sequencing

1 Introduction

The rat was one of the first extensively used mammalian models of human biology and disease, largely owing to the identification of spontaneously occurring genetic mutations resulting in heritable quantitative phenotypic differences between stocks and strains. With the availability of the rat genome sequence and new genome editing technology, hundreds of new genetic models of human disease have been made available in the rat. With the increasing sensitivity of biological assays and techniques, there is a renewed emphasis on understanding sources of biological variation in phenotypes to attempt to control sources of error, leading to over- or misinterpretation of results and lack of reproducibility. Historically, identification and control of pathogens present in laboratory stocks, particularly in the gastrointestinal system, reduced much of the variability of results between institutions and experiments. However, defining these polymicrobial communities has been difficult, as many members of the gut microbiota (GM) have been

G. Thomas Hayman et al. (eds.), *Rat Genomics*, Methods in Molecular Biology, vol. 2018,
https://doi.org/10.1007/978-1-4939-9581-3_9,

refractory to culture using traditional microbiological techniques. While culture methods have progressed tremendously over the last 20 years, it is still not possible to culture all members of the GM. The development of the polymerase chain reaction (PCR) technique in the 1980s revolutionized the field of biomedical research. Coupled with one of several different approaches for high-throughput sequencing of individual strands of DNA, powerful yet relatively affordable platforms for the characterization of polymicrobial communities have emerged. These methods have been applied to various host-associated samples such as feces, greatly enhancing our understanding of the extraordinarily complex and bidirectional relationship between host animals and their cognate microbial communities or "microbiota." Thus, culture-independent amplification of genes or gene regions conserved among all prokaryotes, followed by sequencing of those PCR products, provides a detailed profile of the taxa present in a given sample. While similar methods have been applied to fungi and protozoa, the current chapter focuses on characterization of bacterial (and archaeal) communities.

Eukaryotic and prokaryotic genomes encode distinct small and large ribosomal subunits. In prokaryotes, the small ribosomal RNA (rRNA) subunit is encoded by the 16S rRNA gene, an evolutionarily ancient gene that has undergone mutations at a relatively steady rate over the millennia and thus serves as a molecular clock of sorts and excellent marker for phylogenetic studies. Specifically, the 16S rRNA gene includes nine highly variable regions (termed V1 through V9), separated by highly conserved regions of DNA required to maintain the core structure of the molecule [1]. While the entire 16S rRNA gene is rather long for full-length amplification via standard PCR, primers designed to bind template DNA in conserved regions flanking a variable region will result in amplicon libraries which can then be annotated taxonomically by comparison to a database of 16S rRNA gene sequences. The most commonly used databases are the Ribosomal Database Project (RDP) [2], SILVA [3], and Greengenes [4]. While the taxonomic resolution varies, the V4 region provides the most faithful community phylogeny when compared to whole 16S rRNA assemblies [5], and in our own studies, libraries comprising amplicons roughly 250 nucleotides (nt) in length can generally be annotated to the level of family or genus, and often to species [6]. While the PCR is typically performed in the customary 96-well format, commonly used sequencers hybridize a single PCR library to a single-use "lane" covered in amplicon-binding adaptor molecules. To allow for multiplex sequencing of samples amplified during PCR, sets of 96 (or 384) primer pairs, each containing a specific 8–10 nt index (i.e., "barcode") that will be incorporated into any amplicon generated from a given sample, are available. Thus, 96 different amplicon libraries, each in turn containing a diverse mixture of 16S rRNA sequences, can be combined in a single sequencing run and then deconvoluted during the informatics portion of the analysis, reducing the cost per sample by almost two orders of magnitude.

Here we present a detailed description of the required materials and procedures involved in characterization of the GM of research rats, as well as several powerful informatics tools available for downstream data analysis and prediction of metagenomic capacity and community dynamics. It is fully acknowledged that the informatics approaches associated with these methods are a constantly evolving field and readers are encouraged to review the current literature when analyzing their data. Moreover, there are multiple sequencing platforms available, each with their own advantages and limitations, and this chapter is based on the Illumina MiSeq platform. Details related to sample processing and library preparation are fairly similar between platforms.

2 Materials

2.1 DNA Extraction and 16S rRNA Library Preparation

Feces contain multiple PCR inhibitors, which vary among species and are influenced by diet and other factors. To minimize the amount or effect of such compounds, microbial DNA can be extracted using one of two methods, each requiring only a small portion of a typical rat fecal pellet. The quickest and most effective kit-based method is the PowerFecal kit (Qiagen), performed according to manufacturer's instructions and being careful not to exceed the recommended starting mass (0.25 g) (*see* **Note 1**).

Regardless of the extraction method, DNA needs to be normalized to a standard concentration and volume for 16S rRNA library generation via PCR in a 96-well format. Sequencing via the Illumina MiSeq platform occurs on a single "lane" flow cell, the physical substrate to which individual strands of DNA are hybridized. Thus, to allow for multiplexing of 96 samples on a single lane, libraries are generated using a set of 96 unique primer pairs which incorporate sample-specific indices within all amplicons. Specifically, a common forward primer (U515F) is used in combination with a series of reverse primers with a common DNA-binding sequence (806R) [7, 8] and 96 unique 12-base indices. As a result, all amplicons generated in a given well will contain a unique "barcode" identifier, allowing sequence data to be de-multiplexed during the informatics stage (*see* **Note 2**).

2.2 Sequencing and Informatics

The library sequencing and basic informatics are highly automated and are most commonly performed by a dedicated lab such as an academic institution core facility or fee-for-service partner. Thus, a detailed description of the MiSeq instrument's equipment, reagents, and the primary software likely has minimal utility here. Rather, we describe considerations related to data quality control

(QC), and a number of useful freely available software packages for downstream visualization, analysis, and statistical testing of data.

Depending on the MiSeq chemistry used (i.e., v2 or v3 reagent kit), a maximum of 15 or 25 million reads, respectively, is possible, although quality filtering results in a more realistic total output of 50–60% of those numbers. Thus, an Illumina v2 reagent kit generating eight million high-quality reads would provide a mean of roughly 83,000 reads per sample for 96 amplicon libraries sequenced in a single run. The sufficiency of such coverage can be assessed via metrics such as Good's coverage [9] (an estimate of the percentage of total species detected in a sample) or the Chao estimator [10] (an estimate of the true community richness). The common informatics software packages used to analyze 16S rRNA sequencing data (i.e., Qiime [11] and Mothur [12]) also provide rarefaction curves showing the increase in detected richness as the sequencing data are reiteratively subsampled at increasing read counts. A curve approaching a plateau suggests that sequencing fewer libraries on a single flow cell to increase coverage will provide only an incremental increase in detected richness.

For visualization and analyses of 16S rRNA sequencing data beyond the standard outputs provided via the automated pipelines such as Qiime, the user may possibly use software such as Excel, or several other powerful online tools. Here, we discuss a number of relatively user-friendly desktop-based and online programs capable of generating custom ordination plots and heatmaps, performing relevant statistical tests (e.g., one- and two-way PERMANOVA and ANOSIM, SIMPER), determining microbial interactions in longitudinal datasets, and predicting metagenomic capacity of a polymicrobial community.

2.2.1 PAST

PAleontological **ST**atistics is a free software for scientific data analysis and manipulation, univariate and multivariate statistics including ecological and spatial analyses [13]. In the authors' experience, this tool can be used for many more statistical and empirical tests beyond those described here. The basic utility of this software (*v3.18*, at the time of writing) for the analysis of 16S rRNA sequencing data has only been briefly described and the user is encouraged to explore the various facets of the software at their discretion. The entire potential and methodologies offered by this program designed for Windows and Mac operating systems is well described in the manual available along with the downloaded software, which can be obtained from the following online site: https://folk.uio.no/ohammer/past/

Entering Data, Ordinate Plots, and Multivariate Statistics

The manual accompanying the software, although quite exhaustive, is not necessarily geared toward 16S rRNA sequencing data analysis. We provide a basic protocol for data input and preliminary analysis including ordinate plots and multivariate statistics, including step-by-step instructions. First, the 16S sequencing data is imported into an excel worksheet (*see* Fig. 1), imported into PAST3 (*see* Fig. 2), and grouped for further analysis.

Samples and groups can be assigned various symbols and colors as needed, which may be essential when using ordinate plots to show architectural differences between groups. If the purpose of the analysis is to perform multivariate statistics such as a PERMANOVA (Permutational Multivariate Analysis Of Variance) or

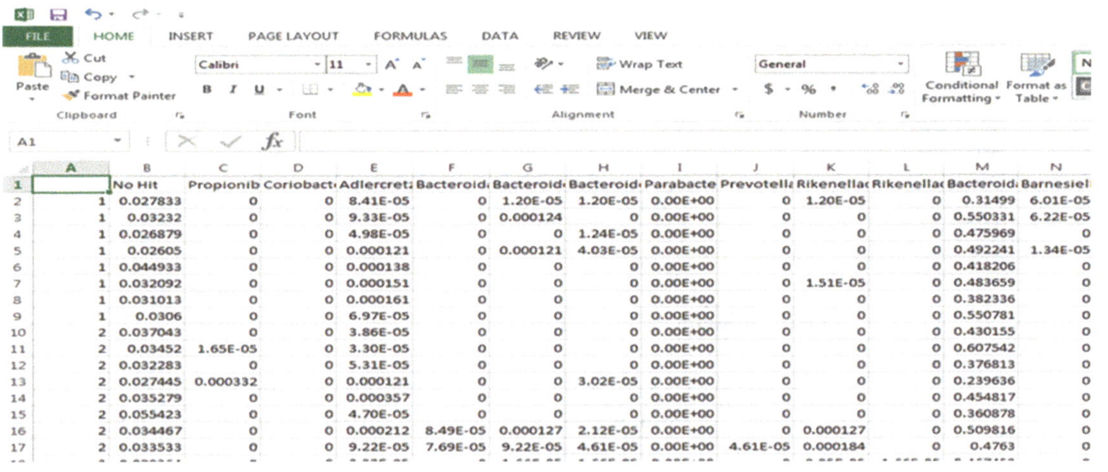

Fig. 1 16S Data organization. Organize and arrange all 16S sequencing data in MS-Excel as shown, where the operational taxonomic units (OTUs) are listed in columns, and individual samples or groups are listed as rows. In the example above, column A contains eight individual samples from group 1 arranged in rows, followed by samples from group 2 etc

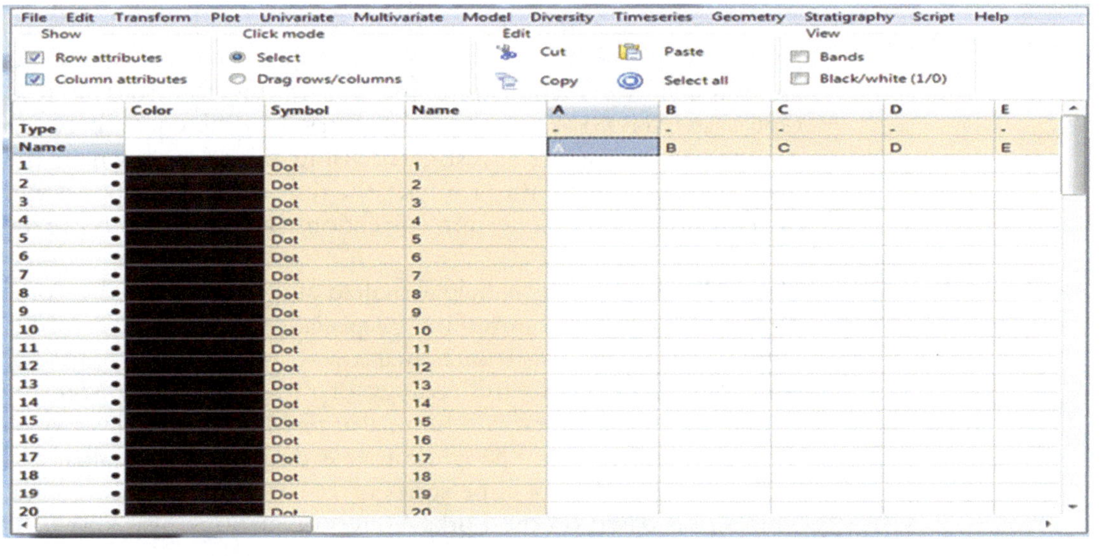

Fig. 2 PAST3 importation. Open PAST3.exe and ensure "Row attributes" and "Column attributes" are checked. Copy the data from Excel and paste into the highlighted box, 'A'

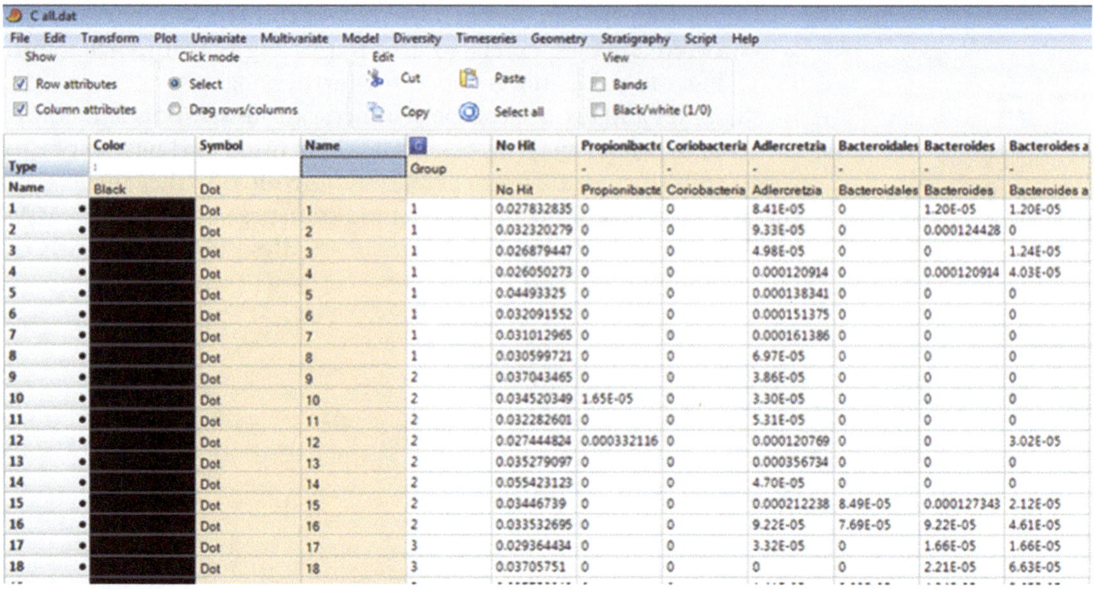

	Color	Symbol	Name		No Hit	Propionibacte	Coriobacteria	Adlercretzia	Bacteroidales	Bacteroides	Bacteroides a
Type	:			Group	-	-	-	-	-	-	-
Name	Black	Dot			No Hit	Propionibacte	Coriobacteria	Adlercretzia	Bacteroidales	Bacteroides	Bacteroides a
1 •		Dot	1	1	0.027832835	0	0	8.41E-05	0	1.20E-05	1.20E-05
2 •		Dot	2	1	0.032320279	0	0	9.33E-05	0	0.000124428	0
3 •		Dot	3	1	0.026879447	0	0	4.98E-05	0	0	1.24E-05
4 •		Dot	4	1	0.026050273	0	0	0.000120914	0	0.000120914	4.03E-05
5 •		Dot	5	1	0.04493325	0	0	0.000138341	0	0	0
6 •		Dot	6	1	0.032091552	0	0	0.000151375	0	0	0
7 •		Dot	7	1	0.031012965	0	0	0.000161386	0	0	0
8 •		Dot	8	1	0.030599721	0	0	6.97E-05	0	0	0
9 •		Dot	9	2	0.037043465	0	0	3.86E-05	0	0	0
10 •		Dot	10	2	0.034520349	1.65E-05	0	3.30E-05	0	0	0
11 •		Dot	11	2	0.032282601	0	0	5.31E-05	0	0	0
12 •		Dot	12	2	0.027444824	0.000332116	0	0.000120769	0	0	3.02E-05
13 •		Dot	13	2	0.035279097	0	0	0.000356734	0	0	0
14 •		Dot	14	2	0.055423123	0	0	4.70E-05	0	0	0
15 •		Dot	15	2	0.03446739	0	0	0.000212238	8.49E-05	0.000127343	2.12E-05
16 •		Dot	16	2	0.033532695	0	0	9.22E-05	7.69E-05	9.22E-05	4.61E-05
17 •		Dot	17	3	0.029364434	0	0	3.32E-05	0	1.66E-05	1.66E-05
18 •		Dot	18	3	0.03705751	0	0	0	0	2.21E-05	6.63E-05

Fig. 3 PAST3 grouping. In PAST click in the "Type" box in column A and change to "Group"

ANOSIM (Analysis Of Similarity), colors and symbols are nonessential as long as the samples are assigned to appropriate groups (*see* Fig. 3).

Data transformation, especially that which is not normally distributed or whose standard deviations are not homogeneous, or both, is a useful method especially when using a parametric test such as an ANOVA or linear regression, which may otherwise give misleading results (*see* **Note 3**). The data can then be visualized by selecting the data as described in Fig. 4.

To perform a univariate or multivariate analysis, similar to the previous step, the data along with its groups should be selected. As an example, we demonstrate the use of the multivariate test, One-Way ANOSIM (*see* Fig. 5). Pairwise differences between groups can be further investigated using the tabs shown in the dialog box generated after running the test, along with application of appropriate multiple testing corrections (*see* Fig. 6 and **Note 4**).

2.2.2 MetaMIS

Network analyses which help visualize and understand the complexity of microbial communities, especially in a dynamic environment such as the gastrointestinal tract, are essential to gauge contributing OTUs to the ecology of the system. The **Meta**genomic **M**icrobial **I**nteraction **S**imulator or MetaMIS [14] is a Windows or Mac OS based program that is designed to visualize and analyze longitudinal or time-series of 16S rRNA sequencing data. This tool is based on the Lotka-Volterra equations, also known as the predator-prey equations [15], frequently used to describe biological system dynamics. These systems of equations, designed

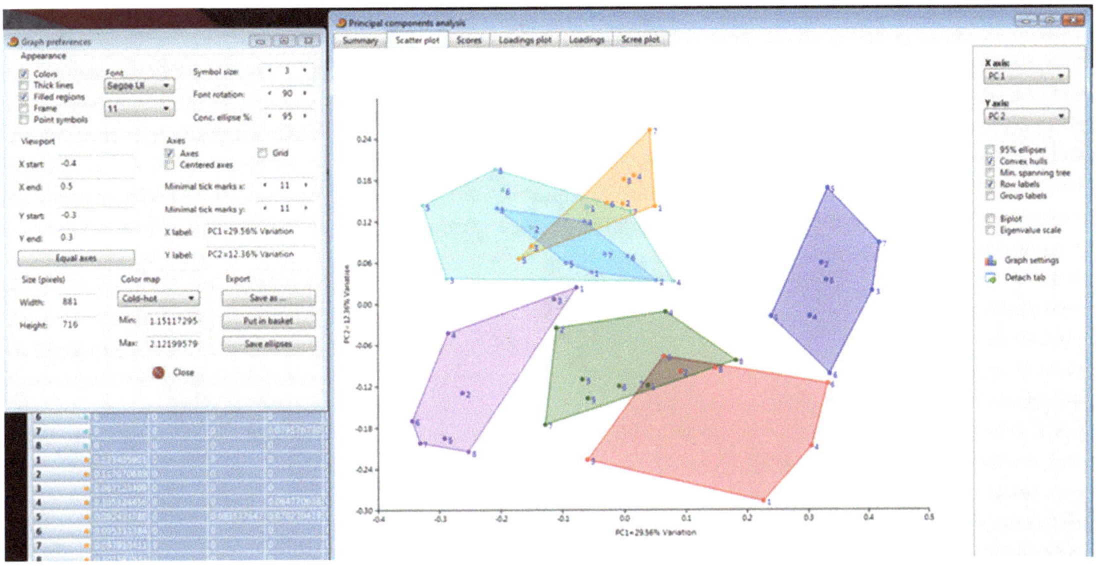

Fig. 4 Ordinate visualization. In order to visualize the data in an ordinate format, select the dataset including the groups and choose an appropriate method from the *Multivariate* → *Ordination* tab in the menu options. An example for a Principal Component Analysis or PCA is shown. The settings of the graph can be altered to the user's preference as well, including re-naming the PCA values and axes

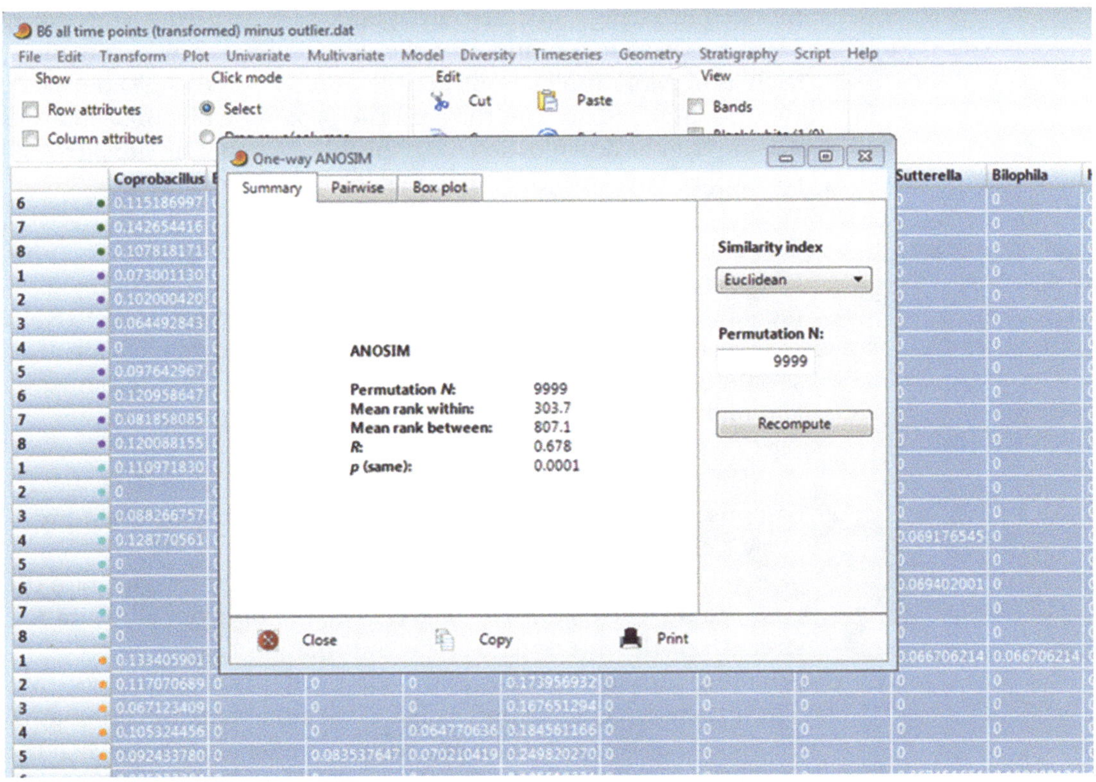

Fig. 5 Multivariant testing. Under the *Multivariate* → *Tests* tab, select the appropriate method depending on the hypothesis being tested and the number of groups contributing to the testing. The summary displays the *p*-value and the similarity index used for the analysis

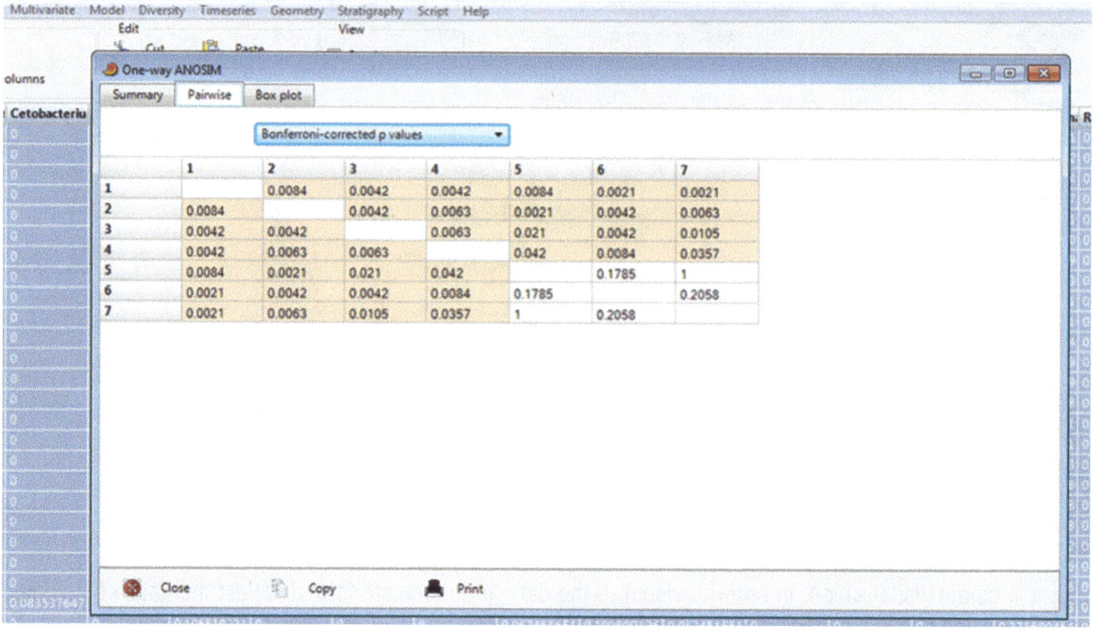

	1	2	3	4	5	6	7
1		0.0084	0.0042	0.0042	0.0084	0.0021	0.0021
2	0.0084		0.0042	0.0063	0.0021	0.0042	0.0063
3	0.0042	0.0042		0.0063	0.021	0.0042	0.0105
4	0.0042	0.0063	0.0063		0.042	0.0084	0.0357
5	0.0084	0.0021	0.021	0.042		0.1785	1
6	0.0021	0.0042	0.0042	0.0084	0.1785		0.2058
7	0.0021	0.0063	0.0105	0.0357	1	0.2058	

Fig. 6 Pairwise comparisons in PAST3

originally to infer animal interactions, are now being applied to model the change in microbial population structure over time. Here we provide a glimpse of the potentially useful information, such as network interactions between the OTUs in a given model, that the user can acquire through the program. The preferred OS versions can be downloaded from the following online site: https://sourceforge.net/projects/metamis/files/?source=navbar

Input Data Structure, Network Plots, and Analyses

The manual provided with the program is fairly exhaustive; however, we provide a basic format of the input file and the requirements for network analysis using longitudinal 16S rRNA sequencing data. Relative abundances from the sequencing data should be saved as a text (.txt) file, with the OTUs listed in rows, and multiple time points from a single sample listed as columns (*see* Fig. 7).

The user interface for MetaMIS is shown with a sample dataset comprised of 3 time points, a subset of the example shown previously (*see* Fig. 8).

Network interactions aside, the program offers insights into the connections between individual OTUs, the directionality and the strength of the interactions. This information can be obtained by browsing through the tabs at the top of the visualization window. The images depicted below (*see* Fig. 9) demonstrate the topology of microbial networks and also the nature of interactions between the OTUs ranging from mutualism to predation to amensalism, among other factors.

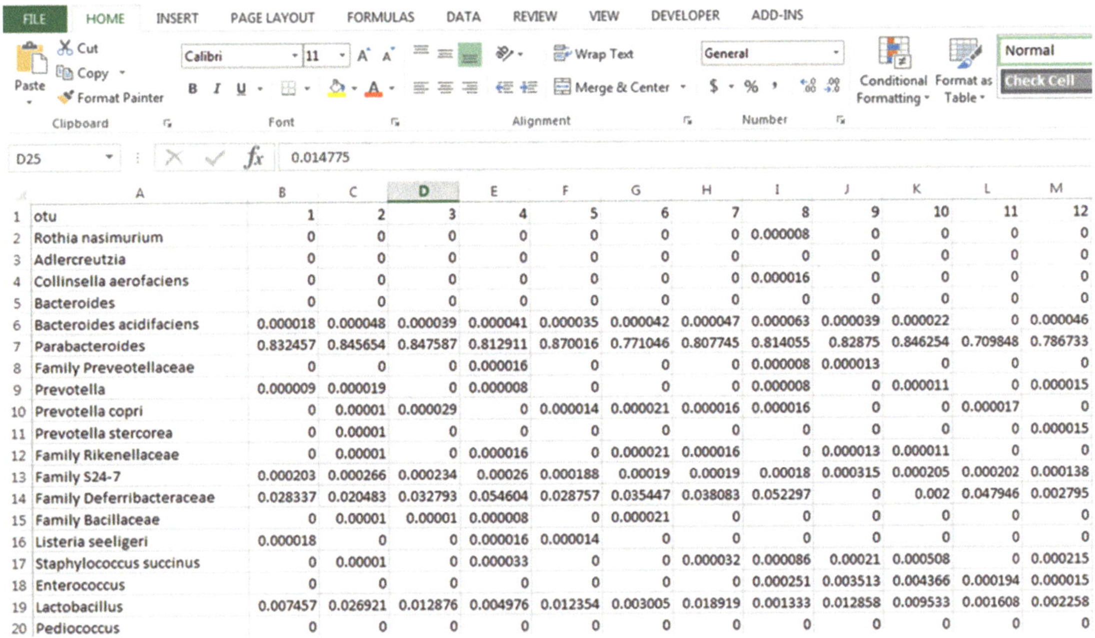

	A	B	C	D	E	F	G	H	I	J	K	L	M
1	otu	1	2	3	4	5	6	7	8	9	10	11	12
2	Rothia nasimurium	0	0	0	0	0	0	0	0.000008	0	0	0	0
3	Adlercreutzia	0	0	0	0	0	0	0	0	0	0	0	0
4	Collinsella aerofaciens	0	0	0	0	0	0	0	0.000016	0	0	0	0
5	Bacteroides	0	0	0	0	0	0	0	0	0	0	0	0
6	Bacteroides acidifaciens	0.000018	0.000048	0.000039	0.000041	0.000035	0.000042	0.000047	0.000063	0.000039	0.000022	0	0.000046
7	Parabacteroides	0.832457	0.845654	0.847587	0.812911	0.870016	0.771046	0.807745	0.814055	0.82875	0.846254	0.709848	0.786733
8	Family Prevoteollaceae	0	0	0	0.000016	0	0	0	0.000008	0.000013	0	0	0
9	Prevotella	0.000009	0.000019	0	0.000008	0	0	0	0.000008	0	0.000011	0	0.000015
10	Prevotella copri	0	0.00001	0.000029	0	0.000014	0.000021	0.000016	0.000016	0	0	0.000017	0
11	Prevotella stercorea	0	0.00001	0	0	0	0	0	0	0	0	0	0.000015
12	Family Rikenellaceae	0	0.00001	0	0.000016	0	0.000021	0.000016	0	0.000013	0.000011	0	0
13	Family S24-7	0.000203	0.000266	0.000234	0.00026	0.000188	0.00019	0.00019	0.00018	0.000315	0.000205	0.000202	0.000138
14	Family Deferribacteraceae	0.028337	0.020483	0.032793	0.054604	0.028757	0.035447	0.038083	0.052297	0	0.002	0.047946	0.002795
15	Family Bacillaceae	0	0.00001	0.00001	0.000008	0	0.000021	0	0	0	0	0	0
16	Listeria seeligeri	0.000018	0	0	0.000016	0.000014	0	0	0	0	0	0	0
17	Staphylococcus succinus	0	0.00001	0	0.000033	0	0	0.000032	0.000086	0.00021	0.000508	0	0.000215
18	Enterococcus	0	0	0	0	0	0	0	0.000251	0.003513	0.004366	0.000194	0.000015
19	Lactobacillus	0.007457	0.026921	0.012876	0.004976	0.012354	0.003005	0.018919	0.001333	0.012858	0.009533	0.001608	0.002258
20	Pediococcus	0	0	0	0	0	0	0	0	0	0	0	0

Fig. 7 Excel data showing relative abundances. In the example, column A contains the OTUs arranged in rows, followed by relative abundances at 12 time points, i.e., samples collected every month from a single animal

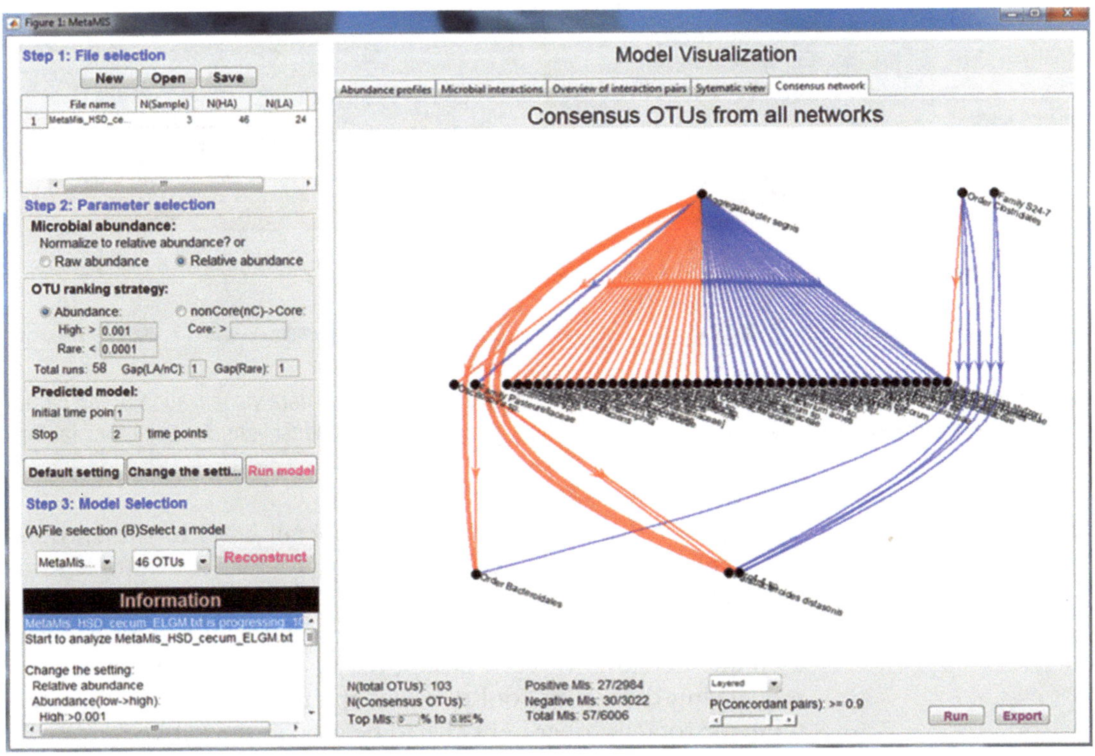

Fig. 8 Visualization of the MetaMIS interface with a sample dataset comprised of 3 time points

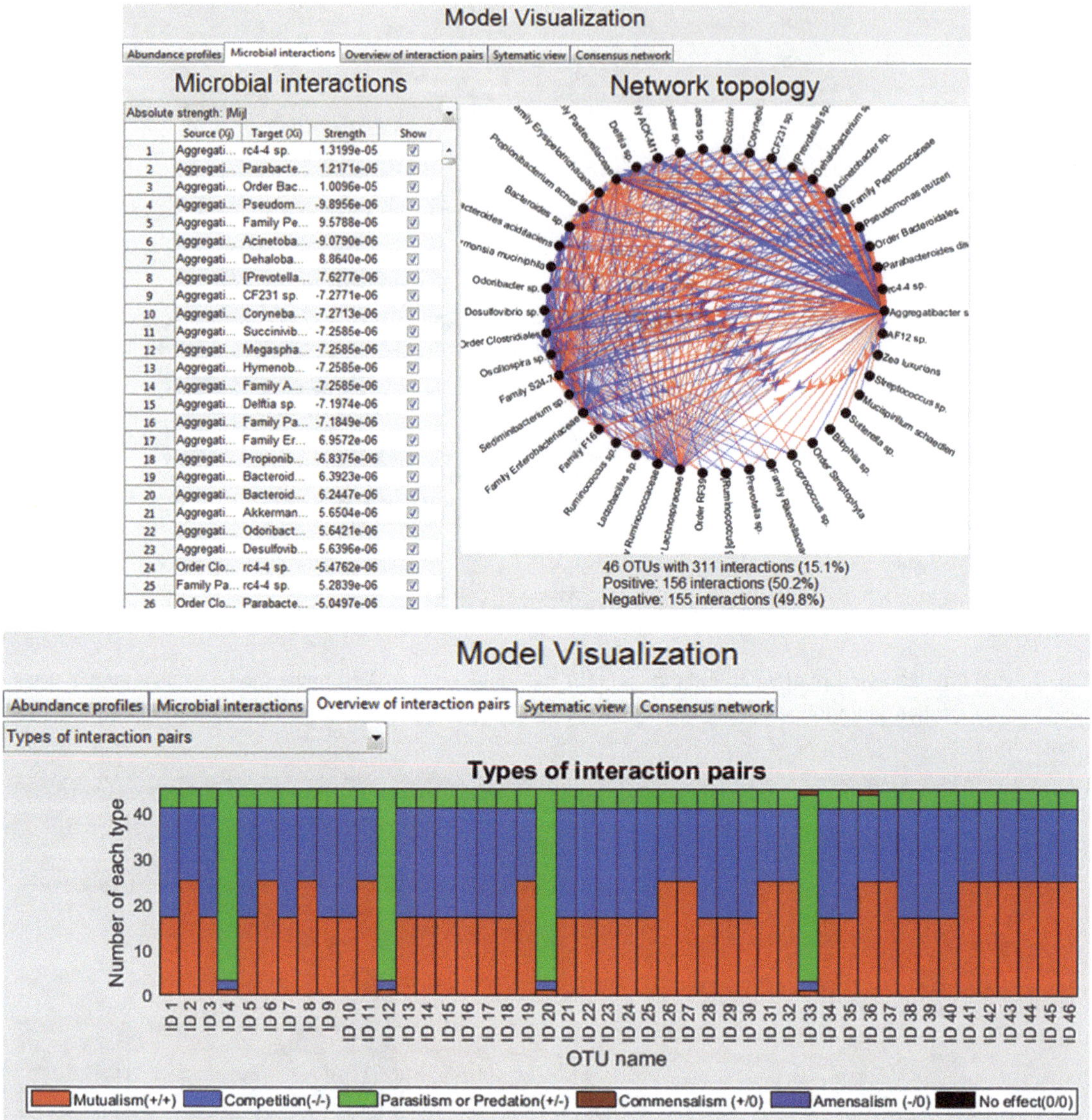

Fig. 9 MetaMIS visualization of the topology of microbial networks that allows data visualization of the nature of interactions between the OTUs ranging from mutualism to predation to amensalism, among other factors

Consensus Networks from Multiple Samples

Though not depicted or described in detail here, in the authors' experience, MetaMIS offers the capacity to analyze multiple samples individually, and then incorporate all of the networks into a "consensus" network. This provides the user the option to establish a unique network for each group, obtained with longitudinal 16S sequencing data. This tool provides the user the option to determine the total number of OTUs to be used in the network analysis, thus offering the advantage of limiting assessment to the highly abundant or selective group of OTUs. Additional abundance profiles, both original and predicted, are also generated based on this information and the Bray-Curtis (BC) dissimilarity matrix, where the lower the BC score, the more similar the two abundance profiles are. This information can be used to assess the validity of a given model, by testing with a synthetic dataset if needed.

2.2.3 MicrobiomeAnalyst

PAST and MetaMIS offer unique visualization and analyses methodologies for the user, especially when working with relative abundance values estimated from 16S rRNA sequencing data. MicrobiomeAnalyst, on the other hand, provides the user with flexibility to not only use relative abundance tables, but also use the raw data acquired from Qiime, MOTHUR and UPARSE [16]. If Shotgun sequencing data are available, MicrobiomeAnalyst is also capable of handling such information to provide data annotation to generate a KEGG [17] Ortholog (KO) count table, providing information about the high-level functions in a biological system, in this case OTUs. MicrobiomeAnalyst is user-friendly and incorporates both statistical analysis and functional interpretation of sequencing data, along with several visualization options including bar plots, diversity analyses, PCAs, and heatmaps.

There are four modules to MicrobiomeAnalyst, each designed for a specific task not limited to 16S rRNA data alone, i.e., the Marker Data Profiling or MDP. Shotgun sequencing data and also processing the users' data in conjunction with publicly available datasets that are similar to the area of study are the other modules. The last module uses the Enrichment analysis method, traditionally applied to gene expression datasets from RNA sequencing methodologies to identify taxonomical composition changes and patterns that are biologically or ecologically relevant. Taxon Set Enrichment Analysis (TSEA) evaluates whether certain taxa or OTUs are represented more often than expected by chance (*see* Fig. 10).

2.2.4 Calypso

With the advent of large datasets due to the advancements in next-generation sequencing, there has been in uptick in the number of tools available for sequencing data analysis. Calypso is another installment of a metagenomic or 16 s rRNA sequencing data [18] analysis software suite available online at the following website: http://cgenome.net/wiki/index.php/Calypso. Similar to the options offered in the previous tools, we have found that Calypso offers a multitude of statistical approaches to identify complex environment-microbiome interactions, apart from generating bubbleplots and heatmaps and providing parametric, non-parametric, and multivariate analyses. In the authors' experience the data transformation and normalization methods offered by Calypso need to be used as per the users' discretion depending on the hypothesis being tested, so no suggestions regarding this can be made at this point. Furthermore, Calypso also offers several "Advanced"

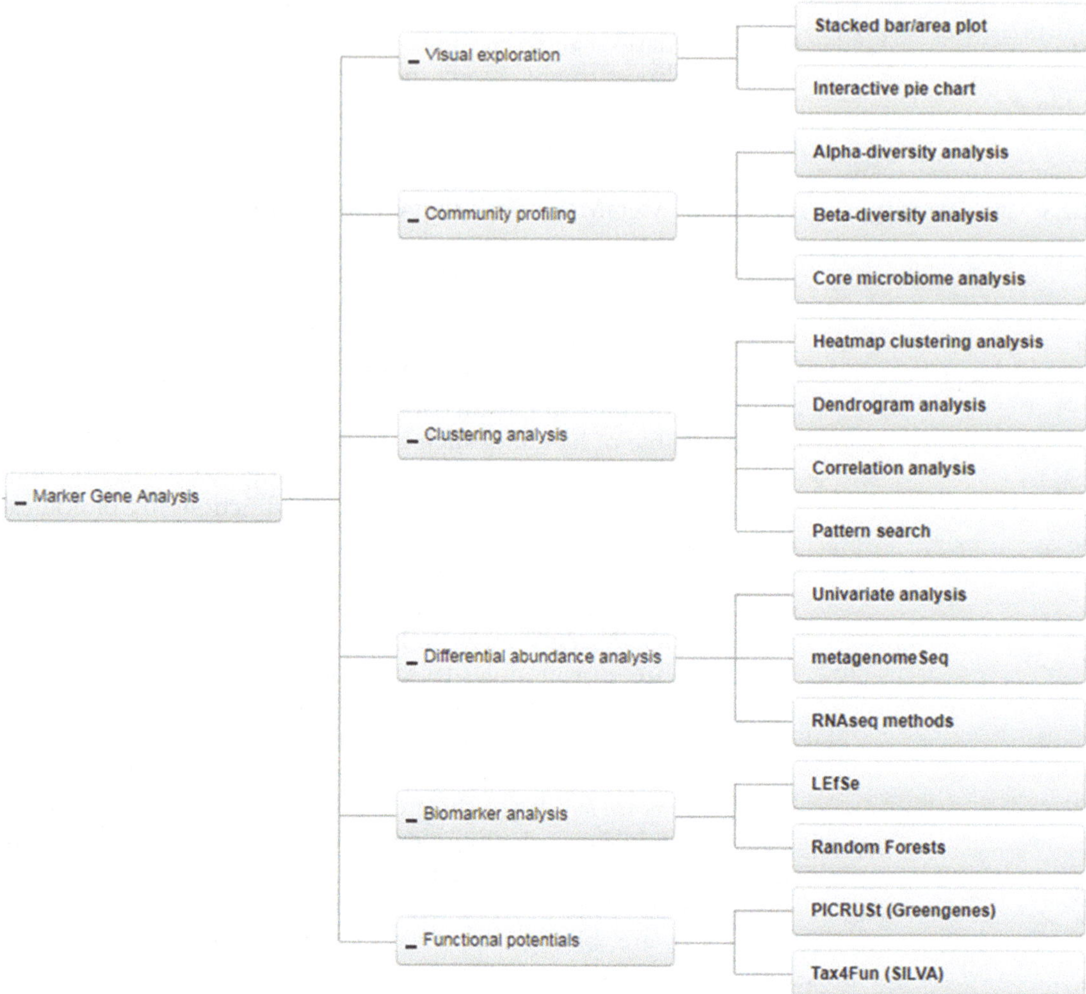

Fig. 10 The outline in the figure was generated on the MicrobiomeAnalyst website, depicting the various tests, visualizations and analyses that can be assessed using this online tool available at http:/www.microbiomeanalyst.ca

analyses which include Biomarker prediction, weighted correlation network analysis, and hierarchical analyses depicted in the form of a Krona [19] plot shown below (*see* Fig. 11 and **Note 5**).

2.2.5 PICRUSt

Phylogenetic **I**nvestigation of **C**ommunities by **R**econstruction of **U**nobserved **St**ates (PICRUSt) is a bioinformatics software tool [20] that requires and assumes that the user has basic working knowledge of Linux programming language and the UNIX system. Setting up and using PICRUSt requires the availability of a high-speed, memory-intensive computer such as a cluster-computing resource, mostly made available through an academic institution's Informatics or Computer Science department. Information

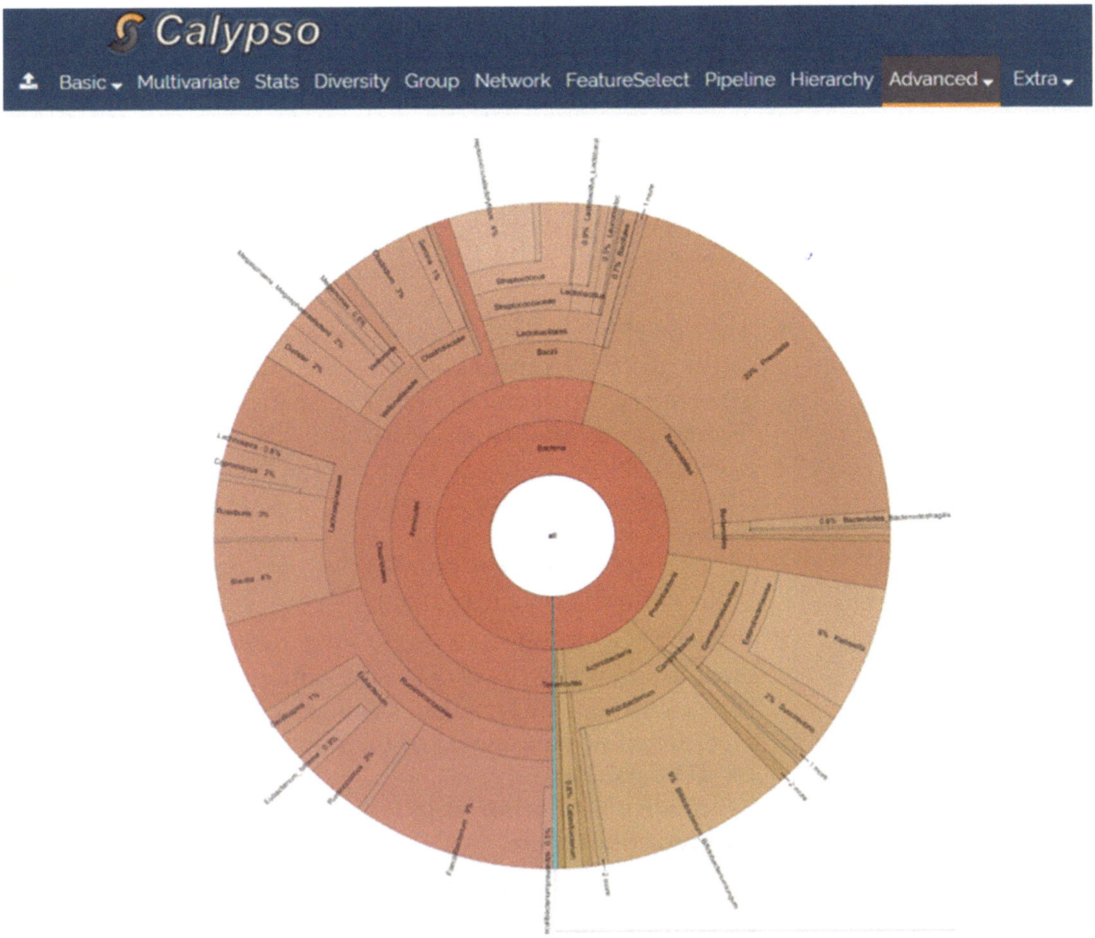

Fig. 11 Calypso web tool with visualization of taxa in Krona plot. These plots are dynamic visual tools that allow for the visualization of changes between parameters such as time or treatment

regarding the installation and utility of this program can be found here: http://picrust.github.io/picrust/.

PICRUSt's output is a matrix with gene-counts, predicting the functional genes in each sample based on a set of known genomes. The ancestral-state reconstruction algorithm used in this tool matches input OTUs against a reference database and obtains a predicted 16S rRNA copy and gene number, and its abundance in the 16S rRNA dataset. This is used to predict the contribution of each OTU to the overall gene content of the samples and their functional profile using the KEGG Orthologs (KO) database. This information is further used to infer the presence, absence, and abundance of microbial pathways using the KEGG pathways information. Here, we provide a generic script for going from 16S rRNA sequencing to PICRUSt outputs.

Required Files, Consideration, and Brief Protocol

PICRUSt currently can only be used with a closed-reference picking of OTUs against the GreenGenes [4] database. For information about closed-reference picking we encourage the users to review current literature along with the methods for clustering 16S rRNA sequences into OTUs by Chen et al. [21]. The protocol described below uses the pooled sequences FASTA file generated from Qiime to obtain a closed-reference OTU table in the *.biom* format. This file serves as the input for the PICRUSt module and downstream analyses.

1. The 16S rRNA sequencing data are first normalized according to the 16S rRNA gene operons since they can vary from 1 to 10 or more copies in bacteria. The relative abundances therefore estimated due to this variation may not be from actual bacterial abundances. We then use the normalize function of PICRUSt to generate a normalized OTU table as shown below:

   normalize_by_copy_number.py -i closed_picked_otus.biom -o normalized_otus.biom.

2. PICRUSt by default uses the KO database for functional predictions; however, the user has the option to choose the COG [22] and Rfam [23] databases alongside the KOs, if required. Here we show an example of the functional predictions based on KO database.

   predict_metagenomes.py -i normalized_otus.biom -o ko_predictions.biom.

3. In the next step, the user can collapse the KOs into KEGG pathways using the categorize function built-in into PICRUSt

   categorize_by_function.py -i ko_predictions.biom -c KEGG_Pathways -l 3 -o pathway_predictions.biom.

4. Since the file is in a *.biom* format, it is necessary to convert the same to a more computer-readable format. For this purpose, we use the following code:

   biom convert -i pathway_predictions.biom -o pathway_predictions.txt --to-tsv --header-key KEGG_Pathways

   The text file generated by this process can then be opened using MS-Excel or other spreadsheet evaluating programs (*see* Fig. 12).

5. If the user is interested in determining which OTUs contribute to each of the pathways, or specific pathways of interest, they may use the "contributions" function of the PICRUSt module as demonstrated. The script requires that the specific KEGG pathway numbers are listed (e.g., K01727, K11049) to generate a table indicating contributions (*see* **Note 6**).

   metagenome_contributions.py -i otus_corrected.biom -l K01727,K01194,K01216,K11049,K00389,K00449 -o metagenome_contributions.txt.

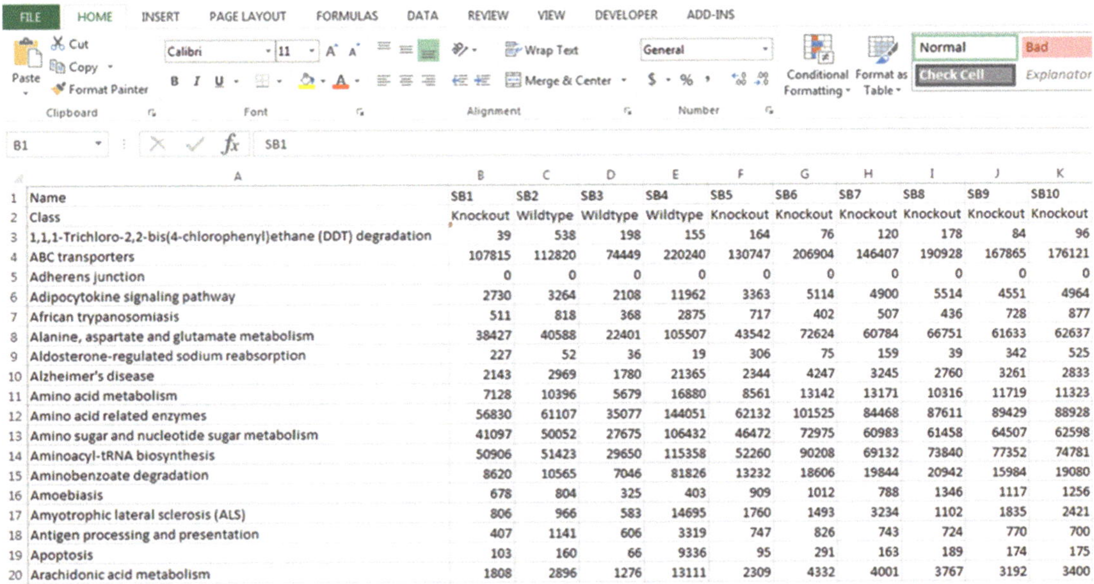

	A	B	C	D	E	F	G	H	I	J	K
1	Name	SB1	SB2	SB3	SB4	SB5	SB6	SB7	SB8	SB9	SB10
2	Class	Knockout	Wildtype	Wildtype	Wildtype	Knockout	Knockout	Knockout	Knockout	Knockout	Knockout
3	1,1,1-Trichloro-2,2-bis(4-chlorophenyl)ethane (DDT) degradation	39	538	198	155	164	76	120	178	84	96
4	ABC transporters	107815	112820	74449	220240	130747	206904	146407	190928	167865	176121
5	Adherens junction	0	0	0	0	0	0	0	0	0	0
6	Adipocytokine signaling pathway	2730	3264	2108	11962	3363	5114	4900	5514	4551	4964
7	African trypanosomiasis	511	818	368	2875	717	402	507	436	728	877
8	Alanine, aspartate and glutamate metabolism	38427	40588	22401	105507	43542	72624	60784	66751	61633	62637
9	Aldosterone-regulated sodium reabsorption	227	52	36	19	306	75	159	39	342	525
10	Alzheimer's disease	2143	2969	1780	21365	2344	4247	3245	2760	3261	2833
11	Amino acid metabolism	7128	10396	5679	16880	8561	13142	13171	10316	11719	11323
12	Amino acid related enzymes	56830	61107	35077	144051	62132	101525	84468	87611	89429	88928
13	Amino sugar and nucleotide sugar metabolism	41097	50052	27675	106432	46472	72975	60983	61458	64507	62598
14	Aminoacyl-tRNA biosynthesis	50906	51423	29650	115358	52260	90208	69132	73840	77352	74781
15	Aminobenzoate degradation	8620	10565	7046	81826	13232	18606	19844	20942	15984	19080
16	Amoebiasis	678	804	325	403	909	1012	788	1346	1117	1256
17	Amyotrophic lateral sclerosis (ALS)	806	966	583	14695	1760	1493	3234	1102	1835	2421
18	Antigen processing and presentation	407	1141	606	3319	747	826	743	724	770	700
19	Apoptosis	103	160	66	9336	95	291	163	189	174	175
20	Arachidonic acid metabolism	1808	2896	1276	13111	2309	4332	4001	3767	3192	3400

Fig. 12 Text file generated by PICRUSt visualized in Excel. The output looks as depicted, where the first column lists the pathways found in the predicted metagenomes, with each of the pathways in rows. Class in the figure refers to the group to which each of the samples belongs, while the samples are listed in individual columns

At this juncture, the authors must acknowledge that the use of PICRUSt's functionality to predict metagenomes and infer the functional profile of samples and groups is fairly controversial. Some researchers believe that this ancestral-state reconstruction does not take into account the vast majority of variation that may exist between known, fully sequenced organisms used in the reference database, compared to the largely unknown populations typically seen in the environment. Simultaneously, the nearly out-dated, i.e., 2012 version of the GreenGenes database serves as the reference for the closed-OTU picking method of the initial PICRUSt input file, leading many to debate the validity of the reference genomes considering the large advances made in sequencing technology and databases since they were last updated.

We present here techniques and tools for the initial identification and characterization of polymicrobial populations. It must be noted that this field is rapidly evolving and there are an abundance of tools and techniques to analyze metagenomic data. What we have presented here is in no way exhaustive. As sequencing costs continue to come down and sequence read lengths increase, whole metagenome and metatranscriptome approaches to characterize the biological capacity of a community will become the standard.

3 Notes

1. The recommended protocol requires that the user purchase a special vortex adaptor for one step of the procedure, although agitation using a Qiagen TissueLyser II (or similar) instrument can be substituted. Using this approach, an experienced user can extract 24 samples in approximately 1.5 h. Alternatively, for scenarios in which the user wishes to maximize yields (e.g., additional DNA needed for other downstream assays, scant sample), an adaptation of the protocol published by Yu et al. [24] consistently results in much higher yields than any kit we have evaluated. In this approach, the samples are homogenized and cells are lysed using a combination of detergent, mechanical forces, and heat, which also serves to denature proteins with potential nuclease activity. Following a brief centrifugation, microbial DNA is then extracted from the supernatant via traditional benchtop nucleic acid precipitation incorporating a chaotropic salt (ammonium acetate) and alcohol (isopropanol). To remove any residual proteins and any RNA that co-precipitated, the resultant DNA pellet is put back into solution and essentially extracted a second time using the DNeasy kit (Qiagen) according to manufacturer's instructions for extraction from cells with one important exception. The elution buffer provided with the DNeasy kits (i.e., AE buffer) must be replaced with an EDTA-free buffer such as EB buffer (Qiagen). In our experience, DNA eluted in AE buffer universally fails to amplify during library preparation. A detailed description of the complete method is provided here [6]. Using this approach, an experienced user can extract 24 samples in approximately 4 h, the advantage being a log-fold increase in DNA yield.
2. All reactions should also be supplemented with 0.4% (wt/vol) bovine serum albumin (BSA) to function as an amplification facilitator. In our experience, fecal samples extracted using either of the methods described above can still occasionally contain problematic PCR inhibitors; addition of highly pure BSA very effectively minimizes the effect of these contaminants.
3. In the authors' experience, a cube-root transformation to the sequencing data can be applied in PAST using the "Evaluate expression" option under the "Transform" tab with the formula (x)^(1/4). Transforming data should only be applied upon careful consideration of distribution and variance of the available dataset.
4. To calculate the contribution of individual OTUs to the overall dissimilarity between the groups, a SIMPER or similarity

percentage analysis can be used which is found under the multivariate tests tab. Simultaneously, the raw-read count table obtained after the automated pipelines can be used to perform diversity analyses, which outputs several alpha diversity indices including Shannon, Chao, and Simpson [25].

5. Calypso also offers the automated pipeline model, where the user can upload their data and submit a job to generate the commonly used plots with respect to microbiome data. The data input formats for Calypso are, however, very specific to the need and design of the experiment, so the authors encourage the user to review the manual and user forums to determine what suits their application the best. Simultaneously it is to be noted that Calypso, among other tools, has the capacity to utilize files generated from the newer version of Qiime, i.e., Qiime2, potentially making it more versatile compared to the other programs and online tools.
6. Tools such as LEfSe [26] (**L**inear discriminant analysis **Ef**fect **S**ize) can be used to determine and visualize the features, i.e., KOs or KEGG pathways that explain differences between the groups/classes can be used to plot and analyze PICRUSt output. Similarly, R [27] packages including the Microbiome Helper [28] Rscript can be utilized to visualize the data using bar plots and perform other statistical tests for identifying differences between groups of interest.

References

1. Van de Peer Y, Chapelle S, De Wachter R (1996) A quantitative map of nucleotide substitution rates in bacterial rRNA. Nucleic Acids Res 24:3381–3391
2. Larsen N, Olsen GJ, Maidak BL, McCaughey MJ, Overbeek R, Macke TJ et al (1993) The ribosomal database project. Nucleic Acids Res 21:3021–3023
3. Pruesse E, Quast C, Knittel K, Fuchs BM, Ludwig W, Peplies J et al (2007) SILVA: a comprehensive online resource for quality checked and aligned ribosomal RNA sequence data compatible with ARB. Nucleic Acids Res 35:7188–7196
4. DeSantis TZ, Hugrnholtz P, Larsen N, Rohas M, Brodie EL, Keller K et al (2006) Greengenes, a chimera-checked 16S rRNA gene database and workbench compatible with ARB. Appl Environ Microbiol 72:5069–5072
5. Yang B, Wang Y, Qian PY (2016) Sensitivity and correlation of hypervariable regions in 16S rRNA genes in phylogenetic analysis. BMC Bioinformatics 17:135
6. Ericsson AC, Akter S, Hanson MM, Busi SB, Parker TW, Schehr RJ et al (2015) Differential susceptibility to colorectal cancer due to naturally occurring gut microbiota. Oncotarget 6 (32):33689–33704. https://doi.org/10.18632/oncotarget.5604
7. Walters WA, Caporaso JG, Lauber CL, Berg-Lyons D, Fierer N, Knight R (2011) Primer-Prospector: de novo design and taxonomic analysis of barcoded polymerase chain reaction primers. Bioinformatics 27:1159–1161
8. Caporaso JG, Lauber CL, Walters WA, Berg-Lyons D, Lozupone CA, Turnbaugh PJ et al (2001) Global patterns of 16S rRNA diversity at a depth of millions of sequences per sample. Proc Natl Acad Sci U S A 108(Suppl 1):4516–4522
9. Good IJ (1953) The population frequencies of species and the estimation of population parameters. Biometrika 40:237–264
10. Chao A, Shen TJ (2003) Nonparametric estimation of Shannon's index of diversity when there are unseen species in sample. Environ Ecol Stat 10:429–433

11. Kuczynski J, Stombaugh J, Walters WA, Gonzalez A, Caporaso JG, Knight R (2011) Using QIIME to analyze 16S rRNA gene sequences from microbial communities. Curr Protoc Bioinformatics 36(1):Unit 10.7
12. Schloss PD, Westcott SL, Ryabin T, Hall JR, Hartmann M, Hollister EB et al (2009) Introducing mothur: open-source, platform-independent, community-supported software for describing and comparing microbial communities. Appl Environ Microbiol 75:7537–7541
13. Hammer Ø, Harper DAT, Ryan PD (2001) PAST: Paleontological statistics software package for education and data analysis. Palaeontol Electron 4(1.1A):1–9
14. Shaw GT, Pao YY, Wang D (2016) MetaMIS: a metagenomic microbial interaction simulator based on microbial community profiles. BMC Bioinformatics 17(1):488
15. Yorke JA, Anderson WN (1973) Predator-prey patterns. Proc Natl Acad Sci U S A 70:2069–2071
16. Edgar RC (2013) UPARSE: highly accurate OTU sequences from microbial amplicon reads. Nat Methods 10:996–998
17. Kanehisa M, Goto S (2000) KEGG: Kyoto Encyclopedia of Genes and Genomes. Nucleic Acids Res 28:27–30
18. Zakrzewski M, Proietti C, Ellis JJ, Hasan S, Brion MJ, Berger B et al (2017) Calypso: a user-friendly web-server for mining and visualizing microbiome–environment interactions. Bioinformatics 33:782–783
19. Ondov BD, Bergman NH, Phillippy AM (2011) Interactive metagenomic visualization in a Web browser. BMC Bioinformatics 12:385
20. Langille MG, Zaneveld J, Caporaso JG, McDonald D, Knights D, Reyes JA et al (2013) Predictive functional profiling of microbial communities using 16S rRNA marker gene sequences. Nat Biotechnol 31:814–821
21. Chen W, Zhang CK, Cheng Y, Zhang S, Zhao H (2013) A comparison of methods for clustering 16S rRNA sequences into OTUs. PLoS One 8:e70837
22. Tatusov RL, Galperin MY, Natale DA, Koonin EV (2000) The COG database: a tool for genome-scale analysis of protein functions and evolution. Nucleic Acids Res 28:33–36
23. Griffiths-Jones S, Bateman A, Marshall M, Khanna A, Eddy SR (2003) Rfam: an RNA family database. Nucleic Acids Res 31:439–441
24. Yu Z, Morrison M (2004) Improved extraction of PCR-quality community DNA from digesta and fecal samples. BioTechniques 36:808–812
25. Morris EK, Caruso T, Buscot F, Fischer M, Hancock C, Maier TS et al (2014) Choosing and using diversity indices: insights for ecological applications from the German Biodiversity Exploratories. Ecol Evol 4:3514–3524
26. Segata N, Izard J, Waldron L, Gevers D, Miropolsky L, Garrett WS et al (2011) Metagenomic biomarker discovery and explanation. Genome Biol 12:R60
27. R Development Core Team (2004) R: a language and environment for statistical computing Version 2.0.1. R Foundation for Statistical Computing, Vienna, Austria
28. Comeau AM, Douglas GM, Langille MGI (2017) Microbiome Helper: a custom and streamlined workflow for microbiome research. mSystems 2(1):e00127–e00116. https://doi.org/10.1128/mSystems.00127-16

Chapter 10

Networking in Biology: The Hybrid Rat Diversity Panel

Boris Tabakoff, Harry Smith, Lauren A. Vanderlinden, Paula L. Hoffman, and Laura M. Saba

Abstract

One of the most fruitful resources for systems genetic studies of nonhuman mammals is a panel of inbred strains that exhibits significant genetic diversity between strains but genetic stability (isogenicity) within strains. These characteristics allow for fine mapping of complex phenotypes (QTLs) and provide statistical power to identify loci which contribute nominally to the phenotype. This type of resource also allows the planning and performance of investigations using the same genetic backgrounds over several generations of the test animals. Often, rats are preferred over mice for physiologic and behavioral studies because of their larger size and more distinguishable anatomy (particularly for their central nervous system). The Hybrid Rat Diversity Panel (HRDP) is a panel of inbred rat strains, which combines two recombinant inbred panels (the HXB/BXH, 30 strains; the LEXF/FXLE, 34 strains and 35 more strains of inbred rats which were selected for genetic diversity, based on their fully sequenced genomes and/or thorough genotyping). The genetic diversity and statistical power of this panel for mapping studies rivals or surpasses currently available panels in mouse. The genetic stability of this panel makes it particularly suitable for collection of high-throughput omics data as relevant technology becomes available for engaging in truly integrative systems biology. The PhenoGen website (http://phenogen.org) is the repository for the initial transcriptome data, making the raw data, the processed data, and the analysis results, e.g., organ-specific protein coding and noncoding transcripts, isoform analysis, expression quantitative trait loci, and co-expression networks, available to the research public. The data sets and tools being developed will complement current efforts to analyze the human transcriptome and its genetic controls (the Genotype-Tissue Expression Project (GTEx)) and allow for dissection of genetic networks that predispose to particular phenotypes and gene-by-environment interactions that are difficult or even impossible to study in humans. The HRDP is an essential population for exploring truly integrative systems genetics.

Key words Hybrid Rat Diversity Panel, Systems genetics, Co-expression networks, Expression quantitative trait loci, Noncoding transcripts, Genetically renewable rodent panel

1 Introduction

More often than not common medical disorders are a result of complex interactions between an individual's genetic characteristics and the environment in which the individual exists. These disorders, along with a plethora of "normal" physiologic functions and behaviors, are referred to as "complex traits." The goal of this

G. Thomas Hayman et al. (eds.), *Rat Genomics*, Methods in Molecular Biology, vol. 2018, https://doi.org/10.1007/978-1-4939-9581-3_10,

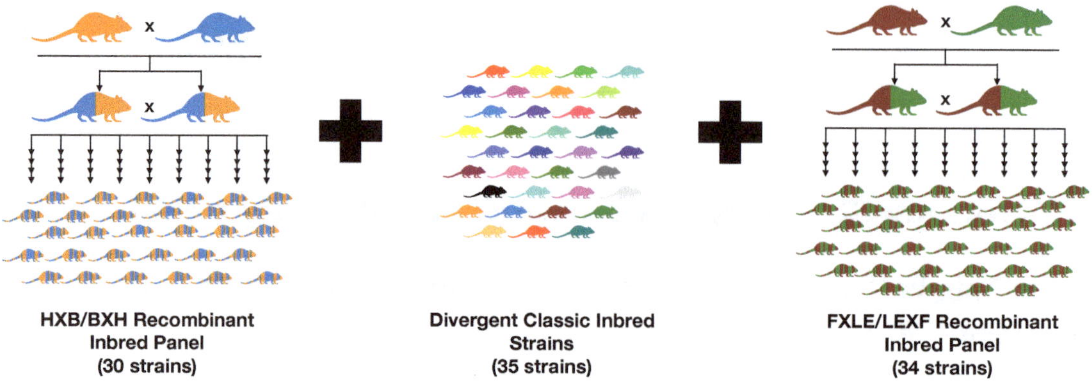

Fig. 1 The Hybrid Rat Diversity Panel (HRDP). The HRDP includes two recombinant inbred (RI) rat panels. The HXB/BXH RI panel is derived from reciprocal crossing of the Spontaneously Hypertensive inbred rat strain (orange) and the normotensive Brown Norway congenic rat strain with polydactyly-luxate syndrome (blue). The FXLE/LEXF is derived from reciprocal crossing of the F344 inbred rat strain (red) and the Long Evans inbred rat strain (green). RI panels are derived from first crossing two divergent inbred strains. Progeny of the F1 cross are mated to form an F2 generation. Rats from the F2 generation are brother-sister mated for many generations to form the final RI panel. The Divergent Classic Inbred Strain panel consists of 35 common inbred laboratory rat strains. This panel includes the four progenitor strains from the two RI panels

chapter is to describe the utility of the Hybrid Rat Diversity Panel (HRDP) for examining the genetic etiology of complex traits using a systems genetics approach (*see* Fig. 1). The HRDP is a panel of rats derived from genetically diverse inbred strains, including two recombinant inbred panels and classic inbred rat strains, which have the power and mapping resolution necessary for a systems approach to understanding genetic predisposition in complex traits, and, in addition, provide the basis for cross-species (including humans) analyses. In humans, genome-wide association studies (GWAS), whole-exome sequencing (WES), and whole-genome sequencing (WGS) have been applied in attempts to identify genetic contributions to many complex diseases [1, 2]. In addition to issues of power and "lost heritability" [3–5], a major problem encountered in current GWAS with humans is the difficulty in translating the genomic findings into an explanation of the biology of the trait being studied. Often the markers associated with disorders fall into intergenic regions, introns, 3′UTR regions, or coding regions of poorly annotated genes, making it difficult to explain the relationship of the polymorphism to the biologic mechanism of the disorder [6–8]. Such results also generate a conclusion that polymorphisms contributing to regulation of gene expression, rather than polymorphisms contributing to mutations in protein sequence, may form the basis of variation predisposing to many complex traits.

1.1 What Is Systems Genetics?

The principles of a systems genetics analysis, and a listing of currently available animal models for such analyses, have been provided by Civelek and Lusis [9] and Moreno-Moral and Petretto [10]. Civelek and Lusis [9] define systems genetics as "A global analysis of molecular factors that underlie variability in physiological or clinical phenotypes across individuals in a population. It considers not only the underlying genetic variation but also intermediate phenotypes such as gene expression, protein levels and metabolite levels, in addition to gene/gene and gene/environment interactions." This definition makes it obvious that a global analysis of all the factors that are mentioned is a massive task and some of the factors, such as protein levels and metabolites, are not yet amenable to truly global and high-throughput technologies. Thus, if one is considering creating a resource for systems genetics analysis in animals, a critical requirement is a genetically stable population which can be studied many times, over many generations, and will produce cumulative data to allow for a complete systems genetics analysis of a complex trait. It should also be stressed that the information on the genome, transcriptome, proteome, and metabolome of an individual can be studied with respect to any complex trait exhibited by the animals that compose the resource established for systems genetics analysis.

1.2 From the Model to Phenotype

One of the fundamental needs in linking genetic variation to phenotype is understanding the biologic repercussions of genomic variation. The majority of disease-associated variants in humans are likely to be involved in regulation of transcription [11, 12]. Transcription is the first step in which environmental variation can interact with genetically defined elements of DNA to produce a quantitative phenotype, i.e., levels or rates of synthesis or degradation of RNA and steady state levels of transcripts. This quantitative information allows for use of analytical approaches to discern the variation in these transcription traits that is due to genetics versus "environment" (i.e., heritability) and to map expression QTLs (i.e., genetic loci contributing to transcriptional control, eQTLs). The knowledge of the quantitative aspects of gene expression can also be used for construction of networks of gene products (*see* below) which provide information on the biological context in which a particular gene product acts to influence phenotype [13]. Such networks of functionally related transcripts can be constructed by use of graph theory to generate co-expression modules (networks of biological relevance; [14, 15]). These modules and individual transcripts can be subjected to genetic association and genetic correlation analysis with higher level phenotypes. The identification and use of network information contributes to a systems genetics approach which can combine studies in humans with studies performed in nonhuman species to understand the relationships between the DNA genetic information and biology [9, 16]. Our efforts to collect gene expression data in the rat species was initiated in 2005, and in a separate effort the Genotype-Tissue Expression Project (GTEx) project for studies of gene expression in human tissues was conceptualized in 2008. The magnitude of the GTEx project is enormous by comparison with our current efforts, but a number of important aspects of the study design for our efforts with the HRDP cannot be reproduced in the design of GTEx. The goals of GTEx are very similar to our goals, which are to establish a resource for accessing gene expression data and an associated tissue bank to assist the scientific community with studies of the relationship between genetic variation and gene expression [17]. The GTEx project is focused on the examination of tissue-specific gene expression in humans and we are focusing on tissues derived from rats. The specific goals of the GTEx project are stated in [17] to be: (1) Create a data resource enabling systematic study of genetic variation and regulation of gene expression, (2) Provide biospecimens to use for determining other molecular phenotypes, (3) Provide results of the study of the ethical, legal, and social issues arising from the GTEx project, (4) Support the development of novel statistical methods for e-QTL analysis, (5) Make data available as fast as possible (*see* gtexportal.org/home/), and (6) Disseminate knowledge, standards, and protocols related to biospecimen collection and analysis methods. The initial portion of the GTEx project produced 11,688 transcriptomes from a median of 17 tissues per individual of 714 individuals, and a follow-up project labeled eGTEx (Enhancing GTEx) is now in progress [18]. The new project will use the tissue samples that provided the aliquots for the original data and expand the database to include information on DNA methylation, histone modification, allele-specific expression, RNA modification, and telomere characteristics [18]. The eGTEx studies are meant to provide more, in-depth, information to understand the impact of various mechanisms of gene regulation. The future goal is to progress the knowledge of transcriptional control to understanding the control features of the proteome and eventually the metabolome. It has to be stressed that the GTEx and eGTEx projects are compiling data on "normal" individuals with no effort currently to link genomic/transcriptomic information to a particular disease, although a significant amount of phenotypic data has been collected on the subjects. Given this wealth of information, what is the benefit of another animal to model the human? The answer comes from the ability to breed the "exact" genomic replicates (both male and female) through inbreeding strategies. Although within an inbred strain there is no genetic variation, one can choose other strains with, a priori, known genetic variations to construct panels, such as the HRDP (99 or more strains) to provide the appropriate level of informative polymorphisms across areas of the genome to maximize mapping power and precision for QTL analysis (*see* below). Construction of panels of inbred mice has been an extensive endeavor but has been hampered by the relatively close genetic origin of laboratory mouse strains [19] and the negative impact on reproduction in mouse strains derived by the introduction of "wild-type" strains into the mix of founders for derivation of recombinant inbred strains [20]. The ability to breed genetically "identical" animals produces a situation akin to having a population of identical human twins, and introduces the ability, not available in GTEx, to study environmental impact on developmental variables or the effect of manipulating the immediate environment (e.g., drug administration, controlled stress) on transcriptional, proteomic, metabolomic variation across *multiple* animals with the same genetic background. With standard recombinant inbred panels of animals, various aspects of twin-related methodology can be utilized as well as current genetic mapping techniques.

The other advantage of a panel of inbred animals over a collection of human individuals is the immortality of the genetic replicates within a strain. Theoretically, except for genetic drift or spontaneous mutations, each generation of inbred animals provides one with genetically identical animals for study. The GTEx collection of tissues from genetically heterogeneous individuals is restrictive. Once the collection of tissue is utilized for a current set of investigations, it is not possible to reproduce the genetically identical tissue for further studies. This is particularly important in the light of technologic progress in omics methods and the continued discovery of novel products of DNA (e.g., circular RNA) and their regulatory mechanisms.

2 Description of the HXB/BXH Panel (The Beginning of HRDP)

Recombinant inbred rodent panels are an extremely valuable resource in systems genetics analysis because of their relatively simple genetic structure (all strains within the panel share 50% of their genetics with every other strain in the panel) and their inbred nature (animals within a strain are genetically identical). The simple genetic structure ensures that minor allele frequencies rarely drop below 25%, thus increasing power to detect association at each locus and the power to detect interactions between loci, i.e., gene-gene interactions. With inbred populations in general, we can accumulate information across generations and across laboratories, and we can explore complicated gene by environmental interaction by essentially holding the genetic background "constant" for different environmental perturbations.

Recombinant inbred panels are generated by first crossing two genetically distinct inbred rodent strains. Often the two progenitor strains are chosen based on known differences in a phenotype of interest. The HXB/BXH recombinant inbred panel was first derived from the spontaneously hypertensive rat strain (SHR) and the normotensive Brown Norway congenic rat strain with polydactyly-luxate syndrome (BN-Lx) in the Czech Republic by Pravenec and Kren [21, 22]. After the F2 generation, animals are mated with siblings over several generations to develop multiple genetically distinct recombinant inbred strains within the panel. The HXB/BXH panel has been breeding for more than 80 generations.

3 Systems Genetics in the HXB/BXH Panel

Early studies with the HXB/BXH panel focused on the quantitative trait of blood pressure to study the genetics of hypertension, e.g., [21]. Since then the HXB/BXH panel has been used to study many additional phenotypes including several alcohol-related traits [23–25], several traits related to metabolic syndrome, e.g., [26, 27], and other phenotypes, e.g., [28–30]. Below we summarize the results of analyses of complex traits using systems genetics approaches. It should be stressed that for these particular studies the genetic and transcriptome data are derived from "normal," *non*-experimentally manipulated animals, and therefore the "gene" networks one derives from the experiments we use as examples are ***predisposing*** a particular phenotype. The other issue to be stressed is that the derived relationships between genes (networks) and the phenotype in the presented examples are obtained by a totally unsupervised approach. In other words, no a priori assumptions are introduced into the analysis, except that the phenotype needs to be heritable. Just as with GTEx, our unsupervised analysis pipeline starts by using DNA sequence information (markers) and RNA expression levels in a tissue of interest to identify e-QTLs and module (eigengene) QTLs for networks of genes. Co-expression networks are typically identified using weighted gene co-expression network analysis (WGCNA; [14]). Briefly, WGCNA uses a mathematical model that first transforms the distribution of co-expression measures between genes, typically correlation coefficients, to resemble a scale-free network (i.e., most genes have few connections and a few genes have many connections) and then the model incorporates both indirect and direct connections between genes to estimate a quantitative "connectivity" between each pair of genes. Groups, or modules, of co-expressed genes are identified based on these connectivity measures. These modules can then be summarized using the module eigengene, the first principal component of the module, and the module eigengene QTL can be identified using standard mapping strategies. The overlap in the location of a phenotypic QTL with the eigengene (module) QTL and the correlation of the eigengene with the phenotype across strains produces a candidate co-expression module for a phenotype. This type of analysis not only identifies genes associated with the trait, but also describes the relationships among the associated genes. This approach gives particular insight when gene products participate in several metabolic networks in the cell, and therefore the context of the other co-expressed genes gives information about which of these multiple networks is relevant to the phenotype. The knowledge of members of a module is also helpful when individual genes within the module are under-annotated. In certain instances, the function of under-annotated gene products can be inferred based on the function of other genes within the module and the relationships of the under-annotated gene with the well-annotated genes.

3.1 Alcohol Consumption

We used this approach to identify co-expression networks in brain [23] and liver [31] of alcohol-naïve animals that were associated with predisposition for alcohol consumption in a two-bottle-choice paradigm. In the brain RNA network study, we combined expression data from both the HXB/BXH recombinant inbred panel and six pairs of selectively bred rat lines that were selected based on alcohol consumption in the same two-bottle-choice paradigm. Interestingly, there was no single gene that was differentially expressed in all six pairs of selectively bred lines, but all pairs had at least one gene from the candidate co-expression module identified in the HXB/BXH panel that was differentially expressed between high and low consuming lines. Furthermore, the hub gene in this candidate module, i.e., the gene with the highest intramodular connectivity, was an unannotated (potentially non-coding) transcript. If the analysis had been done by examining differential expression at the individual gene level rather than the module level, there would be no information to link this transcript to a brain function. However, as the hub gene of an associated co-expression network, we could infer its role through its association with the other "genes" contained in the module. The systems genetics approach allowed us to tie the function of this hub gene to inflammation/immune response and glial/neuronal communication. In the selected lines of rats, the different environmental conditions in different laboratories combined with the selective pressure (the selected lines came from four different countries) can be assumed to have resulted in the differential expression levels of particular genes, but in all cases, the identical network was being affected. Our most recent work (manuscript in preparation) used animals in whom the novel transcript that was the hub gene was deleted. The knockout of the hub gene altered the expression of P2rx4 (another gene in the module) and the expression of a large number of genes outside the module and the knockout also differed in the alcohol consumption phenotype in the direction predicted by the original HXB/BXH analysis. The important lesson from this experiment using the unsupervised approach and our transcriptome dataset was the identification of a module related to immune response and glial/neuronal communication as a predisposing factor for quantitative differences in alcohol consumption. This relationship may seem aberrant, but a plethora of recent data, using completely different approaches, has identified inflammatory mediators and microglial function in rodent brain as drivers for levels of alcohol consumption, e.g., [32]. Our data also uncover the importance of a noncoding RNA which controls transcription of several other genes associated with alcohol consumptive behavior. As compelling as the candidate module in brain was, for linking a biological mechanism to voluntary alcohol consumption, it did not explain all of the genetic variation in alcohol consumption among HXB/BXH strains. To further explore this trait in context of a potential peripheral organ-brain interaction, we used the same system genetics approach to identify a candidate co-expression module in liver of alcohol-naïve animals which was associated with alcohol consumption [31]. Interestingly, the QTL for this module overlapped an alcohol consumption QTL that was *not* associated with any candidate modules in brain. The liver co-expression module was also shown to be related to inflammatory responses. Often signals in response to agents that generate inflammation travel from peripheral organs, such as liver, to inform the brain via cytokines. Such communication can interact with brain-specific mechanisms, mentioned above, for establishing the relation between (neuro) inflammation and alcohol consumption. The interesting lesson learned here, was that a holistic approach can be valuable when one is searching for genomic relationships with behavior. In this particular case, transcriptomic events sub-serving communication between peripheral organs and brain and not just the transcriptomic networks within brain can contribute to a behavioral phenotype.

3.2 Alcohol Metabolism

In addition to alcohol-related behaviors, this panel and approach have been used to examine liver co-expression networks associated with alcohol clearance and circulating acetate levels [25]. Following a single dose of ethanol in male rats of this panel, blood alcohol and acetate levels across a period of 400 min suggest a significant difference in alcohol metabolism between strains. In particular, alcohol clearance was measured via a one-compartment pharmacokinetics model with first-order absorption and elimination in each rat individually and acetate exposure was measured as the area under the curve (AUC) from 0 to 400 min of the acetate concentration-time curves for each individual rat. A single co-expression module was identified, and it was associated with both alcohol clearance and acetate AUC. This module contained two isoforms of alcohol dehydrogenase. The result certainly underscored the validity of the unsupervised pipeline for identifying modules pertinent to the phenotype of interest. However, the result was even more telling about the role of the various isoforms of alcohol dehydrogenase that can contribute to alcohol metabolism in the rat. Alcohol dehydrogenase 1 (ADH1) is the enzyme that most scientists consider to be the most important isoform involved in the metabolism of ethanol. However, our results showed ADH4 to be the hub gene of the module and the mRNA of ADH4 was most highly correlated with the measures of alcohol disappearance across strains. When we reflected on the blood alcohol concentrations reached when we administered ethanol (2 g/kg of body weight) to rats, we, however, noted that these concentrations were certainly in the range that would allow ADH4 (which has a high K_m for ethanol) to be a major player in the alcohol disappearance phenotype. Thus, our results not only pointed to a well-established enzyme system which metabolizes ethanol, but also recognized the nuance of which ADHs come into play at which circulating level of ethanol.

3.3 Brown Adipose Tissue Phenotypes

In addition to alcohol-related traits, several quantitative traits related to metabolic syndrome and the role of brown adipose tissue (BAT) in this syndrome were examined using our methodology and the HXB/BXH panel [26]. In this study, we examined RNA expression in BAT and its association with the amount of BAT in the rat (relative mass) and the amount of glucose incorporation into BAT lipids. The candidate module associated with relative mass of BAT contained several genes associated with intracellular and extracellular signaling including the innervation of BAT by the sympathetic nervous system. This module also included Cd36, a gene known to be associated with long chain fatty acid uptake. Cd36 has a polymorphism in the SHR progenitor strain that renders reduced uptake. A separate module was associated with glucose incorporation. This module had several components with known functions related to glucose uptake and glucose metabolism. In this case, the translational impact of the results obtained in the rat was easily evident. GWAS studies with humans identified loci associated with metabolic syndrome and these loci were syntenic with loci we identified in rat. And, the proteins derived from genes in these regions were critical components of the networks (modules) we identified. Our studies not only substantiate some of the results found in humans but put the role of these genes in the context of a network whose other members may also play significant roles in the genetic predisposition to metabolic syndrome.

4 Definition of HRDP

Although there are several success stories using the HXB/BXH panel, there are phenotypic QTL that are either approaching the significance threshold or that have no associated co-expression module. With 30 strains, we have limited power to identify co-expression networks with more subtle impacts on a quantitative trait and to examine gene-by-gene or gene-by-environment interactions. In addition, with a single RI panel we are limited to the networks/genes/loci that vary in the two progenitor strains. A way to overcome these current limitations is to develop a hybrid panel that contains multiple RI panels and a panel of classic inbred strains.

The best genetically stable population currently available for association mapping is the Hybrid Mouse Diversity Panel (HMDP; [33]), which comprises over 100 inbred strains of mice, including 71 RI strains (for mapping power) and 32 commercially available inbred strains (for mapping resolution). This panel has already shown its benefits in systems genetics studies of diabetes, metabolic anomalies, and cardiovascular disease [33–35]. However, the mouse strains used to create the HMDP are descendants of just two founder sub-strains [19] and this fact limits the genetic variation that can contribute to a trait. The other panel of RI mouse strains which has been initiated is the Collaborative Cross [36, 37]. In this inbred panel, three wild mouse strains were introduced into the breeding to overcome the genetic similarities inherent in the strains of laboratory mice. Recently, we contributed to a study in which 40 strains of rats had their DNA sequenced [38]. This study produced information indicating that these strains are derived from at least nine progenitors, and thus the sequenced rat strains allow for significantly greater genetic diversity if combined into a mapping panel. To produce the HRDP we have included the HXB/BXH RI panel, which currently contains 30 strains but does not, by itself, allow fine mapping of loci contributing to complex traits. We are now adding the LEXF/FXLE RI panel (34 strains; [39]), and 35 more strains of inbred rats (Table 1) which we have carefully selected for genetic diversity.

5 Power/Precision of HRDP

Flint and Eskin [40] provide an excellent set of criteria to be considered for creating a panel or stock of animals for whole genome association studies in mice or rats. These criteria are statistical power, mapping resolution, coverage (extent of genomic polymorphisms between the animals being analyzed), reproducibility (data with inbred or RI strains can be easily replicated), and cost of construction and maintenance of the panel (e.g., in an F2 design or in an outbred population each animal has to be genotyped every time an experiment is started with a new group of animals).

A simple comparison of statistical power, mapping resolution and coverage of the HRDP with the most powerful mouse panels, gives full credibility for the HRDP as a stable resource for systems genetics analysis. For the HRDP, we have estimated statistical power (statistical power is highly dependent on the variance in the

Table 1
List of strains included in the Hybrid Rat Diversity Panel

Classic inbred strains			
ACI/SegHsd	F344/Stm	LN/MavRrrc	RCS/LavRrrc
BDIX/CrCl	FHH/EurMcwi	LOU/MNCr	SHR/NCrl
BN-Lx/Cub	GH/OmrMcwi	LUDW/OlaHsd	SHR/OlaIpcv
BN/NHsdMcwi	HTX/HcjRrrc	M520/N	SHRSP/A3NCrl
BUF/CrCrl	LE/BluGill	MNS/N	SR/JrHsd
COP/CrCrl	LE/Stm	MR/N	SS/JrHsd
DA/OlaHsd	LEW/Crl	MWF/ZtmRrrc	WAG/RijCrl
F344/NCrl	LEW/SsNHsd	PD/Cub	WKY/NCrl
F344/NHsd	LH/MavRrrc	PVG/Seac	
HXB/BXH recombinant inbred panel			
BXH10	BXH8	HXB2	HXB29
BXH11	BXH9	HXB20	HXB3
BXH12	HXB1	HXB21	HXB31
BXH13	HXB10	HXB22	HXB4
BXH2	HXB13	HXB23	HXB5
BXH3	HXB15	HXB24	HXB7
BXH5	HXB17	HXB25	
BXH6	HXB18	HXB27	
FXLE/LEXF recombinant inbred panel			
FXLE12	FXLE21	LEXF11	LEXF6B
FXLE13	FXLE22	LEXF1A	LEXF7A
FXLE14	FXLE23	LEXF1C	LEXF7B
FXLE15	FXLE24	LEXF2A	LEXF7C
FXLE16	FXLE25	LEXF2B	LEXF8A
FXLE17	FXLE26	LEXF2C	LEXF8D
FXLE18	LEXF10A	LEXF3	LEXF9
FXLE19	LEXF10B	LEXF4	
FXLE20	LEXF10C	LEXF5	

trait and the number of animals used) at approximately 80% power to detect loci that account for 12% of phenotypic variance for traits such as alcohol consumption or alcohol metabolism. The median haplotype block size would be 150 Kb. These estimates are similar to the HMDP, although approximately a 2 Mb interval has been estimated for resolution in the mouse panel [40]. If one is considering an experiment based on genetic correlation, the HRDP has the power to detect a significant genetic correlation as small as 0.28.

6 Added Value of HRDP in Transcriptome Characterization

The rat transcriptome is arguably under-annotated. In the human Ensembl transcriptome (v90), there are over 220,000 annotated transcripts compared to the 41,000 transcripts annotated in the Rat Ensembl Transcriptome (v90). Even the Mouse Ensembl Transcriptome (v90) has over 132,000 transcripts, so it is unlikely the difference in the number of transcripts is due to biology. Although human, mouse, and rat have a similar number of annotated protein-coding genes, just over 22,000 in each species, the average number of transcripts (i.e., splice variants) per protein-coding gene is drastically different; 7.2 transcripts per gene in human, 4.3 transcripts per gene in mouse, and 1.3 transcripts per gene in rat. There is also a substantial difference in the number of noncoding transcripts that are annotated across species. For example, in humans 14,841 long intergenic noncoding RNA transcripts are annotated compared to 3,304 in the rat transcriptome. Because of this under-annotation, it is important that de novo transcriptomes are assembled when quantitating RNA-Seq data in rat.

To describe the added value of including classic inbred strains and recombinant inbred strains in the HRDP panel in describing the rat transcriptome, we reconstructed the liver transcriptome in 43 inbred strains of rats including 30 recombinant inbred (HXB/BXH) strains and 13 classic inbred strains. For each strain, we sequenced RNA from livers of three male biological replicates. Total RNA was extracted, and ribosomal RNA was depleted to generate sequencing libraries with both polyadenylated transcripts and non-polyadenylated transcripts. Libraries were sequenced on Illumina's HiSeq machines at a high depth (approximately 65 paired-end reads per sample). Reads were aligned to strain-specific genomes (*see* [25] for methods for developing strain-specific genomes) and strain-specific liver transcriptomes were reconstructed. Individual strain transcriptomes were merged and individual samples were quantified.

In this dataset of quantified expression, individual transcripts, i.e., splice variants/isoforms, were classified as present ($\geq$10 reads in all samples of a strain), absent (<1 read in all samples of a strain), or other (not meeting either of the above criteria) in each of the 43 strains. Each transcript was also classified based on its assignment to one of the three categories: (1) known (i.e., Ensembl

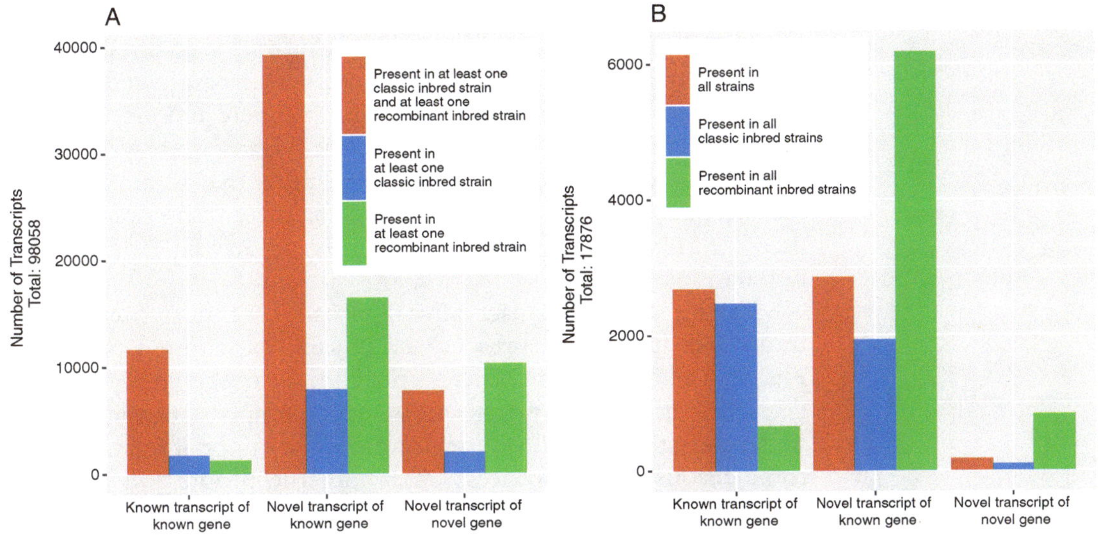

Fig. 2 Transcriptome diversity in the Hybrid Rat Diversity Panel (HRDP). Transcripts are summarized based on their Ensembl annotation: known transcript of known gene, novel transcript of known gene, and novel transcript of novel gene. To compare the recombinant inbred panel and the classic inbred panel, all 11 classic inbred strains were used to represent that panel and the 2 progenitor strains and 9 randomly chosen strains from the recombinant inbred panel were used to represent the recombinant inbred panel. (**a**) Transcripts present in at least one strain. Transcripts had to be present in at least one strain (i.e., >10 reads for each biological replicate). Red represents transcripts present in at least one strain from both the recombinant inbred panel and the classic inbred panel. Blue represents transcripts that were present in at least one classic inbred strain and not present in any recombinant inbred strains. Green represents transcripts that were in at least one recombinant inbred strain and not in any classic inbred strains. (**b**) Transcripts present in all strains of at least one panel. Transcripts had to be present in all strains of a given inbred panel. Red represents transcripts present in all inbred strains. Blue represents transcripts that were present in all classic inbred strains and not present in all recombinant inbred strains. Green represents transcripts that were in all recombinant inbred strain and not in all classic inbred strains

annotated) transcript of a known gene, (2) novel transcript of a known gene, and (3) novel transcript of a novel gene.

The reconstructed liver transcriptome contained 328,096 transcripts. Using the above criteria, 114,982 transcripts were present in at least one of the 43 strains. Of these, 15,621 (14%) were known transcripts of known genes. 73,855 (64%) were novel transcripts of known genes, which can represent anything from unprocessed transcripts to novel splice variants of a gene, and 25,506 (22%) were novel transcripts of novel genes.

To examine the impact of combining a RI panel with a classic inbred panel, we identified transcripts that were only expressed either in the classic inbred panel or in the recombinant inbred panel (*see* Fig. 2). In order to reduce bias due to the greater number of recombinant inbred strains compared to classic inbred strains, we created a smaller dataset to represent the recombinant inbred panel that included the two progenitor strains and nine randomly sampled recombinant inbred strains. Transcripts were included in Fig. 2a if they were present in at least one strain of either the classic inbred panel or the reduced set of recombinant inbred strains. This captures transcripts that may be expressed in only one of the classic inbred strains. As expected, most transcripts are expressed in both panels. However, about 40% of transcripts were unique to either panel. Furthermore, the "panel-specific" transcripts were more likely to be unannotated than the transcripts that were expressed in both panels. The lack of annotation for these transcripts may be due to this heterogeneity in expression across different rat populations.

Taking a more conservative approach, we next focused on transcripts that were "present" in all samples within a panel (*see* Fig. 2b). Even with this stricter criterion, we observed more novel transcripts than known transcripts, which emphasizes the current under-annotation of the rat genome and the utility of the transcriptome reconstruction process. Although the proportion of novel transcripts remained high, the majority of the novel transcripts were derived from known (i.e., annotated) genes. Only a small proportion of transcripts were classified as novel transcripts of novel genes. Two possible explanations for this are: (1) the novel transcripts of novel genes are more likely to be noise, i.e., a false positive, and (2) genetic diversity has a much more dramatic impact

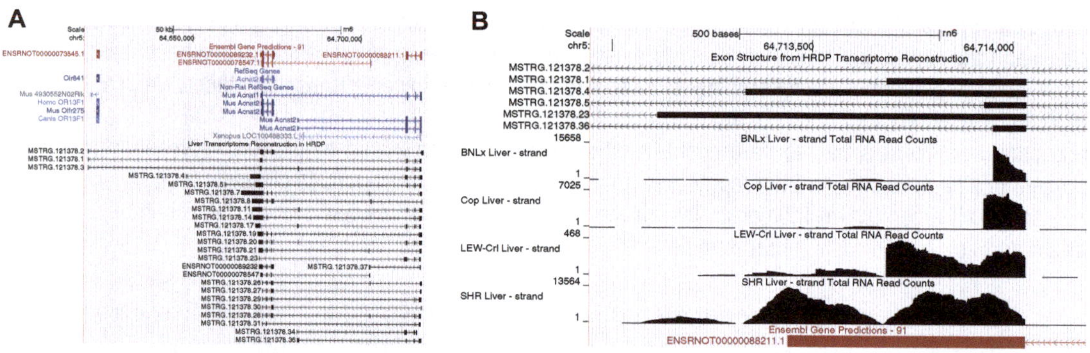

Fig. 3 Demonstration of strain-specific differences in exon structure of Acnat1 and Acnat2. (**a**) Liver transcriptome reconstruction of Acnat1 and Acnat2. The first track (red) indicates the Ensembl transcript annotation in the rat region on Chromosome 5 from 64,617,992 to 64,719,930 base pairs. Translated regions of the transcript are indicated by the tallest vertical bars, the transcribed, but untranslated regions of the transcripts have the shorter vertical bars, and the thin horizontal lines indicate intronic regions with the arrow heads pointing in the direction of transcription (i.e., 5′–3′). The next track (light blue) includes the RefSeq annotation in this region. The third track (blue) indicates the orthologous RefSeq transcripts within this region. The final track (black) includes 25 transcripts identified in the reconstruction of the liver transcriptome from 43 strains of the Hybrid Rat Diversity Panel and present in at least one strain (greater than 10 reads in all biological replicates of that strain). In the transcriptome reconstruction the translated and untranslated regions are not distinguished from one another. (**b**) Comparison of read coverage by strain to alternative exon structures within the reconstructed Acnat1/Acnat2 liver transcripts. The first track (black) includes six of the reconstructed transcripts that represent the five different versions of this exon. The next four tracks represent the RNA-Seq reads that align to this region by strain. The four strains represent the general patterns observed in this region. The last track (red) is the Ensembl transcriptome annotation for Acnat1

on the expression levels of these potentially noncoding transcripts. Also, with this stricter criterion, we still observed a substantial number of transcripts (more than 60%) that were only expressed in one of the two panels.

To demonstrate an observed striking strain difference in transcript structure, we examined the exonic structure of acyl-coenzyme A amino acid *N*-acyltransferase 1/2 (Acnat1/Acnat2) in liver (*see* Fig. 3). The Ensembl annotation of these two genes includes two transcripts for Acnat2 and one transcript for Acnat1. In the transcriptome reconstruction, several isoforms overlap coding regions and untranslated regions of both annotated genes. The "fusion" was also observed in the ortholog RefSeq genes from mouse. Our liver transcriptome reconstruction identified 25 different isoforms among the two genes that were highly expressed in at least one strain of the 43 HRDP strains included with many isoforms overlapping transcribed regions of both genes. Two of the transcripts (MSTRG.121378.36 and MSTRG.121378.5 in Fig. 3) were highly expressed in at least one of the classic inbred samples (>1000 reads) and had little expression in the RI panel (<10 reads in every sample). One region of the gene(s) with the most heterogeneity among transcripts was the exon that overlapped the last exon of the Ensembl transcript for Acnat1 (ENSRNOT00000088211.1 in Fig. 3). In this region, five different versions of the exon were included in transcripts identified in the reconstruction (*see* Fig. 3b). Many of the classic inbred strains differed dramatically based on which version(s) of the exon were expressed. In BN-Lx strain, only the shortest version of this exon was expressed. In the LEW/Crl strain, at least two versions of the exon are expressed if not more. These differences in exon structure represent differences in the protein-coding area of the transcript and could result in differences in the protein expressed. This example demonstrates strain-specific difference in transcript structure that could have a dramatic impact on protein function.

In summary, the liver transcriptome is much more diverse than what is currently annotated within the Ensembl database. Furthermore, genetic diversity contributes substantially to the diversity of the transcriptome.

7 Overview of PhenoGen: Your Site for Rat RNA Information

The PhenoGen website was designed as a systems genetics tool box and repository for transcriptome data on renewable animal models such as the HRDP (*see* Fig. 4). Currently, the PhenoGen database contains RNA-Seq data from both small RNAs (<200 nucleotides) and long RNAs (>200 nucleotides, including both polyadenylated transcripts and non-polyadenylated transcripts) from brain and liver of 43 strains of the HRDP and full genome sequencing of 12 of the

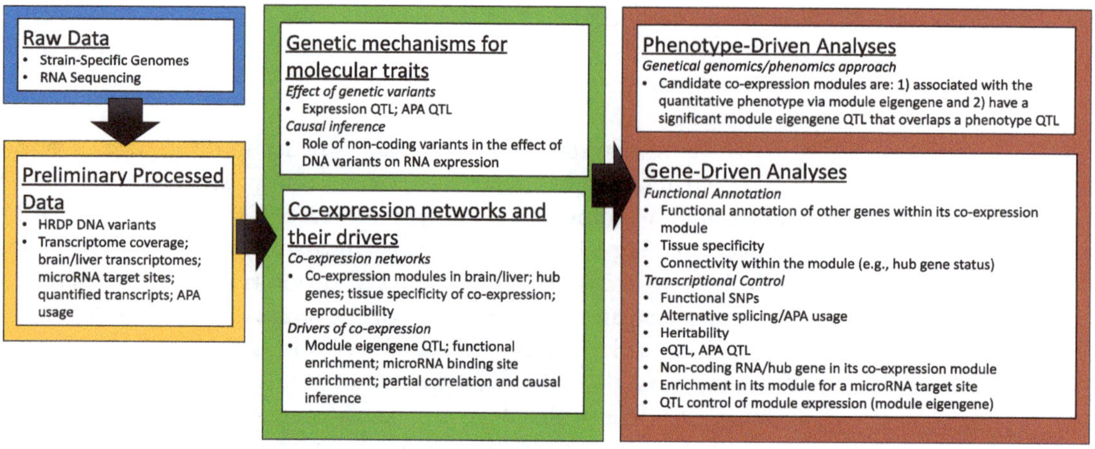

Fig. 4 Overview of transcriptomic and genomic information from the Hybrid Rat Diversity Panel (HRDP) either currently available or projected to be available on the PhenoGen website (http:/phenogen.org) by 2020. The goal of the PhenoGen website is to provide a sophisticated and intuitive interface for the high-throughput sequencing information gathered on the genetically renewable HRDP. The level of data includes the *Raw Data* (blue box), e.g., DNA sequence and RNA sequence reads. We recognize that analytical methods and our perspective on RNA biology are rapidly evolving and that access to the original data will provide researchers with the flexibility to interrogate novel hypotheses in an efficient manner. The next level of data is the *Preliminary Processed Data* (yellow box). These data distill the terabytes of information in the raw data to the position and character of DNA variants, quantitative levels of expression, and DNA signals for microRNA binding, polyadenylation, etc. Providing this level of data allows researchers without the bioinformatics expertise and resources to explore the data for both hypothesis generation and hypothesis testing. The third level of data (green box) includes both *Genetic mechanisms for molecular traits* and *Co-expression networks and their driver*. This level of data represents a statistically rigorous analysis of how DNA sequence influences RNA expression in both structure of transcripts expressed and quantitative expression levels and how and why transcripts interact with one another. The final level of data (red box) is the products of *Phenotype-Driven Analyses* and *Gene-Driven Analyses*. These are co-expression modules associated with a genetic predisposition for a phenotypic trait, e.g., alcohol consumption, and detailed information about an individual gene including both functional annotation and information about transcription control

13 classic inbred strains and dense genotype data for the 30 recombinant inbred strains. Data are preprocessed and made available to researchers through an interactive genome browser based on physical location, through co-expression networks with interactive features to further explore the relationships between genes, and through eQTL integrated through both the genome browser interface and the co-expression networks. The PhenoGen data reflect a basal state of the animal and should be considered for use when one is interested in a predisposition to a complex trait. The data can be used to either interrogate the genetic factors that predispose to disease or other quantitative phenotypes using the genetical genomic/phenomic approach described earlier or it can be used to explore the biological context under which a candidate gene functions within a surveyed tissue. For example, if you had a candidate gene from a GWAS of alcohol use disorder, you could explore its

co-expression network in rat brain and its drivers of expression. This can bring insight about the role of this gene in predisposition to alcohol use disorders, what its function is in the absence of disease and alcohol exposure, what molecular features drive differences in expression, and how current animal models can be manipulated to ask specific biological questions about this gene and its related networks.

8 The Future of the HRDP

Panels like the HRDP represent the future of systems genetics and will facilitate exploration of gene by environment interactions. As omics technologies continue to develop and mature, we will have a stable population that we can continue to interrogate from different prospectives. We expect that many of the past studies of disease, behavior, physiology, and toxicology that used a single strain or a standard outbred rat (e.g., Sprague Dawley) will be expanded in genetically diverse panels to explore the impact of genetic variability. For instance, instead of exploring the general toxicity of a substance within a single strain of rats, we can now explore the role of genetic contributions to susceptibility to the toxin or to a medication. The uncovered susceptibility factors (which can also include the susceptibility to the positive effects of a medication) can certainly inform the efforts to establish precision medicine in humans.

The major advantage of focusing on a renewable rat model, such as the HRDP, and creating a public repository, such as PhenoGen, is that data can be accumulated across generations and laboratories. This means that our PhenoGen databases will continue to grow and that other omics data can be integrated to give a broader view of the complex intracellular, intercellular, and trans-tissue networks. As our databases grow, our statistical methodologies will continue to advance. We will move from co-expression networks to directed predictive models. We will explore the many genetic features that contribute and interact to produce disease. These tools will drive research by allowing individual researchers to build better hypotheses by taking full advantage of the plethora of data already available and will foster new collaborations and new ways of applying science to benefit humanity.

Acknowledgments

This work is supported by NIH/NIAAA (R24 AA013162), NIH/NIDA (P30 DA044223), and the Banbury Fund.

References

1. Hancock DB, Markunas CA, Bierut LJ, Johnson EO (2018) Human genetics of addiction: new insights and future directions. Curr Psychiatry Rep 20(2):8. https://doi.org/10.1007/s11920-018-0873-3
2. Dodoo SN, Benjamin IJ (2017) Genomic approaches to hypertension. Cardiol Clin 35 (2):185–196. https://doi.org/10.1016/j.ccl.2016.12.001
3. Heath AC, Whitfield JB, Martin NG, Pergadia ML, Goate AM, Lind PA et al (2011) A quantitative-trait genome-wide association study of alcoholism risk in the community: findings and implications. Biol Psychiatry 70 (6):513–518. https://doi.org/10.1016/j.biopsych.2011.02.028
4. Hong S, Kim Y, Park T (2014) Practical issues in screening and variable selection in genome-wide association analysis. Cancer Informat 13 (Suppl 7):55–65. https://doi.org/10.4137/CIN.S16350
5. Eichler EE, Flint J, Gibson G, Kong A, Leal SM, Moore JH et al (2010) Missing heritability and strategies for finding the underlying causes of complex disease. Nat Rev Genet 11 (6):446–450. https://doi.org/10.1038/nrg2809
6. Schierding W, Antony J, Cutfield WS, Horsfield JA, O'Sullivan JM (2016) Intergenic GWAS SNPs are key components of the spatial and regulatory network for human growth. Hum Mol Genet 25(15):3372–3382. https://doi.org/10.1093/hmg/ddw165
7. Chen J, Tian W (2016) Explaining the disease phenotype of intergenic SNP through predicted long range regulation. Nucleic Acids Res 44(18):8641–8654. https://doi.org/10.1093/nar/gkw519
8. Ballantyne RL, Zhang X, Nunez S, Xue C, Zhao W, Reed E et al (2016) Genome-wide interrogation reveals hundreds of long intergenic noncoding RNAs that associate with cardiometabolic traits. Hum Mol Genet 25 (14):3125–3141. https://doi.org/10.1093/hmg/ddw154
9. Civelek M, Lusis AJ (2014) Systems genetics approaches to understand complex traits. Nat Rev Genet 15(1):34–48. https://doi.org/10.1038/nrg3575
10. Moreno-Moral A, Petretto E (2016) From integrative genomics to systems genetics in the rat to link genotypes to phenotypes. Dis Model Mech 9(10):1097–1110. https://doi.org/10.1242/dmm.026104
11. GTEx Consortium (2015) Human genomics. The Genotype-Tissue Expression (GTEx) pilot analysis: multitissue gene regulation in humans. Science 348(6235):648–660. https://doi.org/10.1126/science.1262110
12. Nicolae DL, Gamazon E, Zhang W, Duan S, Dolan ME, Cox NJ (2010) Trait-associated SNPs are more likely to be eQTLs: annotation to enhance discovery from GWAS. PLoS Genet 6(4):e1000888. https://doi.org/10.1371/journal.pgen.1000888
13. Nica AC, Dermitzakis ET (2013) Expression quantitative trait loci: present and future. Philos Trans R Soc Lond Ser B Biol Sci 368 (1620):20120362. https://doi.org/10.1098/rstb.2012.0362
14. Zhang B, Horvath S (2005) A general framework for weighted gene co-expression network analysis. Stat Appl Genet Mol Biol 4:Article17. https://doi.org/10.2202/1544-6115.1128
15. Dong J, Horvath S (2007) Understanding network concepts in modules. BMC Syst Biol 1:24. https://doi.org/10.1186/1752-0509-1-24
16. Farber CR, Lusis AJ (2008) Integrating global gene expression analysis and genetics. Adv Genet 60:571–601. https://doi.org/10.1016/S0065-2660(07)00420-8
17. GTEx Consortium (2013) The Genotype-Tissue Expression (GTEx) project. Nat Genet 45(6):580–585. https://doi.org/10.1038/ng.2653
18. eGTEx Project (2017) Enhancing GTEx by bridging the gaps between genotype, gene expression, and disease. Nat Genet 49 (12):1664–1670. https://doi.org/10.1038/ng.3969
19. Reuveni E, Birney E, Gross CT (2010) The consequence of natural selection on genetic variation in the mouse. Genomics 95 (4):196–202. https://doi.org/10.1016/j.ygeno.2010.02.004
20. Odet F, Pan W, Bell TA, Goodson SG, Stevans AM, Yun Z et al (2015) The founder strains of the collaborative cross express a complex combination of advantageous and deleterious traits for male reproduction. G3 5(12):2671–2683. https://doi.org/10.1534/g3.115.020172
21. Pravenec M, Klir P, Kren V, Zicha J, Kunes J (1989) An analysis of spontaneous hypertension in spontaneously hypertensive rats by means of new recombinant inbred strains. J Hypertens 7(3):217–221
22. Pravenec M, Kren V, Landa V, Mlejnek P, Musilova A, Silhavy J et al (2014) Recent progress in the genetics of spontaneously hypertensive rats. Physiol Res 63(Suppl 1):S1–S8
23. Saba LM, Flink SC, Vanderlinden LA, Israel Y, Tampier L, Colombo G et al (2015) The

sequenced rat brain transcriptome—its use in identifying networks predisposing alcohol consumption. FEBS J 282(18):3556–3578. https://doi.org/10.1111/febs.13358

24. Vanderlinden LA, Saba LM, Printz MP, Flodman P, Koob G, Richardson HN et al (2014) Is the alcohol deprivation effect genetically mediated? Studies with HXB/BXH recombinant inbred rat strains. Alcohol Clin Exp Res 38(7):2148–2157. https://doi.org/10.1111/acer.12471
25. Lusk R, Saba LM, Vanderlinden LA, Zidek V, Silhavy J, Pravenec M et al (2018) Unsupervised, statistically based systems biology approach for unraveling the genetics of complex traits: a demonstration with ethanol metabolism. Alcohol Clin Exp Res 42 (7):1177–1191
26. Pravenec M, Saba LM, Zidek V, Landa V, Mlejnek P, Silhavy J et al (2018) Systems genetic analysis of brown adipose tissue function. Physiol Genomics 50(1):52–66. https://doi.org/10.1152/physiolgenomics.00091.2017
27. Morrissey C, Grieve IC, Heinig M, Atanur S, Petretto E, Pravenec M et al (2011) Integrated genomic approaches to identification of candidate genes underlying metabolic and cardiovascular phenotypes in the spontaneously hypertensive rat. Physiol Genomics 43 (21):1207–1218. https://doi.org/10.1152/physiolgenomics.00210.2010
28. Abdelmagid N, Bereczky-Veress B, Atanur S, Musilova A, Zidek V, Saba L et al (2016) Von Willebrand factor gene variants associate with herpes simplex encephalitis. PLoS One 11(5): e0155832. https://doi.org/10.1371/journal.pone.0155832
29. Pravenec M, Kozich V, Krijt J, Sokolova J, Zidek V, Landa V et al (2016) Genetic variation in renal expression of folate receptor 1 (Folr1) gene predisposes spontaneously hypertensive rats to metabolic syndrome. Hypertension 67 (2):335–341. https://doi.org/10.1161/HYPERTENSIONAHA.115.06158
30. Kuda O, Brezinova M, Silhavy J, Landa V, Zidek V, Dodia C et al (2018) Nrf2-mediated antioxidant defense and peroxiredoxin 6 are linked to biosynthesis of palmitic acid ester of 9-hydroxystearic acid. Diabetes 67 (6):1190–1199. https://doi.org/10.2337/db17-1087
31. Hoffman PL, Saba LM, Vanderlinden LA, Tabakoff B (2017) Voluntary exposure to a toxin: the genetic influence on ethanol consumption. Mamm Genome 29 (1–2):128–140. https://doi.org/10.1007/s00335-017-9726-3
32. Mayfield J, Ferguson L, Harris RA (2013) Neuroimmune signaling: a key component of alcohol abuse. Curr Opin Neurobiol 23 (4):513–520. https://doi.org/10.1016/j.conb.2013.01.024
33. Bennett BJ, Farber CR, Orozco L, Kang HM, Ghazalpour A, Siemers N et al (2010) A high-resolution association mapping panel for the dissection of complex traits in mice. Genome Res 20(2):281–290. https://doi.org/10.1101/gr.099234.109
34. Ghazalpour A, Rau CD, Farber CR, Bennett BJ, Orozco LD, van Nas A et al (2012) Hybrid mouse diversity panel: a panel of inbred mouse strains suitable for analysis of complex genetic traits. Mamm Genome 23(9–10):680–692. https://doi.org/10.1007/s00335-012-9411-5
35. Lusis AJ, Seldin MM, Allayee H, Bennett BJ, Civelek M, Davis RC et al (2016) The Hybrid Mouse Diversity Panel: a resource for systems genetics analyses of metabolic and cardiovascular traits. J Lipid Res 57(6):925–942. https://doi.org/10.1194/jlr.R066944
36. Valdar W, Flint J, Mott R (2006) Simulating the collaborative cross: power of quantitative trait loci detection and mapping resolution in large sets of recombinant inbred strains of mice. Genetics 172(3):1783–1797. https://doi.org/10.1534/genetics.104.039313
37. Chesler EJ (2014) Out of the bottleneck: the Diversity Outcross and Collaborative Cross mouse populations in behavioral genetics research. Mamm Genome 25(1–2):3–11. https://doi.org/10.1007/s00335-013-9492-9
38. Hermsen R, de Ligt J, Spee W, Blokzijl F, Schafer S, Adami E et al (2015) Genomic landscape of rat strain and substrain variation. BMC Genomics 16:357. https://doi.org/10.1186/s12864-015-1594-1
39. Voigt B, Kuramoto T, Mashimo T, Tsurumi T, Sasaki Y, Hokao R et al (2008) Evaluation of LEXF/FXLE rat recombinant inbred strains for genetic dissection of complex traits. Physiol Genomics 32(3):335–342. https://doi.org/10.1152/physiolgenomics.00158.2007
40. Flint J, Eskin E (2012) Genome-wide association studies in mice. Nat Rev Genet 13 (11):807–817. https://doi.org/10.1038/nrg3335

Chapter 11

Using Heterogeneous Stocks for Fine-Mapping Genetically Complex Traits

Leah C. Solberg Woods and Abraham A. Palmer

Abstract

In this chapter we will review both the rationale and experimental design for using Heterogeneous Stock (HS) populations for fine-mapping of complex traits in mice and rats. We define an HS as an outbred population derived from an intercross between two or more inbred strains. HS have been used to perform genome-wide association studies (GWAS) for multiple behavioral, physiological, and gene expression traits. GWAS using HS require four key steps, which we review: selection of an appropriate HS population, phenotyping, genotyping, and statistical analysis. We provide advice on the selection of an HS, comment on key issues related to phenotyping, discuss genotyping methods relevant to these populations, and describe statistical genetic analyses that are applicable to genetic analyses that use HS.

Key words Genetic mapping, Outbred, Advanced intercross, Heterogeneous stock, Diversity outbred

1 Introduction

In the past 10 years, heterogeneous stock (HS) rodents have been used for fine-mapping and identification of underlying causal genes and genetic variants of multiple complex traits. HS are highly recombinant rodent populations that are created from two or more (most commonly eight) inbred founder strains. Examples of HS populations include advanced intercross lines (AIL), HS mice and rats, HS-CC, and the diversity outbred (DO) mouse. The original purpose of the early HS populations was to serve as a source of genetic diversity for selection studies, whereas more recently created HS (HS-CC and DO) were intended primarily for genome-wide association studies (GWAS). In each case, a collection of inbred strains were chosen to capture high levels of genetic diversity. Breeding schemes for these populations require that many families are maintained in each generation and that familial relationships among breeding pairs are minimized. With each generation of outbreeding, the distance between

G. Thomas Hayman et al. (eds.), *Rat Genomics*, Methods in Molecular Biology, vol. 2018,
https://doi.org/10.1007/978-1-4939-9581-3_11,

recombination events is reduced, allowing for genetic mapping to increasingly smaller intervals. The individuals within an HS represent a random mosaic of haplotypes of the inbred founder strains, with each animal being genetically and phenotypically distinct. Thus, HS more closely resemble the variation found in natural populations (including humans) when compared to "reference populations" such as recombinant inbred (RI; [1]), collaborative cross (CC; [2]), and other inbred panels. New mutations within HS populations are generally assumed to be negligible, although future work is needed to test this assumption.

There are several advantages to using HS populations for genetic mapping relative to traditional mapping strategies such as an F_2 intercross or backcross. The first is that they allow mapping of complex traits to small regions, addressing a critical weakness of traditional mapping strategies [3–5], and decreasing the number of candidate genes within each interval. A second advantage is that the founders of all extant HS have been fully sequenced and these data are available in public databases, making it possible to identify informative markers, potentially causal coding variants and more complex repetitive or structural differences. Ancestral haplotype information can also be used to inform which founder allele (s) contribute to the phenotypic trait (*see* [6, 7]). Third, the use of HS models for genetic mapping minimizes the chance that mapped alleles will be highly dependent on a specific genetic background/genetic context (*see* [8]), because loci are identified on a mix of genetic backgrounds. Finally, due to the genetic diversity within HS populations, there is often a high degree of phenotypic variability, ranging from physiological to behavioral traits, allowing genetic mapping of many different traits in the same cohort. Advantages and disadvantages of HS rats relative to other mapping strategies have been previously discussed [3–5].

For genetic mapping studies in HS populations, four key steps are necessary: selection of an appropriate HS population, phenotyping, genotyping, and statistical analysis. All four are discussed in more detail below.

2 Selection of an Appropriate HS Population

Below we review available rat and mouse HS populations that can be used for genetic studies. In deciding whether to use HS rats or mice, the main consideration should be the phenotype to be studied. Rats are better suited for certain behavioral phenotypes that do not work well in the mouse [9]. Rats are also larger and thus may be more amenable to phenotypes involving surgery. Mice are less expensive to maintain and benefit from the availability of a greater diversity of genetic tools and resources. Because genetic mapping is dependent on the heritability of the trait, it is important to establish heritability of a phenotype before beginning a large GWAS. This can sometimes be accomplished by measuring differences among the inbred founder strains; however for HS rats, the original founder strains are no longer available; therefore, heritability must be established using the most closely related inbred strains or by pedigree or marker-based relatedness (e.g. [10]).

2.1 HS Mice

The oldest HS mouse population, which is now extinct (personal communication, Dr. Jerry Stitzel 6-30-18), was the Boulder HS which was created in UC Berkeley but thereafter maintained at the Institute for Behavioral Genetics in Boulder, CO (aka HS/IBG; [11]). The HS/IBG was created by intercrossing eight inbred strains: A, AKR, BALB/c, C3H/2, C57BL, DBA/2, Is/Bi, RIII (note that the exact strain/vendor designations are unknown). A second HS population, called the Northport HS (aka HS/NPT; [12]), which is alive in the laboratory of Dr. Robert Hitzemann as of July 1, 2018, was created using six of the same inbred strains: A/J, AKR/NCrl, BALB/cAnNCrl, C3H/HeNCrl, C57BL6/NCrl, DBA/2J, but replaced Is/Bi and RIII with CBA/J, and LP/J. Both the HS/IBG and HS/NPT were originally created to provide a genetically diverse population for selection studies, but were later used for genetic fine-mapping. The HS/NPT was one of the earlier stocks used for genetic fine-mapping, with the goal of narrowing previously identified loci discovered using an F_2 cross [12, 13]. Using HS for genome-wide analysis, in which discovery of genetic loci and fine-mapping for multiple phenotypic traits were done in the same cohort, occured several years later [14]. HS mice have since been used to map multiple traits including fear behavior [15], ethanol consumption [16], and arthritis [17].

An advanced intercross line (AIL) is an HS that is composed of only two inbred strains [18]. At least two AIL have been created using C57BL/6J and DBA/2J [19, 20], but the only currently extant AIL was created by crossing the LG/J and SM/J inbred strains and has been used for numerous mapping studies (e.g., [19, 21–34]).

More recently, several HS mouse populations that were specifically intended for genetic fine-mapping have been developed. The HS-CC was created by intercrossing eight inbred strains: A/J, C57BL/6J, 129S1/SvImJ, NOD/LtJ, NZO/HlLtJ, CAST/EiJ, PWK/PhJ, and WSB/EiJ [35]. Those strains were selected because they were the same eight strains used to make the collaborative cross (CC; for further discussion of the CC *see* [36]). Another HS termed the HS4, which is now extinct, was similarly created by intercrossing only four inbred strains: C57BL/6J, DBA/2J, BALB/cJ, and LP/J [35]. Using the partially inbred founders of the CC, another HS was created at Jackson Labs and is termed the diversity outbred (DO; [37]). The DO are currently being maintained using 175 breeding pairs [7], a much larger number than used for other HS [38], which should significantly reduce inbreeding. The DO have been used for a wide variety of GWAS and related projects (e.g., [7, 39–44]). Both the HS-CC and DO are made up of the same eight inbred strains. That said, the HS-CC were started frm the original inbred strains whereas the DO were started from the partially inbred CC lines. For more information about the similarities and differences between the HS-CC and the DO, *see* [45].

2.2 HS Rats

The N/NIH HS is the only rat HS population that we are aware of; it was first established at the National Institute of Health (NIH) in 1984 [46] and was derived from the following eight inbred progenitor (founder) strains: ACI/N, BN/SsN, BUF/N, F344/N, M520/N, MR/N, WKY/N, and WN/N. The HS rat colony was maintained using a rotational breeding strategy and 60 breeder pairs by Dr. Carl Hansen at NIH until 2003. From 2003 to 2006, the colony was transferred to the laboratory of Dr. Eva Redei at Northwestern University and the number of breeder pairs decreased to 25. In 2006, the HS colony was transferred to two locations: Dr. Solberg Woods at the Medical College of Wisconsin and Dr. Alberto Fernando Teruel at the Autonomous University of Barcelona in Spain. At that time, the colony had been through 55 generations of breeding (50 at the NIH and 5 at Northwestern University). At MCW the number of breeding pairs was increased to 46 and the colony was maintained using a rotational breeding strategy [1, 23]. The HS animals at the Medical College of Wisconsin were named NMcwi:HS (Rat Genome Database identification number: 2314009). In 2013, after 15 generations of breeding at MCW (70 total), the colony was expanded to include 64 breeder pairs per generation and have been maintained in this way since that time. At that time a new breeding strategy that uses kinship coefficients was employed (https://github.com/pcarbo/breedail). As of early 2018, the colony had been through 81 generations of breeding. This HS rat colony is currently a national resource funded through a NIDA Center of Excellence for genome-wide association studies (GWAS) in outbred rats (P50DA037844) which seeks to identify genetic loci underlying drug abuse behaviors (*see* www.ratgenes.org). Dr. Solberg Woods' laboratory currently ships rats to several investigators throughout the United States, on a cost-recovery basis (enquiries to: lsolberg@wakehealth.edu). As of this writing, the HS colony is being maintained at both the Medical College of Wisconsin (MCW) and Wake Forest School of Medicine (WFSM); however, the MCW colony will be phased out by the end of 2019.

HS rats have also been used for GWAS. They were first used to fine-map a previously identified locus for glucose tolerance [47, 48], which led to identification of *Tpcn2* as the likely underlying causal gene [49]. They were subsequently used to map multiple behavioral and physiological traits [50]. Since then they have been used for genetic mapping of adiposity [6] and studies show that the HS rat will be a promising model for mapping kidney-related traits [51], bone fragility [52], drug abuse behavior [53–55], depression-like behavior [10], as well as behavioral and physiological responses to stress [56–58] and ethanol [59–61].

2.3 Breeding Considerations

In many cases, HS mice or rats can be obtained from an existing colony for much less than the cost of maintaining an independent colony. When maintaining a colony, both the breeding methods (i.e., circular vs. random) and the number of families per generation are critical parameters. Rockman and Kruglyak [62] conducted a series of simulation studies that compare several outbred designs. They concluded that randomized breeding, using equal contributions from each breeder pair for the next generation (one male and one female from each breeder pair, no sib matings), is one of the most effective designs. Random breeding results in an expanded genetic map, decreased bin sizes between recombination breakpoints and controls genetic drift. An added advantage is that it is relatively easy to implement. Both inbreeding and genetic drift are also dependent on population size, with smaller populations more vulnerable to genetic drift [63]. In this regard, it is best to maintain the largest colony that is economically feasible. Rockman and Kruglyak [62] found that 64 breeder pairs increased map expansion, decreased bin size, and controlled genetic drift relative to a population of only 16 breeder pairs. In general, more breeder pairs are always better for achieving these goals. Dr. Palmer's lab has created software that implements a more sophisticated breeding scheme that uses the pedigree to minimize relatedness across all pairs (https://github.com/pcarbo/breedail).

3 Phenotyping

We will not discuss any particular phenotype as these are beyond the scope of this chapter. Phenotypic traits studied in HS mice or rats, however, should be amenable to high-throughput phenotyping, as large numbers of animals are needed to achieve sufficient statistical power (*see* Subheading 3.3 below). Because each HS animal is genetically and phenotypically unique, it is not possible to have biological replicates when working with HS populations. For this reason, it is important to use phenotypes that are robust, stable, and replicable—heritable traits should fulfill these criteria. Due to the cost of genotyping, it is advantageous to phenotype multiple traits in each animal. Examples of studies with multiple high-throughput phenotypes include Valdar et al. [14], Baud et al. [50], and Gonzales et al. [25].

3.1 Tracking Animals and Covariates

Record keeping is critical to all scientific work, and is especially so for a large GWAS with HS. Each animal must be assigned a unique ID. At the time of weaning, all animals should receive some indelible identifier, typically an ear, neck tag or a transponder ID chip, secondary identifiers, like ear punches can also be helpful. Pertinent information that should be recorded includes sex, coat color (convenient for identifying the animal while alive and for checking genetically predicted coat color versus recorded coat color, which can identify sample mix-ups), date of birth, date of death, and the ID of the mother and the father. Phenotypic data that are collected should always refer to the individual by their unique ID; abbreviating this ID is sometimes tempting, but always unwise. Covariates collected at the time of phenotyping such as batch, experimenter, and time of day should also be carefully recorded; a list of possible confounding environmental covariates has been presented by [64]. As discussed in Subheading 5, covariates can then be regressed out in advance or included in the statistical model.

3.2 Data Checking

Once all phenotype data have been collected, it is important to remove erroneous data and to locate missing data; in practice, identifying outliers is a good first step. Once outliers are identified, it can be determined whether the outlier is legitimate or represents some type of error. Often batches of data (e.g., data collected on a given day) can be compared to data collected in all other batches, to determine whether batch-wide errors have occurred.

3.3 Statistical Power

The number of animals to be used depends greatly on the heritability of the trait, which can be estimated in advance. However, power also depends on the effect sizes of the QTLs, which are not knowable in advance. Power also depends on the degree of recombination among the HS founders, with greater recombination increasing mapping precision but also decreasing statistical power, thus increasing the sample size required (*see* [4]). Previous power calculations have been run in multi-founder populations and suggest that 1000–1500 HS animals provide sufficient power for mapping QTL explaining 5% of the variance [65–67]. These simulations do not account for the confounding effects of relatedness (e.g., [22, 68]), or marker ascertainment (e.g., [69]). That said, previous studies using >1000 rats or mice have had high degrees of success in mapping many traits [14, 50]. Indeed, we have had success mapping adiposity traits with only 742 HS rats, although the QTL identified just reached the significance level despite explaining >5% of the variance [6]. Expression QTL (eQTL) mapping generally requires far fewer animals because eQTL explain a greater percentage of the variance.

4 Genotyping

A critical consideration for any quantitative trait locus (QTL) mapping study is the genotyping platform. The number of genetic markers needed is a function of the recombination in the HS population; in general, more generations of breeding will require more markers. Prior studies have used anywhere between a few thousand to a few million markers. If marker density is not sufficient, QTLs may not be detected. For example, in a study conducted a decade ago, we used a single nucleotide polymorphism (SNP) genotyping microarray that contained ~4,500 informative markers to study a 34th generation AIL [22]; subsequent reanalysis with a denser set of markers showed that some apparently true loci were not discovered due to inadequate coverage (https://www.biorxiv.org/content/10.1101/387613v2). Similarly, in one of our recent HS rat studies, we used a 10,000 SNP array, containing 8218 informative markers in the HS rat population [6]. Although successful for genetic mapping, we found that this relatively low number led to uncertainty when we tried to impute founder haplotypes (*see* Subheading 5 below).

Many recent mouse studies have used the Mouse Universal Genotyping Array (MUGA) which contains about 8000 SNPs [7], the MegaMUGA which contains about 78,000 SNPs and the more recent GigaMUGA, which contains about 143,000 SNPs; these arrays are available from Neogen (http://genomics.neogen.com/en/mouse-universal-genotyping-array). Previous studies in the HS rat have used an 800K array [50]; however this array is no longer commercially available.

Microarrays are typically designed with a particular population in mind, which means they may not provide satisfactory coverage for other populations. For example, the above mentioned MUGA arrays, despite their name, were predominantly designed for the DO mouse and may be less informative in other populations (e.g., *see* [70, 71]). Designing a new array incurs a significant cost. As the price of next generation sequencing has decreased, strategies have been developed to obtain genotypes via sequencing rather than from microarrays. The two most prominent sequencing-based strategies are genotyping-by-sequencing and low-coverage whole-genome sequencing.

For low-coverage whole-genome sequencing, the entire genome of each rat or mouse is sequenced at very low coverage (~0.2X). An imputation algorithm is then used to call genotypes. Davies et al. [72] recently developed a genotype imputation algorithm, Sequencing to Imputation Through Constructing Haplotypes (STITCH), that is tailored to low-coverage sequence data for which no reference haplotypes are available. This method has recently been applied to commercially available outbred mice

[70]. In the case of HS, for which reference haplotypes are available, conventional imputation software such as Beagle [73] may perform equally well [72].

Genotype-by-sequencing (GBS), which was originally developed for use in plants, has been adapted for use in both mice and rats [25, 71, 74]. In this method, the DNA is cut using one or more restriction enzymes, ligated to bar-coded adapters, pooled, and sequenced. As the cost of sequencing continues to decrease, these methods may supplant the use of microarrays; however, it is important to note that the bioinformatic analysis required for these sequencing-based methods is more burdensome than for microarrays.

Genotype data can be used to perform several important quality checks. Markers on the X chromosome should be homozygous (actually hemizygous) in males, the presence of too many heterozygous genotypes in a male, or too few in a female, suggests that the sex was recorded incorrectly or that samples were inadvertently switched. Such errors should be corrected or the affected samples should be excluded. When sequencing data are available it is also possible to compute the fraction of reads that map to the X chromosome; females are expected to have twice as many such reads. As with genotype information, deviations from this expectation indicate some sort of error. Genotype information can also be used to determine whether the genotypes are consistent with the recorded pedigree; this can be done in many ways, including by performing "Mendelization checks" or by comparing identical by descent statistics to kinship as calculated from the pedigree. Finally, the genotype at known coat color alleles can be compared to the recorded coat color; deviations from expectation may indicate genotyping error at that locus, errors in the recording of coat color, or sample mix-ups.

5 Statistical Analysis

Once phenotype and genotype data have been quality checked, the data are statistically analyzed. Association analyses can be conducted on SNP genotypes or on ancestral haplotypes, as described below.

GWAS is widely used in human genetics and is a general term for many similar methods that seek to estimate the association between a genetic marker (typically a SNP) and a phenotype of interest. Rodent populations have relatively small effective population sizes, meaning that there will be variable degrees of relatedness among subjects (siblings, cousins, etc.), which must be accounted for when performing a GWAS [22, 68, 75–78]. The most commonly used approach to account for relatedness is to employ a linear mixed model (LMM *see* [6]). Various software packages are available for this purpose (reviewed in [78]) that can include covariates such as experimenter, time of day, and sex. The only LMM software packages specifically targeted at model organisms are QTLRel [7, 75], DOQTL [79], and R/qtl2 [80], all of which have been used to analyze multi-founder populations like the HS. In addition, GEMMA, while originally designed for use in human GWAS, has been widely used for both human and rodent GWAS (e.g., [71, 81]). Non-normally distributed phenotypes may require transformations as with other statistical tests that assume normality. Web-based tools are also available to allow dynamic analyses with historical or user-supplied datasets (*see* www.genenetwork.org).

An advantage of HS populations is the ability to impute founder haplotypes. Imputation is commonly used in human genetics to obtain genotypes at markers that are not directly genotyped. However because an HS has a finite number of haplotypes and because parental lines have been sequenced, ancestral haplotype probabilities can be used for genetic mapping, potentially providing information beyond that obtained using only SNP genotypes [66]. There are several methods for haplotype imputation, including HAPPY, originally developed by Mott and colleagues [66], DOQTL, developed by Gatti and Churchill [79] or QTL2Geno, developed by Broman [80, 82]. All methods employ a Hidden Markov model to determine the probability of each possible haplotype across the genome. To date, a comparison of the three methods has not been published. Once genetic loci are identified, founder haplotype effects can be determined using tools within DOQTL [79] or using the Diploeffect model (*see* [6, 83]) or simply by comparing strain distribution patterns for the lead SNP and adjacent SNPs.

5.1 Significance Thresholds

In human GWAS, 5×10^{-8} is an almost universally accepted threshold for significance [84]. In HS rodent populations, where the degree of linkage disequilibrium is highly variable [4], it is more common to use empirically derived thresholds. Genome-wide significance thresholds can be determined using Bonferroni thresholds (which are highly conservative), estimated by parametric bootstrap samples from the fitted null [47, 68], or estimated using permutation, if and only if an LMM was used to account for relatedness [76]. There are also more recently proposed methods (e.g., multiTrans), and our experience suggests that thresholds obtained using multiTrans are very similar to those obtained using permutation (unpublished observation).

5.2 Confidence Intervals

To determine confidence intervals, LD intervals for the detected QTL can be defined by including neighboring markers that meet a set level of LD, measured with the squared correlation coefficient r^2 as previously described [6]. An r^2 threshold of 0.4 is widely used in human genetics.

6 Post-GWAS: How to Find Causal Genes and Variants

Genetic loci identified using HS rats or mice typically span just a few Mb or less (there are several examples of loci that are less than 1 Mb), depending on many factors including the population used, the effect size of the locus, the density of the markers, and the local LD structure. In rare cases these loci contain only one gene, however it is common to identify loci that include many genes. It is therefore important to follow up genetic mapping with additional strategies to identify the causal genes and/or variants. Non-synonymous variants within the QTL that are both highly conserved and predicted to be damaging by SIFT (http://sift.jcvi.org/) or Polyphen (http://genetics.bwh.harvard.edu/pph/) should be considered, if and only if they match the strain distribution pattern of the QTL. As demonstrated in our lab [6] and by others [50], this can be followed up by using protein modeling strategies to demonstrate a functional effect of the amino acid change. Although this strategy has been successful, human GWAS suggests that most causal variants are likely to be regulatory instead of coding. Although there are still challenges to identify the causal regulatory variant, RNA expression, eQTL mapping (e.g., [49, 71]), co-localization [85], and mediation analysis [6] can be used to identify candidate genes that underlie these loci. When combined, we have demonstrated that these strategies allow identification of multiple candidate genes underlying a single locus [6]. Merge analysis [86] has also been used to narrow the number of potentially causal variants within a QTL, thus further refining candidate genes within a locus.

7 Conclusions

HS populations allow genetic fine-mapping of complex disease traits to relatively small intervals of the genome. Due to genetic and phenotypic diversity, these model systems are useful for mapping multiple behavioral and physiological traits. We have outlined the general methods used for genetic mapping using outbred rodent resources, outlining considerations for phenotyping, genotyping, and statistical analysis.

Acknowledgments

L.S.W. and A.A.P. were both supported by P50DA037844. L.S.W. is also supported by R01 DK106386.

References

1. Peirce JL, Lu L, Gu J, Silver LM, Williams RW (2004) A new set of BXD recombinant inbred lines from advanced intercross populations in mice. BMC Genet 5:7. https://doi.org/10.1186/1471-2156-5-7
2. Churchill GA, Airey DC, Allayee H, Angel JM, Attie AD, Beatty J et al (2004) The Collaborative Cross, a community resource for the genetic analysis of complex traits. Nat Genet 36(11):1133–1137. https://doi.org/10.1038/ng1104-1133
3. Mott R, Flint J (2013) Dissecting quantitative traits in mice. Annu Rev Genomics Hum Genet 14:421–439. https://doi.org/10.1146/annurev-genom-091212-153419
4. Parker CC, Palmer AA (2011) Dark matter: are mice the solution to missing heritability? Front Genet 2:32. https://doi.org/10.3389/fgene.2011.00032
5. Solberg Woods LC (2014) QTL mapping in outbred populations: successes and challenges. Physiol Genomics 46(3):81–90. https://doi.org/10.1152/physiolgenomics.00127.2013
6. Keele GR, Prokop JW, He H, Holl K, Littrell J, Deal A et al (2018) Genetic fine-mapping and identification of candidate genes and variants for adiposity traits in outbred rats. Obesity (Silver Spring) 26(1):213–222. https://doi.org/10.1002/oby.22075
7. Svenson KL, Gatti DM, Valdar W, Welsh CE, Cheng R, Chesler EJ et al (2012) High-resolution genetic mapping using the Mouse Diversity outbred population. Genetics 190(2):437–447. https://doi.org/10.1534/genetics.111.132597
8. Sittig LJ, Carbonetto P, Engel KA, Krauss KS, Barrios-Camacho CM, Palmer AA (2016) Genetic background limits generalizability of genotype-phenotype relationships. Neuron 91(6):1253–1259. https://doi.org/10.1016/j.neuron.2016.08.013
9. Parker CC, Chen H, Flagel SB, Geurts AM, Richards JB, Robinson TE et al (2014) Rats are the smart choice: rationale for a renewed focus on rats in behavioral genetics. Neuropharmacology 76(Pt B):250–258. https://doi.org/10.1016/j.neuropharm.2013.05.047
10. Holl K, He H, Wedemeyer M, Clopton L, Wert S, Meckes JK et al (2018) Heterogeneous stock rats: a model to study the genetics of despair-like behavior in adolescence. Genes Brain Behav 17(2):139–148. https://doi.org/10.1111/gbb.12410
11. McClearn GE, Wilson JR, Meredith W (1970) The use of isogenic and heterogenic mouse stocks in behavioral research. In: Lindzey G, Thiessen D (eds) Contributions to behavior-genetic analysis: the mouse as a prototype. Appleton Century Crofts, New York, pp 3–22
12. Demarest K, Koyner J, McCaughran J Jr, Cipp L, Hitzemann R (2001) Further characterization and high-resolution mapping of quantitative trait loci for ethanol-induced locomotor activity. Behav Genet 31(1):79–91
13. Talbot CJ, Nicod A, Cherny SS, Fulker DW, Collins AC, Flint J (1999) High-resolution mapping of quantitative trait loci in outbred mice. Nat Genet 21(3):305–308
14. Valdar W, Solberg LC, Gauguier D, Burnett S, Klenerman P, Cookson WO et al (2006) Genome-wide genetic association of complex traits in heterogeneous stock mice. Nat Genet 38(8):879–887
15. Talbot CJ, Radcliffe RA, Fullerton J, Hitzemann R, Wehner JM, Flint J (2003) Fine scale mapping of a genetic locus for conditioned fear. Mamm Genome 14(4):223–230
16. Hitzemann R, Edmunds S, Wu W, Malmanger B, Walter N, Belknap J et al (2009) Detection of reciprocal quantitative trait loci for acute ethanol withdrawal and ethanol consumption in heterogeneous stock mice. Psychopharmacology 203(4):713–722. https://doi.org/10.1007/s00213-008-1418-y
17. Ahlqvist E, Ekman D, Lindvall T, Popovic M, Forster M, Hultqvist M et al (2011) High-resolution mapping of a complex disease, a model for rheumatoid arthritis, using heterogeneous stock mice. Hum Mol Genet 20(15):3031–3041. https://doi.org/10.1093/hmg/ddr206
18. Darvasi A, Soller M (1995) Advanced intercross lines, an experimental population for fine genetic mapping. Genetics 141(3):1199–1207
19. Parker CC, Cheng R, Sokoloff G, Palmer AA (2012) Genome-wide association for methamphetamine sensitivity in an advanced intercross mouse line. Genes Brain Behav 11(1):52–61. https://doi.org/10.1111/j.1601-183X.2011.00747.x
20. Peirce JL, Broman KW, Lu L, Chesler EJ, Zhou G, Airey DC et al (2008) Genome Reshuffling for Advanced Intercross Permutation (GRAIP): simulation and permutation for advanced intercross population analysis. PLoS One 3(4):e1977. https://doi.org/10.1371/journal.pone.0001977

21. Carroll AM, Cheng R, Collie-Duguid ES, Meharg C, Scholz ME, Fiering S et al (2017) Fine-mapping of genes determining extrafusal fiber properties in murine soleus muscle. Physiol Genomics 49(3):141–150. https://doi.org/10.1152/physiolgenomics.00092.2016
22. Cheng R, Lim JE, Samocha KE, Sokoloff G, Abney M, Skol AD et al (2010) Genome-wide association studies and the problem of relatedness among advanced intercross lines and other highly recombinant populations. Genetics 185 (3):1033–1044. https://doi.org/10.1534/genetics.110.116863
23. Cheverud JM, Lawson HA, Bouckaert K, Kossenkov AV, Showe LC, Cort L et al (2014) Fine-mapping quantitative trait loci affecting murine external ear tissue regeneration in the LG/J by SM/J advanced intercross line. Heredity (Edinb) 112(5):508–518. https://doi.org/10.1038/hdy.2013.133
24. Ehrich TH, Hrbek T, Kenney-Hunt JP, Pletscher LS, Wang B, Semenkovich CF et al (2005) Fine-mapping gene-by-diet interactions on chromosome 13 in a LG/J x SM/J murine model of obesity. Diabetes 54 (6):1863–1872
25. Gonzales NM, Seo J, Hernandez-Cordero AI, St. Pierre CL, Gregory JS, Distler MG et al (2018) Genome wide association study of behavioral, physiological and gene expression traits in a multigenerational mouse intercross. BioRxiv. https://doi.org/10.1101/230920
26. Hernandez Cordero AI, Carbonetto P, Riboni Verri G, Gregory JS, Vandenbergh DJ, Gyekis JP et al (2018) Replication and discovery of musculoskeletal QTLs in LG/J and SM/J advanced intercross lines. Phys Rep 6(4). https://doi.org/10.14814/phy2.13561
27. Lawson HA, Lee A, Fawcett GL, Wang B, Pletscher LS, Maxwell TJ et al (2011) The importance of context to the genetic architecture of diabetes-related traits is revealed in a genome-wide scan of a LG/J x SM/J murine model. Mamm Genome 22(3–4):197–208. https://doi.org/10.1007/s00335-010-9313-3
28. Lawson HA, Zelle KM, Fawcett GL, Wang B, Pletscher LS, Maxwell TJ et al (2010) Genetic, epigenetic, and gene-by-diet interaction effects underlie variation in serum lipids in a LG/JxSM/J murine model. J Lipid Res 51 (10):2976–2984. https://doi.org/10.1194/jlr.M006957
29. Norgard EA, Lawson HA, Pletscher LS, Wang B, Brooks VR, Wolf JB et al (2011) Genetic factors and diet affect long-bone length in the F34 LG,SM advanced intercross. Mamm Genome 22(3–4):178–196. https://doi.org/10.1007/s00335-010-9311-5
30. Parker CC, Carbonetto P, Sokoloff G, Park YJ, Abney M, Palmer AA (2014) High-resolution genetic mapping of complex traits from a combined analysis of F2 and advanced intercross mice. Genetics 198(1):103–116. https://doi.org/10.1534/genetics.114.167056
31. Parker CC, Cheng R, Sokoloff G, Lim JE, Skol AD, Abney M et al (2011) Fine-mapping alleles for body weight in LG/J x SM/J F(2) and F(34) advanced intercross lines. Mamm Genome 22(9–10):563–571. https://doi.org/10.1007/s00335-011-9349-z
32. Parker CC, Sokoloff G, Cheng R, Palmer AA (2012) Genome-wide association for fear conditioning in an advanced intercross mouse line. Behav Genet 42(3):437–448. https://doi.org/10.1007/s10519-011-9524-8
33. Rai MF, Schmidt EJ, Hashimoto S, Cheverud JM, Sandell LJ (2015) Genetic loci that regulate ectopic calcification in response to knee trauma in LG/J by SM/J advanced intercross mice. J Orthop Res 33(10):1412–1423. https://doi.org/10.1002/jor.22944
34. Samocha KE, Lim JE, Cheng R, Sokoloff G, Palmer AA (2010) Fine mapping of QTL for prepulse inhibition in LG/J and SM/J mice using F(2) and advanced intercross lines. Genes Brain Behav 9(7):759–767. https://doi.org/10.1111/j.1601-183X.2010.00613.x
35. Iancu OD, Darakjian P, Walter NA, Malmanger B, Oberbeck D, Belknap J et al (2010) Genetic diversity and striatal gene networks: focus on the heterogeneous stock-collaborative cross (HS-CC) mouse. BMC Genomics 11:585. https://doi.org/10.1186/1471-2164-11-585
36. Consortium CC (2012) The genome architecture of the Collaborative Cross mouse genetic reference population. Genetics 190 (2):389–401. https://doi.org/10.1534/genetics.111.132639
37. Matsumoto Y, Goto T, Nishino J, Nakaoka H, Tanave A, Takano-Shimizu T et al (2017) Selective breeding and selection mapping using a novel wild-derived heterogeneous stock of mice revealed two closely-linked loci for tameness. Sci Rep 7(1):4607. https://doi.org/10.1038/s41598-017-04869-1
38. Yalcin B, Flint J (2012) Association studies in outbred mice in a new era of full-genome sequencing. Mamm Genome 23 (9–10):719–726. https://doi.org/10.1007/s00335-012-9409-z

39. Gatti D, French JE, Schughart K (2017) QTL Mapping and identification of candidate genes in DO mice: a use case model derived from a benzene toxicity experiment. Methods Mol Biol 1488:265–281. https://doi.org/10.1007/978-1-4939-6427-7_12
40. Logan RW, Robledo RF, Recla JM, Philip VM, Bubier JA, Jay JJ et al (2013) High-precision genetic mapping of behavioral traits in the diversity outbred mouse population. Genes Brain Behav 12(4):424–437. https://doi.org/10.1111/gbb.12029
41. Recla JM, Robledo RF, Gatti DM, Bult CJ, Churchill GA, Chesler EJ (2014) Precise genetic mapping and integrative bioinformatics in Diversity Outbred mice reveals Hydin as a novel pain gene. Mamm Genome 25 (5–6):211–222. https://doi.org/10.1007/s00335-014-9508-0
42. Shorter JR, Huang W, Beak JY, Hua K, Gatti DM, de Villena FP et al (2018) Quantitative trait mapping in Diversity Outbred mice identifies two genomic regions associated with heart size. Mamm Genome 29(1–2):80–89. https://doi.org/10.1007/s00335-017-9730-7
43. Smallwood TL, Gatti DM, Quizon P, Weinstock GM, Jung KC, Zhao L et al (2014) High-resolution genetic mapping in the diversity outbred mouse population identifies Apobec1 as a candidate gene for atherosclerosis. G3 4(12):2353–2363. https://doi.org/10.1534/g3.114.014704
44. Winter JM, Gildea DE, Andreas JP, Gatti DM, Williams KA, Lee M et al (2017) Mapping complex traits in a diversity outbred F1 mouse population identifies germline modifiers of metastasis in human prostate cancer. Cell Syst 4(1):31–45 e36. https://doi.org/10.1016/j.cels.2016.10.018
45. Chesler EJ, Gatti DM, Morgan AP, Strobel M, Trepanier L, Oberbeck D et al (2016) Diversity Outbred mice at 21: maintaining allelic variation in the face of selection. G3 6 (12):3893–3902. https://doi.org/10.1534/g3.116.035527
46. Hansen C, Spuhler K (1984) Development of the National Institutes of Health genetically heterogeneous rat stock. Alcohol Clin Exp Res 8(5):477–479
47. Solberg Woods LC, Holl K, Tschannen M, Valdar W (2010) Fine-mapping a locus for glucose tolerance using heterogeneous stock rats. Physiol Genomics 41(1):102–108. https://doi.org/10.1152/physiolgenomics.00178.2009
48. Solberg Woods LC, Holl KL, Oreper D, Xie Y, Tsaih SW, Valdar W (2012) Fine-mapping diabetes-related traits, including insulin resistance, in heterogeneous stock rats. Physiol Genomics 44(21):1013–1026. https://doi.org/10.1152/physiolgenomics.00040.2012
49. Tsaih SW, Holl K, Jia S, Kaldunski M, Tschannen M, He H et al (2014) Identification of a novel gene for diabetic traits in rats, mice, and humans. Genetics 198(1):17–29. https://doi.org/10.1534/genetics.114.162982
50. Baud A, Hermsen R, Guryev V, Stridh P, Graham D, McBride MW et al (2013) Combined sequence-based and genetic mapping analysis of complex traits in outbred rats. Nat Genet 45(7):767–775. https://doi.org/10.1038/ng.2644
51. Solberg Woods LC, Stelloh C, Regner KR, Schwabe T, Eisenhauer J, Garrett MR (2010) Heterogeneous stock rats: a new model to study the genetics of renal phenotypes. Am J Physiol Ren Physiol 298(6):F1484–F1491. https://doi.org/10.1152/ajprenal.00002.2010
52. Alam I, Koller DL, Sun Q, Roeder RK, Canete T, Blazquez G et al (2011) Heterogeneous stock rat: a unique animal model for mapping genes influencing bone fragility. Bone 48(5):1169–1177. https://doi.org/10.1016/j.bone.2011.02.009
53. King CP, Palmer AA, Woods LC, Hawk LW, Richards JB, Meyer PJ (2016) Premature responding is associated with approach to a food cue in male and female heterogeneous stock rats. Psychopharmacology 233 (13):2593–2605. https://doi.org/10.1007/s00213-016-4306-x
54. Richards JB, Lloyd DR, Kuehlewind B, Militello L, Paredez M, Solberg Woods L et al (2013) Strong genetic influences on measures of behavioral-regulation among inbred rat strains. Genes Brain Behav 12(5):490–502. https://doi.org/10.1111/gbb.12050
55. Wang T, Han W, Wang B, Jiang Q, Solberg-Woods LC, Palmer AA et al (2014) Propensity for social interaction predicts nicotine-reinforced behaviors in outbred rats. Genes Brain Behav 13(2):202–212. https://doi.org/10.1111/gbb.12112
56. Diaz-Moran S, Palencia M, Mont-Cardona C, Canete T, Blazquez G, Martinez-Membrives E et al (2012) Coping style and stress hormone responses in genetically heterogeneous rats: comparison with the Roman rat strains. Behav Brain Res 228(1):203–210. https://doi.org/10.1016/j.bbr.2011.12.002
57. Lopez-Aumatell R, Guitart-Masip M, Vicens-Costa E, Gimenez-Llort L, Valdar W, Johannesson M et al (2008) Fearfulness in a large N/Nih genetically heterogeneous rat stock:

differential profiles of timidity and defensive flight in males and females. Behav Brain Res 188(1):41–55. https://doi.org/10.1016/j.bbr.2007.10.015

58. Lopez-Aumatell R, Vicens-Costa E, Guitart-Masip M, Martinez-Membrives E, Valdar W, Johannesson M et al (2009) Unlearned anxiety predicts learned fear: a comparison among heterogeneous rats and the Roman rat strains. Behav Brain Res 202(1):92–101. https://doi.org/10.1016/j.bbr.2009.03.024
59. Bice PJ, Liang T, Zhang L, Graves TJ, Carr LG, Lai D et al (2010) Fine mapping and expression of candidate genes within the chromosome 10 QTL region of the high and low alcohol-drinking rats. Alcohol 44 (6):477–485. https://doi.org/10.1016/j.alcohol.2010.06.004
60. Foroud T, Bice P, Castelluccio P, Bo R, Miller L, Ritchotte A et al (2000) Identification of quantitative trait loci influencing alcohol consumption in the high alcohol drinking and low alcohol drinking rat lines. Behav Genet 30 (2):131–140
61. Spuhler K, Deitrich RA (1984) Correlative analysis of ethanol-related phenotypes in rat inbred strains. Alcohol Clin Exp Res 8 (5):480–484
62. Rockman MV, Kruglyak L (2008) Breeding designs for recombinant inbred advanced intercross lines. Genetics 179(2):1069–1078. https://doi.org/10.1534/genetics.107.083873
63. Falconer DS (1960) Introduction to quantitative genetics. Ronald Press, New York
64. Wurbel H (2002) Behavioral phenotyping enhanced—beyond (environmental) standardization. Genes Brain Behav 1(1):3–8
65. Mott R, Flint J (2002) Simultaneous detection and fine mapping of quantitative trait loci in mice using heterogeneous stocks. Genetics 160 (4):1609–1618
66. Mott R, Talbot CJ, Turri MG, Collins AC, Flint J (2000) A method for fine mapping quantitative trait loci in outbred animal stocks. Proc Natl Acad Sci U S A 97 (23):12649–12654
67. Valdar WS, Flint J, Mott R (2003) QTL fine-mapping with recombinant-inbred heterogeneous stocks and in vitro heterogeneous stocks. Mamm Genome 14(12):830–838. https://doi.org/10.1007/s00335-003-3021-1
68. Valdar W, Holmes CC, Mott R, Flint J (2009) Mapping in structured populations by resample model averaging. Genetics 182 (4):1263–1277. https://doi.org/10.1534/genetics.109.100727
69. Valdar W, Flint J, Mott R (2006) Simulating the collaborative cross: power of quantitative trait loci detection and mapping resolution in large sets of recombinant inbred strains of mice. Genetics 172(3):1783–1797. https://doi.org/10.1534/genetics.104.039313
70. Nicod J, Davies RW, Cai N, Hassett C, Goodstadt L, Cosgrove C et al (2016) Genome-wide association of multiple complex traits in outbred mice by ultra-low-coverage sequencing. Nat Genet 48(8):912–918. https://doi.org/10.1038/ng.3595
71. Parker CC, Gopalakrishnan S, Carbonetto P, Gonzales NM, Leung E, Park YJ et al (2016) Genome-wide association study of behavioral, physiological and gene expression traits in outbred CFW mice. Nat Genet 48 (8):919–926. https://doi.org/10.1038/ng.3609
72. Davies RW, Flint J, Myers S, Mott R (2016) Rapid genotype imputation from sequence without reference panels. Nat Genet 48 (8):965–969. https://doi.org/10.1038/ng.3594
73. Browning BL, Browning SR (2016) Genotype imputation with millions of reference samples. Am J Hum Genet 98(1):116–126. https://doi.org/10.1016/j.ajhg.2015.11.020
74. Fitzpatrick CJ, Gopalakrishnan S, Cogan ES, Yager LM, Meyer PJ, Lovic V et al (2013) Variation in the form of Pavlovian conditioned approach behavior among outbred male Sprague-Dawley rats from different vendors and colonies: sign-tracking vs. goal-tracking. PLoS One 8(10):e75042. https://doi.org/10.1371/journal.pone.0075042
75. Cheng R, Abney M, Palmer AA, Skol AD (2011) QTLRel: an R package for genome-wide association studies in which relatedness is a concern. BMC Genet 12:66. https://doi.org/10.1186/1471-2156-12-66
76. Cheng R, Palmer AA (2013) A simulation study of permutation, bootstrap, and gene dropping for assessing statistical significance in the case of unequal relatedness. Genetics 193(3):1015–1018. https://doi.org/10.1534/genetics.112.146332
77. Cheng R, Parker CC, Abney M, Palmer AA (2013) Practical considerations regarding the use of genotype and pedigree data to model relatedness in the context of genome-wide association studies. G3 3(10):1861–1867. https://doi.org/10.1534/g3.113.007948
78. Gonzales NM, Palmer AA (2014) Fine-mapping QTLs in advanced intercross lines and other outbred populations. Mamm Genome 25(7–8):271–292. https://doi.org/10.1007/s00335-014-9523-1

79. Gatti DM, Svenson KL, Shabalin A, Wu LY, Valdar W, Simecek P et al (2014) Quantitative trait locus mapping methods for diversity outbred mice. G3 4(9):1623–1633. https://doi.org/10.1534/g3.114.013748
80. Broman KW, Gatti DM, Simecek P, Furlotte NA, Prins P, Sen Ś, Yandell BS, Churchill GA (2018) R/qtl2: software for mapping quantitative trait loci with high-dimensional data and multi-parent populations. Genetics 211:495–502. https://doi.org/10.1534/genetics.118.301595. PMID: 30591514
81. Pallares LF, Carbonetto P, Gopalakrishnan S, Parker CC, Ackert-Bicknell CL, Palmer AA et al (2015) Mapping of craniofacial traits in outbred mice identifies major developmental genes involved in shape determination. PLoS Genet 11(11):e1005607. https://doi.org/10.1371/journal.pgen.1005607
82. Broman KW (2016) qtl2geno: treatment of marker genotypes for QTL experiments. R package version 0.4-21. http://kbroman.org/qtl2
83. Zhang Z, Wang W, Valdar W (2014) Bayesian modeling of haplotype effects in multiparent populations. Genetics 198(1):139–156. https://doi.org/10.1534/genetics.114.166249
84. Risch N, Merikangas K (1996) The future of genetic studies of complex human diseases. Science 273(5281):1516–1517
85. Hormozdiari F, van de Bunt M, Segre AV, Li X, Joo JWJ, Bilow M et al (2016) Colocalization of GWAS and eQTL signals detects target genes. Am J Hum Genet 99(6):1245–1260. https://doi.org/10.1016/j.ajhg.2016.10.003
86. Yalcin B, Flint J, Mott R (2005) Using progenitor strain information to identify quantitative trait nucleotides in outbred mice. Genetics 171(2):673–681

Chapter 12

Mapping Mammary Tumor Traits in the Rat

Michael J. Flister, Amit Joshi, Carmen Bergom, and Hallgeir Rui

Abstract

For nearly a century, the rat has served as a key model for studying the pathophysiology and genetic risk modifiers of breast cancer. Rat mammary tumors that initiate after exposure to carcinogens or estrogens closely resemble the etiological, histopathological, and genomic features of human breast cancer. Recent developments in genome-editing techniques in the rat have also enabled the development of sophisticated models for identifying the genetic modifiers of the nonmalignant tumor microenvironment that contribute to the formation, progression, and outcome of breast cancer. In this protocol review, we discuss the current methodologies for the three genetic mapping techniques in the rat that are widely used for identifying and testing the heritable genetic modifiers of breast cancer.

Key words Rat, Breast cancer, Genetics, Tumor microenvironment

1 Introduction

Each year, 1.7 million cases of breast cancer are diagnosed worldwide and nearly 600,000 women die from the disease [1–3]. Approximately 30% of breast cancer risk is heritable [4], of which less than 10% is attributable to relatively rare alleles of moderate and high penetrance (e.g., BRCA1, BRCA2, PTEN, TP53, ATM, BRIP1, CHEK2, and PALB2) [5, 6]. To date, a much larger group of more than 170 candidate loci have been identified by genome-wide association studies (GWAS) [7], which confer lower relative risks (RR) of breast cancer (<1.5-fold RR) compared with the more penetrant risk modifiers listed above [5, 6]. Although some rare yet highly penetrant alleles convey a very high relative risk (>5-fold RR) to a small pool of affected individuals, the overall population risk of breast cancer attributed to these high-risk alleles is far less than that attributed to common risk alleles with lower relative risk [8]. Thus, defining the common genetic modifiers of breast cancer is likely to have a major impact on our understanding of the heritability of breast cancer.

G. Thomas Hayman et al. (eds.), *Rat Genomics*, Methods in Molecular Biology, vol. 2018,
https://doi.org/10.1007/978-1-4939-9581-3_12,

Despite what is known about the heritability of breast cancer, only a small fraction of breast cancer risk modifiers have been fully characterized [9] and this is potentially due to the diverse nature and complexity of common breast cancer risk alleles with relatively weak phenotypic effects. Unlike the highly/moderately penetrant modifier alleles that are typically linked with the malignant transformation of breast epithelial cells through disruption of pathways regulating DNA damage repair, cell cycle, and apoptosis, GWAS candidate genes span a diverse range of molecular and cellular pathways [10]. Additionally, the majority of risk loci contain multiple candidate genes in linkage disequilibrium (LD), which are "co-inherited" and therefore considered equally plausible candidates, until proven otherwise [11]. This is further complicated by the evidence that a single underlying genetic cause of breast cancer in most cases cannot be found, but rather multiple genetic and environmental factors likely contribute to overall susceptibility to breast cancer. Collectively, these challenges combined with complex genetic interactions in both the pre-malignant epithelial cells and the emerging nonmalignant tumor microenvironment (TME) pose significant challenges in defining the contributions of the many genetic variants that likely contribute to breast cancer risk.

One strategy to characterize modifiers of heritable breast cancer risk is by genetic mapping using rat strains with varying susceptibility to the formation of mammary tumors. More than ten commonly used inbred rat strains have been characterized for their susceptibility or resistance to mammary cancer formation, and heritable susceptibility to mammary tumor formation in these rat strains is polygenic, which resembles the risk conferred by common alleles that are linked to human breast cancer [8, 12]. To date, at least 32 mammary cancer risk loci have been identified in the rat (rgd.mcw.edu) and, more importantly, the overlap of these loci with human breast cancer risk loci is significantly higher than expected by chance [13], suggesting that heritable modifiers of breast cancer are conserved between human and model species, such as the rat. In addition to conservation of risk modifiers, spontaneous rat mammary tumors share many etiological and pathophysiological characteristics with human breast cancer [8, 12]. Thus, the rat offers many advantages to understanding the complex genetic modifiers of breast cancer. Here, we discuss the basic methodology for mapping the polygenic risk factors in the rat that modify the susceptibility to breast cancer in the malignant cancer cells and in the nonmalignant TME.

2 Methodologies for Spontaneous Mammary Tumor Models in the Rat

Multiple established methods for eliciting carcinogen-induced mammary tumors in the rat exist, including administration of the carcinogens, DMBA (7,12-dimethylbenz[*a*]anthracene) [12, 14] or NMU (*N*-methylnitrosourea) [12, 15, 16], or following a continuous physiological dose of the hormone, 17β-Estradiol (E2) [8, 17]. Despite the different mechanisms by which DMBA/NMU and E2 elicit mammary tumor formation (detailed below), several common characteristics are shared between these tumor models. Most notably, all three models typically form various stages of benign hyperplastic lesions, ductal carcinoma in situ, and invasive carcinoma, and predominantly form the ER^+/PR^+ luminal A and B subtypes [8, 12]. The majority of the carcinogen- and E2-induced mammary tumors are also responsive to the anti-estrogen therapy, tamoxifen [8, 12, 18], further demonstrating the importance of ERα signaling in these models. Compared with the ~20% incidence of $ER^-/PR^-/HER2^-$ triple-negative breast cancer (TNBC) cases in human [19], it appears that far fewer TNBC tumors are initially induced by carcinogen or E2 treatment in the rat [8, 18]. Interestingly, however, we have found that serial passaging of carcinogen-induced mammary tumors or derived cell lines ultimately results in the selection of aggressive TNBC tumors. As a whole, mammary tumors induced by carcinogens or prolonged estrogen exposure rarely undergo distant metastasis, which could be due in part to the relatively shorter follow-up time of most rat experiments compared with human patients [18]. Thus, the current inducible rat mammary tumor models are likely best suited for studying the mechanisms and risk factors driving early tumor initiation, proliferation, survival, and local invasiveness at the primary site.

2.1 Carcinogen-Induced Mammary Tumor Models

DMBA and NMU are the most frequently used chemical carcinogens for inducing mammary tumors in the rat [12]. DMBA is a polynuclear aromatic hydrocarbon (PAH) that requires metabolic activation by mammary epithelium to form polar epoxides that generate random DNA-damaging adducts [12]. In contrast, NMU is a metabolically active carcinogen that forms O^6-methylguanine in the DNA and causes G-A transitions during replication [20, 21]. An activating mutation in H-Ras (G-A transition in codon 12) is characteristic of >85% of NMU-induced mammary tumors [22, 23], whereas the genomic instability induced by DMBA is less defined [24]. Despite the distinct mechanisms of action of DMBA and NMU, the other etiological characteristics and histopathological features of mammary tumors induced by either carcinogen appear to be similar. The cellular targets of DMBA and NMU within the mammary gland are predominantly the epithelial cells of the terminal end bud (TEB) and some terminal ducts (TD) that arise from the TEB [12]. The other structures arising from the TEB, namely the alveolar buds (AB) and lobules, are largely resistant to malignant transformation and instead will infrequently form benign ductal hyperplasia or tubular adenomas [12]. The number of TEB peaks at 21 days-of-age and is followed by differentiation into AB that begins from 30 to 42 days of age [12]. The incidence of mammary tumors is 100% in susceptible strains when DMBA or NMU is administered between 30 to 55 days of age, yet the greatest number of tumors per rat (i.e., multiplicity) occurs at 40 to 46 days of age, when TEB density has already reached its maximum and they are beginning to differentiate to AB [12, 25, 26]. At this stage, the TEB and the TD (to a slightly lesser extent) contain epithelium with the highest proliferative index, the shortest cell cycle, and the least cell turnover within the mammary gland [12, 27], all of which likely contribute to the high rate of malignant transformation by DMBA or NMU. As might be expected, this susceptibility decreases considerably with age and following pregnancy, when the number of highly proliferative TEB and TD are dramatically reduced [27]. In young virgin rats, the ratio of AB to TEB progressively increases with each estrous cycle and the majority of TEB differentiate to AB, and subsequently to lobules upon first pregnancy [28, 29]. Although a few remaining TEB exist in older virgin rats (>180 days-of-age), the proliferative capacity of the epithelial cells within the TEB is markedly diminished and the incidence of tumor formation is reduced [27, 29, 30].

Following exposure to DMBA or NMU, female rats will typically develop a wide spectrum of mammary neoplasia subtypes, including benign fibroadenomas (FA), intraductal carcinomas, or invasive carcinomas, and the incidence of different neoplasia varies between rat genetic backgrounds, ranging from as high as ~68% FA in the Sprague Dawley (SD) [18] to as low as ~10% FA in the SS rat [31]. Intraductal carcinomas of the rat closely resemble ductal carcinoma in situ (DCIS), which represents roughly 30% of human breast cancer cases [32]. Of the rat tumors that are classified as ER^+ invasive carcinomas, multiple subtypes with reported incidences (in a cohort of 584 rats) have been classified, including invasive papillary carcinoma (IPC) types I and II (~55%), invasive cribriform carcinoma (~27%), invasive comedo carcinoma (~1%), invasive tubular carcinoma (~7%), and adenoid cystic carcinoma (~10%) [18]. The ER^+ invasive carcinomas in 59 rats treated with DMBA were further subclassified into luminal A (~75%; ER^+/PR^+, $Her2^-$, low Ki67) and luminal B (~25%; ER^+/PR^+, $Her2^+$, high Ki67) subtypes by molecular immunophenotyping [18].

The susceptibility to carcinogen-induced mammary tumors varies between rat genetic backgrounds [33]. The SD [12, 18, 27, 34], LEW [34], SS [31, 35], NSD [12, 34], and WF [12, 34] rat strains show varying levels of susceptibility to DMBA-induced mammary tumors, whereas the BN [31, 36], WKY [37, 38], ACI [12], F344 [34], and COP [12, 34] rat strains are resistant or only moderately susceptible to DMBA-induced mammary tumors. In comparison, the SD [15, 39, 40], NSD [34], F344 [15, 41], and BUF/N [15] rat strains have been shown to be susceptible to NMU-induced mammary tumors, whereas LE [39], WKY [42], and COP [34, 42] rat strains are largely resistant. To date,

22 DMBA-induced mammary cancer QTL have been mapped via F2 intercrosses of WF × COP (Mcs1) [43], WF × COP (Mcs2–4) [44], WF × WKY (Mcs5–9 and Mcsm1) [37], and SPRD/Ztm × WKY (Mcs16–22) [38]. Congenic mapping has been used to refine multiple QTL to the candidate gene level, including Mcs1 (*Mier3*, *Nr2f1*) [33, 45, 46], Mcs3 (*Ilk*, *Il18bp*, *Emsy*, *Pak1*, *Rsf1*, *Gab2*, *Cyp2c19*) [47], Mcs2/6 (*Igf1*, *Socs2*) [48], and Mcs5 (*Fbxo10*, *Pappa*) [49, 50]. Compared with DMBA, far less is known about the genetic modifiers of NMU-induced mammary cancer, as only one QTL by a single [(F344 × Cop) F1 × F344] N2 intercross has been identified (Mcs30) [51]. However, due to evidence that F344 is only moderately susceptible to NMU-induced mammary tumors [52], it is likely that multiple other QTL exist and might be detected in a larger study using other parental genetic backgrounds with more divergent phenotypes. At least one QTL interval (Mcs5a) has been shown to modify mammary tumor risk to DMBA and NMU [49], yet it remains unclear whether other QTL overlap between the DMBA and NMU models.

2.1.1 Experimental Procedures

As mentioned above, the susceptibility of mammary tumors induced by DMBA or NMU is influenced by multiple factors, including gender, genetic background, age, and parity. To achieve optimal and reproducible results, we typically use virgin female rats that are 49–55 days of age. DMBA that has been freshly reconstituted in sesame seed oil is administered by a single oral gavage (65 mg/kg) (*see* ref. 53 for a video demonstrating proper oral gavage technique). NMU is typically reconstituted in saline and a single dose (50 mg/kg) can be administered by multiple injection routes (intravenous, intraperitoneal, or subcutaneous). However, the route of administration should be considered an experimental variable that will potentially confound tumor phenotypes when comparing different routes of NMU administration. In susceptible rat genetic backgrounds (e.g., SD and SS), mammary tumors can arise as early as several weeks after DMBA or NMU administration and typically reach 100% incidence within 200 days [34] and far sooner in highly susceptible strains (e.g., SS) [31]. Following the carcinogen exposure, tumor formation is monitored weekly by manual palpation of each mammary fat pad (MFP) by a blinded observer. In the cases of questionable diagnosis, it is typical to query a second blinded observer for confirmation of the tumor diagnosis or to wait for a second confirmation by the same blinded observer in the following week, in which case the date of initial diagnosis would be considered the tumor onset. In large genetic mapping studies, the standard weekly recorded tumor phenotypes include the anatomical location of each tumor, the overall incidence, the latency, and the multiplicity (i.e., number of tumors per rat). At experimental endpoint, mammary tumors are collected and weighed to record overall tumor burden per rat, followed by tissue processing for downstream analyses. As noted previously [12], the incidence of different benign and invasive mammary neoplastic subtypes might vary between rat genetic backgrounds, which should be taken into consideration and, if necessary, quantified histologically at experimental endpoint.

Various strategies for genetic mapping of mammary tumor risk modifiers have been used in the carcinogen-induced mammary tumor models, including linkage analysis [37, 38, 43, 44], consomic mapping [31], and congenic mapping [31, 45, 48, 54, 55]. A typical linkage analysis for susceptibility to mammary tumors induced by DMBA or NMU has consisted of ~200 female rats and ~100 genotype markers [37, 38, 43]. For genotyping, previous mapping studies have made use of microsatellite markers, such as simple sequence length polymorphisms (SSLP), which continues to serve as a cost-effective and efficient method for genome-wide mapping. Many polymorphic SSLP markers have been characterized for commonly used rat strains and are currently available through the Rat Genome Database (https://rgd.mcw.edu/rgdweb/search/markers.html?100). The statistical framework for QTL interval mapping was first established by Lander and Botstein [56] and the methodology for QTL mapping using the R/qtl statistical package has been reviewed elsewhere, and we refer readers to the authoritative work of the program's developers [57].

Following the localization of carcinogen-induced mammary tumor QTL by linkage analysis [37, 38, 43, 44], fine-mapping studies are used to positionally clone the causative gene (s) [46, 50, 54, 55] and functional testing is necessary for establishing the causative mechanism(s) [46, 50]. Positional cloning entails backcrossing one of the parental strains onto the other, followed by repeated backcrossing of the recombinant offspring that contain the QTL and ultimately brother-sister mating to generate homozygous congenic strains (*see* ref. 58 for greater detail). By comparison to the original parental strains, the congenic strain can be used to confirm the QTL position and phenotypes, and further localize the causative gene(s) by reducing the congenic interval using smaller subcongenics [59]. Ultimately, to established causality of a candidate gene, one must reduce the congenic interval until only the causative allele(s) have been isolated, which becomes progressively more challenging as the interval size decreases. Alternatively, methods for gene-editing and creating transgenics for candidate genes are now established in the rat and can be used to directly interrogate the most likely candidate(s) within the region of interest [60].

An alternative strategy for mapping the genetic modifiers of mammary tumor risk loci is to first localize QTL to a chromosome using chromosome substitution strains (i.e., consomics). The methodology for constructing consomic panels has been described elsewhere [58, 61] and currently two full consomic rat panels exist, which consist of individual BN chromosomes introgressed onto the SS or FHH genetic backgrounds [61]. Although the susceptibility of the FHH rat strain to DMBA- or NMU-induced mammary tumors is currently unknown, the divergent susceptibility to DMBA-induced mammary tumors in the SS (highly susceptible) and BN (highly resistant) is well established [36]. Moreover, the full SS.BN consomic panel has been used for mapping the susceptibility to DMBA-induced mammary tumors [31]. Compared with genome-wide linkage analysis, genetic mapping using consomic strains offers several advantages: (1) variability is reduced and phenotypes can be rapidly reproduced due to the fixed genetics of the inbred consomic strains and (2) congenic intervals can also be more rapidly isolated from consomic strains in as few as two generations, whereas isolation of a congenic interval directly from the parental background can take as many as eight generations [61]. However, once QTL have been mapped to congenic intervals, many of the same challenges exist to localize and functionally test the underlying causative allele(s).

2.2 Estrogen-Induced Mammary Tumor Models

Prolonged exposure to endogenous E2 or exogenous synthetic estrogens has long been correlated with increased risk of luminal breast cancers [8, 12, 18, 62, 63] and blockade of the E2 signaling reduces breast cancer incidence in high-risk patient cohorts [64–66]. Prolonged exposure to E2 also induces mammary tumors in some rat genetic backgrounds [17, 67, 68], which is primarily dependent on estrogen receptor-α (ERα) [69] and progesterone-mediated activation of the progesterone receptor (PR) [70, 71]. The majority of E2-induced mammary tumors are $ER\alpha^+/PR^+$ and proceed from lobuloalveolar hyperplasia (i.e., sustained luminal epithelial proliferation) to atypical focal hyperplastic lesions, ductal carcinoma in situ (DCIS), and invasive ductal carcinoma [67, 72]. Some E2 metabolites can form DNA adducts (e.g., estrogen quinones) [73], yet DNA damage through this mechanism does not appear to be a major contributor to E2-induced mammary tumor formation [74]. Nonetheless, E2-induced mammary tumors frequently display aneuploidy and chromosome instability, including multiple frequently amplified or deleted regions that are homologous to copy number variations in human breast cancer [75, 76].

The susceptibility to E2-induced mammary tumors varies between rat genetic backgrounds. The ACI [17, 67, 68] and DA rat strains [8] are susceptible to E2-induced mammary tumors, whereas the BN [77] and COP [67, 68] rat strains are resistant. Nine E2-induced mammary cancer QTL have been mapped via F2 linkage analysis of ACI × COP (Emca1–2) [78] and ACI × BN (Emca3–9) [79]. Congenic mapping has been used to confirm Emca QTL and nominate candidate genes for Emca3, Emca4, Emca5, Emca6, Emca7, and Emca9 [80]. Notably, all nine Emca

QTL overlap with human breast cancer GWAS loci [8], albeit the identity of the causative mechanisms within these loci and whether they are conserved between human and rat has yet to be definitively established.

In addition to mammary tumors, anterior pituitary lactotroph hyperplasia/adenoma [81, 82] and hyperprolactinemia [17, 83] are uniformly induced in the ACI rat strain, whereas the DA, BN, and COP rat strains are largely resistant to these traits [82, 84–88]. Prolactin (Prl) is an E2-responsive hormone that stimulates mammary ductal side branching during puberty and proliferative expansion of the secretory alveolar cells during pregnancy [89]. Prl is also critical for ERα signaling in breast cancer [90] and elevated circulating Prl levels are associated with increased risk of breast cancer in human [91, 92] and rat [93]. Chronic E2 administration is associated with prolactin-producing pituitary adenomas in the ACI rat, and two F2 linkage analyses, ACI × BN [82] and ACI × COP [87], have identified eight E2-induced pituitary tumor (Ept) QTL that largely do not overlap the Emca QTL mapped in the same strains [8]. Thus, although Prl likely contributes to E2-dependent mammary tumor formation, the genetic modifiers of the pituitary tumors and mammary tumors appear to be distinct and it has yet to be determined whether epistasis between these modifiers exists. Of note, an ACI substrain was recently developed by the Shull laboratory (ACWi) that is susceptible to the E2-induced mammary tumors exhibited by the ACI parental strain, but lacks the morbidity associated with pituitary lactotroph hyperplasia/adenoma [94]. This unique model is likely to aid in disentangling the genetic and environmental modifiers of E2-dependent mammary tumor formation in the rat, without the confounding phenotypes associated with pituitary lactotroph hyperplasia/adenoma and hyperprolactinemia.

2.2.1 Experimental Procedures

The induction of mammary tumors by prolonged exposure to estrogen is influenced by multiple factors, including the type of estrogen [17, 95], ovary function [17], gender [95], and the genetic background [8, 95]. E2 treatment of ovary-intact virgin ACI female rats typically begins between 55 to 65 days of age [17, 67, 68]. Historically, silastic medical tubing containing crystalline E2 (27.5 mg) has been implanted subcutaneously into the interscapular region [17, 96], which results in continuous circulating levels of E2 that recapitulate the periovulatory phase of the menstrual cycle or during pregnancy [17]. In the highly susceptible ACI strain, palpable tumors have been detected as early as 99 days posttreatment and 100% of rats developed tumors by 198 days posttreatment [17]. Compared with E2, other forms of estrogen (e.g., estrone and diethylstilbestrol) induce mammary tumors in the ACI rat, yet are less effective [17, 68, 95]. Beginning at 5–7 weeks posttreatment with E2, tumor formation is monitored weekly by manual palpation of the MFP by a blinded observer. If diagnosis of a tumor is unclear, it is typical to query a second blinded observer for confirmation of the tumor diagnosis or to wait for a second confirmation by the same blinded observer in the following week, in which case the date of initial diagnosis would be considered the tumor onset. In large genetic mapping studies [78, 79], the standard weekly recorded tumor phenotypes include the anatomical location of each tumor, the overall incidence, the latency, and the multiplicity. At experimental endpoint, mammary tumors are typically collected and weighed to record overall tumor burden per rat. The strategies for mapping genetic modifiers of E2-induced mammary tumor risk are identical to the methodology used for mapping genetic modifiers of carcinogen-induced mammary tumors (*see* Subheading 2.1.1).

3 Methodologies for Mapping Host TME Modifiers in the Rat

The TME comprises multiple nonmalignant cell types that interact with malignant tumor cells at all disease stages [97–101]. Moreover, outcome has been correlated with stromal expression signatures in the TME of human breast cancer patients, demonstrating the importance of the TME to the pathophysiological characteristics of breast tumors [102–104]. The TME includes cancer-associated fibroblasts (CAF) [105], blood vasculature [106, 107], lymphatic vasculature [108], and infiltrating leukocytes [109]. CAF are a major component of the breast TME and modulate most aspects of tumor pathophysiology, including malignant progression of cancer cells, fibrosis, tumor-associated immunity, and the formation of tumor-associated vasculature (i.e., tumor angiogenesis) [105]. In addition to CAF, tumor angiogenesis is mediated by endothelial cells, infiltrating leukocytes, and by growth factors secreted by malignant cancer cells [106, 107]. A denser tumor vasculature is typically correlated with adverse outcome, which is due to enhanced oxygen supply, nutrients, and routes for metastatic dissemination of tumor cells [110]. Likewise, tumor lymphatic vessels provide routes for tumor cell metastasis and tumor lymphangiogenesis also correlates with poor clinical outcomes [107, 108]. Finally, tumor-associated vasculature is also the primary route for trafficking leukocytes, which play both pro- and anti-tumorigenic roles during tumor initiation, progression, and response to therapy [109].

Although many breast cancer risk modifiers undoubtedly impact cancer cells directly, some modifier alleles likely impact breast cancer risk through the TME. To date, at least two host TME modifier loci of breast cancer have been experimentally validated in the rat, and evidence of several more TME modifier loci exists. In one example, the Mcs5a rat mammary tumor risk locus was shown to modify mammary carcinoma progression via the immune system, which was driven by *Fbxo10* and was dependent upon T lymphocytes [49]. A homologous *Fbxo10*-dependent mechanism has been associated with human breast cancer risk [55, 111]. In another example, we identified a host TME modifier QTL, which appears to be DLL4-dependent and impacts breast cancer growth and hematogenous metastasis in the rat by inducing dysfunctional angiogenesis [35, 112, 113]. Despite the evidence that host TME modifiers of breast cancer exist, the majority of heritable genetic variants that influence human breast cancer through the host TME remain poorly understood.

To discover host TME modifiers, we recently developed a genetic mapping strategy called the Consomic/Congenic Xenograft Model (CXM) [35, 112, 113]. In CXM, human breast cancer cells are orthotopically implanted into genetically modified consomic and congenic xenograft host strains, which are derived from two parental strains with different susceptibilities to breast cancer. Because the host strain backgrounds are different, whereas the inoculated tumor cells are the same, any phenotypic variation is due to TME modifier(s) on the substituted chromosome (i.e., consomic) or subchromosomal region (i.e., congenic) of the host's germline DNA. CXM utilizes transgenically tagged human cancer cells with defined clinical properties (e.g., triple-negative, pro-metastatic). Thus, it enables testing of clinically relevant cancer models in strain backgrounds with varying genetic predispositions to breast cancer. An important distinction between CXM and the traditional mammary mapping studies is that modifier alleles in CXM are only able to exert their effects through the TME, whereas in mammary tumors induced by exposure to carcinogens or E2, the same alleles are present in both the malignant cancer cells and the TME.

3.1 Experimental Design and Procedures

Multiple gene-editing and breeding strategies can be used to generate a CXM host strain in the rat [60]. One strategy is to first create SCID status in two parental strains of varying susceptibility to mammary cancer or another trait of interest, which can then be used to generate SCID consomic and congenic strains. A strength of this strategy is the ability to use genome-wide mapping strategies (e.g., F2 linkage analysis or consomic panel phenotyping) for discovery of de novo host TME modifiers. A second strategy is to create SCID status directly in consomic or congenic strains that have already revealed mammary tumor traits via other methods (e.g., carcinogen- or E2-induced mammary tumor models). An advantage of this strategy is the ability to leverage existing consomic panels and congenic strains, which reduces cost and speeds up discovery.

An important consideration for either strategy is whether the target rat strains for creating SCID status are amenable to genome-engineering strategies, including Zinc-Finger Nucleases (ZFN), Transcription Activator-Like Effector Nucleases (TALEN), and Clustered Regularly Interspaced Short Palindromic Repeats (CRISPR)/Cas9 [60, 114]. In addition to identifying an effective genome-engineering strategy, choosing the optimal target gene (s) and germline mutation(s) for inducing SCID status are also key to success. We and others have found that using standard genome-editing techniques to target the IL2Rγ locus is sufficient for creating SCID status in the rat, which should be confirmed by analysis of circulating peripheral blood leukocytes [35, 115]. Although there is no quantitative definition of SCID status in the rat, we typically observe >10-fold reduction in circulating T-cells, B-cells, and NK cells [35]. Several other gene mutations have been reported to induce immunodeficiency, including *Foxn1* (nude rat and mouse) [116], *Prkdc* (SCID mouse) [117], or the combination of *Il2rg* and *Prkdc* mutation (NSG mouse) [118]. Although any SCID-inducing mutation would be suitable for generating CXM strains, we recommend targeting *Il2rg* for two main reasons: (1) *Il2rg* deficiency alone is sufficient for SCID and *Il2rg* deficient rats will accept human tumors with >95% take rate, based on our experience [35, 112, 113] and (2) *Il2rg* is X-linked, which enables gene mapping on any autosome without interference by introgression of the mutant allele and achieving homozygous maintenance colonies more quickly through male germline transmission.

Since different mutations within the same gene can cause different functional effects, it is also important that the same or very similar gene mutation(s) be introduced into the rat strains being compared. It is foreseeable that a complete ablation of gene activity (e.g., large deletion or frameshift) would likely result in the same severity of SCID in two rat strains, albeit not guaranteed. A safer alternative strategy is to mutate the *Il2rg* locus in one parental strain and introgress the mutant allele into the recipient background, whether it be another parental strain or consomic/congenic strain. For example, in generating the first CXM strain, a germline mutation of *Il2rg* was introduced into the susceptible SS/JrHsdMcwi genetic background (SS^{Il2rg}) and then intercrossed with a resistant SS-Chr3^{BN}/Mcwi consomic male rat to yield heterozygous SS.BN3^{Il2rg} offspring (F_1), followed by marker-assisted brother-sister mating to yield homozygous SS.BN3^{Il2rg} offspring by the F_3 generation [35]. The advantage of this strategy is twofold: (1) gene-editing is required in only one strain and (2) the SCID mutation in both strains is identical, thus removing the potentially confounding effects of two different gene mutations. In either strategy, we routinely confirm genotypes of the target gene and within the chromosome of interest, following standard methodology (e.g., Sanger sequencing and fluorescent genotyping, respectively).

Upon generation of homozygous CXM and parental SCID strains, standard tumor xenografting techniques can be used to test whether host TME modifier(s) influence breast cancer traits (e.g., growth, metastasis) [35, 112, 113]. The observation of phenotypic differences between the CXM and parental strain demonstrates that host TME modifier(s) of the measured trait exist on the chromosome of interest (i.e., consomic) [35, 113] or the subchromosomal region of interest (i.e., congenic) [112]. The methodology for tumor xenograft implantation and phenotypic measurement resembles that of standard xenograft techniques. We typically use luciferase-tagged transgenic cell lines that can be readily used to track tumor growth and metastatic dissemination by luminescence; however, non-tagged parental cell lines or patient-derived xenografts (PDX) can also be used without luminescent measurements. A typical CXM experiment entails orthotopic implantation of 6–8 $\times$ 10^6 human breast cancer cells suspended in 50% matrigel into the MFP. One week following tumor implantation, growth measurements are typically recorded every 2–3 days by in vivo bioluminescence or caliper measurements, as described previously [119–121]. An added advantage of in vivo bioluminescence imaging is the ability to detect and quantify distant metastatic lesions in real time. However, caution should be exercised in quantifying and interpreting bioluminescence data, as our previous CXM studies have shown that host TME modifiers of vascular perfusion are likely to impact the in vivo measurements of bioluminescence [112, 113]. An alternative that circumvents this issue is to collect organs at experimental endpoint and measure luciferase activity in tissue lysate ex vivo [119–121]. Such measurements are both highly quantitative and cost effective, typically requiring far less luciferin substrate than analysis in vivo.

4 Conclusions and Future Directions

In this protocol review, we have summarized three widely established models for identifying and studying the genetic modifiers of breast cancer in the rat. Exposure to carcinogens or estrogens induces the formation of mammary tumors and enables the study of germline variants that impact tumor formation through malignant epithelial cells and the nonmalignant TME, whereas the novel CXM strategy is specifically designed to map genetic modifiers in the host TME only. In several cases, the genetic modifiers of breast cancer that were discovered via these methods have been replicated in humans [54, 111] or show strong overlap with candidate regions in human breast cancer association studies [13, 47, 80]. Thus, these studies are likely to yield high translational impact on the current understanding of the genetic modifiers that affect breast cancer risk.

References

1. Jemal A, Bray F, Center MM, Ferlay J, Ward E, Forman D (2011) Global cancer statistics. CA Cancer J Clin 61(2):69–90. https://doi.org/10.3322/caac.20107
2. Youlden DR, Cramb SM, Dunn NA, Muller JM, Pyke CM, Baade PD (2012) The descriptive epidemiology of female breast cancer: an international comparison of screening, incidence, survival and mortality. Cancer Epidemiol 36(3):237–248. https://doi.org/10.1016/j.canep.2012.02.007
3. Stewart BW, Wild C, International Agency for Research on Cancer, World Health Organization (2014) World cancer report 2014. International Agency for Research on Cancer WHO Press, Lyon
4. Moller S, Mucci LA, Harris JR, Scheike T, Holst K, Halekoh U et al (2016) The heritability of breast cancer among women in the Nordic Twin Study of Cancer. Cancer Epidemiol Biomark Prev 25(1):145–150. https://doi.org/10.1158/1055-9965.EPI-15-0913
5. Stratton MR, Rahman N (2008) The emerging landscape of breast cancer susceptibility. Nat Genet 40(1):17–22. https://doi.org/10.1038/ng.2007.53
6. Foulkes WD (2008) Inherited susceptibility to common cancers. N Engl J Med 359 (20):2143–2153. https://doi.org/10.1056/NEJMra0802968
7. Michailidou K, Lindstrom S, Dennis J, Beesley J, Hui S, Kar S et al (2017) Association analysis identifies 65 new breast cancer risk loci. Nature 551(7678):92–94. https://doi.org/10.1038/nature24284
8. Shull J, Dennison K, Chack A, Trentham-Dietz A (2018) Rat models of 17beta-estradiol-induced mammary cancer reveal novel insights into breast cancer etiology and prevention. Physiol Genomics 50:215–234. https://doi.org/10.1152/physiolgenomics.00105.2017
9. Michailidou K, Hall P, Gonzalez-Neira A, Ghoussaini M, Dennis J, Milne RL et al (2013) Large-scale genotyping identifies 41 new loci associated with breast cancer risk. Nat Genet 45(4):353–361, 361e1–2. https://doi.org/10.1038/ng.2563
10. Nielsen FC, van Overeem Hansen T, Sorensen CS (2016) Hereditary breast and ovarian cancer: new genes in confined pathways. Nat Rev Cancer 16(9):599–612. https://doi.org/10.1038/nrc.2016.72
11. Flister MJ, Tsaih SW, O'Meara CC, Endres B, Hoffman MJ, Geurts AM et al (2013) Identifying multiple causative genes at a single GWAS locus. Genome Res 23 (12):1996–2002. https://doi.org/10.1101/gr.160283.113
12. Russo IH, Russo J (1996) Mammary gland neoplasia in long-term rodent studies. Environ Health Perspect 104(9):938–967
13. Sanders J, Samuelson DJ (2014) Significant overlap between human genome-wide association-study nominated breast cancer risk alleles and rat mammary cancer susceptibility loci. Breast Cancer Res 16(1):R14. https://doi.org/10.1186/bcr3607
14. Huggins C, Grand LC, Brillantes FP (1961) Mammary cancer induced by a single feeding of polynuclear hydrocarbons, and its suppression. Nature 189:204. https://doi.org/10.1038/189204a0
15. Gullino PM, Pettigrew HM, Grantham FH (1975) N-Nitrosomethylurea as mammary gland carcinogen in rats. J Natl Cancer Inst 54(2):401–414. https://doi.org/10.1093/jnci/54.2.401
16. Druckrey H, Steinhoff D, Preussmann R, Ivankovic S (1964) [Induction of cancer by a single dose of methylnitroso-urea and various dialkylnitrosamines in rats]. Z Krebsforsch 66:1–10
17. Shull JD, Spady TJ, Snyder MC, Johansson SL, Pennington KL (1997) Ovary-intact, but not ovariectomized female ACI rats treated with 17beta-estradiol rapidly develop mammary carcinoma. Carcinogenesis 18 (8):1595–1601
18. Russo J (2015) Significance of rat mammary tumors for human risk assessment. Toxicol Pathol 43(2):145–170. https://doi.org/10.1177/0192623314532036
19. DeSantis C, Ma J, Bryan L, Jemal A (2014) Breast cancer statistics, 2013. CA Cancer J Clin 64(1):52–62. https://doi.org/10.3322/caac.21203
20. Richardson KK, Richardson FC, Crosby RM, Swenberg JA, Skopek TR (1987) DNA base changes and alkylation following in vivo exposure of *Escherichia coli* to N-methyl-N-nitrosourea or N-ethyl-N-nitrosourea. Proc Natl Acad Sci U S A 84(2):344–348
21. Burns PA, Gordon AJ, Glickman BW (1988) Mutational specificity of N-methyl-N-nitrosourea in the lacI gene of *Escherichia coli*. Carcinogenesis 9(9):1607–1610
22. Sukumar S, Notario V, Martin-Zanca D, Barbacid M (1983) Induction of mammary carcinomas in rats by nitroso-methylurea involves

malignant activation of H-ras-1 locus by single point mutations. Nature 306 (5944):658–661

23. Lu SJ, Archer MC (1992) Ha-ras oncogene activation in mammary glands of N-methyl-N-nitrosourea-treated rats genetically resistant to mammary adenocarcinogenesis. Proc Natl Acad Sci U S A 89(3):1001–1005
24. Adamovic T, McAllister D, Guryev V, Wang X, Andrae JW, Cuppen E et al (2009) Microalterations of inherently unstable genomic regions in rat mammary carcinomas as revealed by long oligonucleotide array-based comparative genomic hybridization. Cancer Res 69(12):5159–5167. https://doi.org/10.1158/0008-5472.CAN-08-4038
25. Russo J, Saby J, Isenberg WM, Russo IH (1977) Pathogenesis of mammary carcinomas induced in rats by 7,12-dimethylbenz[a] anthracene. J Natl Cancer Inst 59 (2):435–445. https://doi.org/10.1093/jnci/59.2.435
26. Dai ML, Huang XF, Wang QF, Cai WJ, Jin ZB, Wang Y (2016) CFI-rs7356506 polymorphisms associated with Vogt-Koyanagi-Harada syndrome. Mol Vis 22:9–17
27. Russo J, Wilgus G, Russo IH (1979) Susceptibility of the mammary gland to carcinogenesis: I. Differentiation of the mammary gland as determinant of tumor incidence and type of lesion. Am J Pathol 96(3):721–736
28. Russo IH, Medado J, Russo J (1989) Endocrine influences on the mammary gland. In: Jones TC, Mohr U, Hunt RD (eds) Integument and mammary glands. Springer, Berlin, pp 252–266. https://doi.org/10.1007/978-3-642-83749-4_39
29. Russo IH, Russo J (1978) Developmental stage of the rat mammary gland as determinant of its susceptibility to 7,12-dimethylbenz [a]anthracene. J Natl Cancer Inst 61 (6):1439–1449
30. Russo J, Russo IH (1978) DNA labeling index and structure of the rat mammary gland as determinants of its susceptibility to carcinogenesis. J Natl Cancer Inst 61 (6):1451–1459
31. Adamovic T, McAllister D, Wang T, Adamovic D, Rowe JJ, Moreno C et al (2010) Identification of novel carcinogen-mediated mammary tumor susceptibility loci in the rat using the chromosome substitution technique. Genes Chromosomes Cancer 49 (11):1035–1045. https://doi.org/10.1002/gcc.20812
32. Kerlikowske K (2010) Epidemiology of ductal carcinoma in situ. J Natl Cancer Inst Monogr 2010(41):139–141. https://doi.org/10.1093/jncimonographs/lgq027
33. denDekker AD, Xu X, Vaughn MD, Puckett AH, Gardner LL, Lambring CJ et al (2012) Rat Mcs1b is concordant to the genome-wide association-identified breast cancer risk locus at human 5q11.2 and MIER3 is a candidate cancer susceptibility gene. Cancer Res 72 (22):6002–6012. https://doi.org/10.1158/0008-5472.CAN-12-0748
34. Isaacs JT (1986) Genetic control of resistance to chemically induced mammary adenocarcinogenesis in the rat. Cancer Res 46 (8):3958–3963
35. Flister MJ, Endres BT, Rudemiller N, Sarkis AB, Santarriaga S, Lemke A et al (2014) CXM—a new tool for mapping breast cancer risk in the tumor microenvironment. Cancer Res 74(22):6419–6429. https://doi.org/10.1158/0008-5472.CAN-13-3212
36. Adamovic T, McAllister D, Rowe JJ, Wang T, Jacob HJ, Sugg SL (2008) Genetic mapping of mammary tumor traits to rat chromosome 10 using a novel panel of consomic rats. Cancer Genet Cytogenet 186(1):41–48. https://doi.org/10.1016/j.cancergencyto.2008.05.014
37. Lan H, Kendziorski CM, Haag JD, Shepel LA, Newton MA, Gould MN (2001) Genetic loci controlling breast cancer susceptibility in the Wistar-Kyoto rat. Genetics 157 (1):331–339
38. Quan X, Laes JF, Stieber D, Riviere M, Russo J, Wedekind D et al (2006) Genetic identification of distinct loci controlling mammary tumor multiplicity, latency, and aggressiveness in the rat. Mamm Genome 17 (4):310–321. https://doi.org/10.1007/s00335-005-0125-9
39. Huggins CB, Ueda N, Wiessler M (1981) N-Nitroso-N-methylurea elicits mammary cancer in resistant and sensitive rat strains. Proc Natl Acad Sci U S A 78(2):1185–1188
40. Thompson HJ, Meeker LD (1983) Induction of mammary gland carcinomas by the subcutaneous injection of 1-methyl-1-nitrosourea. Cancer Res 43(4):1628–1629
41. Cha RS, Thilly WG, Zarbl H (1994) N-Nitroso-N-methylurea-induced rat mammary tumors arise from cells with preexisting oncogenic Hras1 gene mutations. Proc Natl Acad Sci U S A 91(9):3749–3753
42. Haag JD, Newton MA, Gould MN (1992) Mammary carcinoma suppressor and susceptibility genes in the Wistar-Kyoto rat. Carcinogenesis 13(10):1933–1935
43. Hsu LC, Kennan WS, Shepel LA, Jacob HJ, Szpirer C, Szpirer J et al (1994) Genetic identification of Mcs-1, a rat mammary carcinoma suppressor gene. Cancer Res 54 (10):2765–2770

44. Shepel LA, Lan H, Haag JD, Brasic GM, Gheen ME, Simon JS et al (1998) Genetic identification of multiple loci that control breast cancer susceptibility in the rat. Genetics 149(1):289–299
45. Haag JD, Shepel LA, Kolman BD, Monson DM, Benton ME, Watts KT et al (2003) Congenic rats reveal three independent Copenhagen alleles within the Mcs1 quantitative trait locus that confer resistance to mammary cancer. Cancer Res 63(18):5808–5812
46. Smits BM, Haag JD, Rissman AI, Sharma D, Tran A, Schoenborn AA et al (2013) The gene desert mammary carcinoma susceptibility locus Mcs1a regulates Nr2f1 modifying mammary epithelial cell differentiation and proliferation. PLoS Genet 9(6):e1003549. https://doi.org/10.1371/journal.pgen.1003549
47. Le S, Martin ZC, Samuelson DJ (2017) Physical confirmation and comparative genomics of the rat mammary carcinoma susceptibility 3 quantitative trait locus. G3 7 (6):1767–1773. https://doi.org/10.1534/g3.117.039388
48. Sanders J, Haag JD, Samuelson DJ (2011) Physical confirmation and mapping of overlapping rat mammary carcinoma susceptibility QTLs, Mcs2 and Mcs6. PLoS One 6(5): e19891. https://doi.org/10.1371/journal.pone.0019891
49. Smits BM, Sharma D, Samuelson DJ, Woditschka S, Mau B, Haag JD et al (2011) The non-protein coding breast cancer susceptibility locus Mcs5a acts in a non-mammary cell-autonomous fashion through the immune system and modulates T-cell homeostasis and functions. Breast Cancer Res 13(4): R81. https://doi.org/10.1186/bcr2933
50. Henning AN, Haag JD, Smits BM, Gould MN (2016) The non-coding mammary carcinoma susceptibility locus, Mcs5c, regulates pappa expression via age-specific chromatin folding and allele-dependent DNA methylation. PLoS Genet 12(8):e1006261. https://doi.org/10.1371/journal.pgen.1006261
51. Ren X, Graham JC, Jing L, Mikheev AM, Gao Y, Lew JP et al (2013) Mapping of Mcs30, a new mammary carcinoma susceptibility quantitative trait locus (QTL30) on rat chromosome 12: identification of *Fry* as a candidate Mcs gene. PLoS One 8(9): e70930. https://doi.org/10.1371/journal.pone.0070930
52. Gould MN, Zhang R (1991) Genetic regulation of mammary carcinogenesis in the rat by susceptibility and suppressor genes. Environ Health Perspect 93:161–167
53. Machholz E, Mulder G, Ruiz C, Corning BF, Pritchett-Corning KR (2012) Manual restraint and common compound administration routes in mice and rats. J Vis Exp (67):2771. https://doi.org/10.3791/2771
54. Samuelson DJ, Aperavich BA, Haag JD, Gould MN (2005) Fine mapping reveals multiple loci and a possible epistatic interaction within the mammary carcinoma susceptibility quantitative trait locus, Mcs5. Cancer Res 65 (21):9637–9642. https://doi.org/10.1158/0008-5472.CAN-05-1498
55. Samuelson DJ, Hesselson SE, Aperavich BA, Zan Y, Haag JD, Trentham-Dietz A et al (2007) Rat Mcs5a is a compound quantitative trait locus with orthologous human loci that associate with breast cancer risk. Proc Natl Acad Sci U S A 104(15):6299–6304. https://doi.org/10.1073/pnas.0701687104
56. Lander ES, Botstein D (1989) Mapping mendelian factors underlying quantitative traits using RFLP linkage maps. Genetics 121 (1):185–199
57. Broman KW, Sen S (2009) A guide to QTL mapping with R/qtl. Statistics for biology and health. Springer, Dordrecht
58. Lagrange D, Fournie GJ (2010) Generation of congenic and consomic rat strains. Methods Mol Biol 597:243–266. https://doi.org/10.1007/978-1-60327-389-3_17
59. Aitman TJ, Petretto E, Behmoaras J (2010) Genetic mapping and positional cloning. Methods Mol Biol 597:13–32. https://doi.org/10.1007/978-1-60327-389-3_2
60. Flister MJ, Prokop JW, Lazar J, Shimoyama M, Dwinell M, Geurts A (2015) 2015 Guidelines for establishing genetically modified rat models for cardiovascular research. J Cardiovasc Transl Res 8 (4):269–277. https://doi.org/10.1007/s12265-015-9626-4
61. Cowley AW Jr, Roman RJ, Jacob HJ (2004) Application of chromosomal substitution techniques in gene-function discovery. J Physiol 554(Pt 1):46–55. https://doi.org/10.1113/jphysiol.2003.052613
62. Titus-Ernstoff L, Hatch EE, Hoover RN, Palmer J, Greenberg ER, Ricker W et al (2001) Long-term cancer risk in women given diethylstilbestrol (DES) during pregnancy. Br J Cancer 84(1):126–133. https://doi.org/10.1054/bjoc.2000.1521
63. Colton T, Greenberg ER, Noller K, Resseguie L, Van Bennekom C, Heeren T et al (1993) Breast cancer in mothers

prescribed diethylstilbestrol in pregnancy. Further follow-up. JAMA 269 (16):2096–2100

64. Fisher B, Costantino JP, Wickerham DL, Cecchini RS, Cronin WM, Robidoux A et al (2005) Tamoxifen for the prevention of breast cancer: current status of the National Surgical Adjuvant Breast and Bowel Project P-1 study. J Natl Cancer Inst 97 (22):1652–1662. https://doi.org/10.1093/jnci/dji372
65. Goss PE, Ingle JN, Ales-Martinez JE, Cheung AM, Chlebowski RT, Wactawski-Wende J et al (2011) Exemestane for breast-cancer prevention in postmenopausal women. N Engl J Med 364(25):2381–2391. https://doi.org/10.1056/NEJMoa1103507
66. Vogel VG, Costantino JP, Wickerham DL, Cronin WM, Cecchini RS, Atkins JN et al (2006) Effects of tamoxifen vs raloxifene on the risk of developing invasive breast cancer and other disease outcomes: the NSABP Study of Tamoxifen and Raloxifene (STAR) P-2 trial. JAMA 295(23):2727–2741. https://doi.org/10.1001/jama.295.23.joc60074
67. Harvell DM, Strecker TE, Tochacek M, Xie B, Pennington KL, McComb RD et al (2000) Rat strain-specific actions of 17beta-estradiol in the mammary gland: correlation between estrogen-induced lobuloalveolar hyperplasia and susceptibility to estrogen-induced mammary cancers. Proc Natl Acad Sci U S A 97 (6):2779–2784. https://doi.org/10.1073/pnas.050569097
68. Dunning WF, Curtis MR (1952) The incidence of diethylstilbestrol-induced cancer in reciprocal F hybrids obtained from crosses between rats of inbred lines that are susceptible and resistant to the induction of mammary cancer by this agent. Cancer Res 12 (10):702–706
69. Li SA, Weroha SJ, Tawfik O, Li JJ (2002) Prevention of solely estrogen-induced mammary tumors in female aci rats by tamoxifen: evidence for estrogen receptor mediation. J Endocrinol 175(2):297–305
70. Blank EW, Wong PY, Lakshmanaswamy R, Guzman R, Nandi S (2008) Both ovarian hormones estrogen and progesterone are necessary for hormonal mammary carcinogenesis in ovariectomized ACI rats. Proc Natl Acad Sci U S A 105(9):3527–3532. https://doi.org/10.1073/pnas.0710535105
71. Boopalan T, Arumugam A, Parada J, Saltzstein E, Lakshmanaswamy R (2015) Receptor activator for nuclear factor-kappaB ligand signaling promotes progesterone-mediated estrogen-induced mammary carcinogenesis. Cancer Sci 106(1):25–33. https://doi.org/10.1111/cas.12571
72. Ding L, Zhao Y, Warren CL, Sullivan R, Eliceiri KW, Shull JD (2013) Association of cellular and molecular responses in the rat mammary gland to 17beta-estradiol with susceptibility to mammary cancer. BMC Cancer 13:573. https://doi.org/10.1186/1471-2407-13-573
73. Cavalieri EL, Rogan EG, Zahid M (2017) Critical depurinating DNA adducts: estrogen adducts in the etiology and prevention of cancer and dopamine adducts in the etiology and prevention of Parkinson's disease. Int J Cancer 141(6):1078–1090. https://doi.org/10.1002/ijc.30728
74. el-Bayoumy K, Ji BY, Upadhyaya P, Chae YH, Kurtzke C, Rivenson A et al (1996) Lack of tumorigenicity of cholesterol epoxides and estrone-3,4-quinone in the rat mammary gland. Cancer Res 56(9):1970–1973
75. Adamovic T, Roshani L, Chen L, Schaffer BS, Helou K, Levan G et al (2007) Nonrandom pattern of chromosome aberrations in 17beta-estradiol-induced rat mammary tumors: indications of distinct pathways for tumor development. Genes Chromosomes Cancer 46(5):459–469. https://doi.org/10.1002/gcc.20428
76. Li JJ, Papa D, Davis MF, Weroha SJ, Aldaz CM, El-Bayoumy K et al (2002) Ploidy differences between hormone- and chemical carcinogen-induced rat mammary neoplasms: comparison to invasive human ductal breast cancer. Mol Carcinog 33(1):56–65
77. Schaffer BS, Leland-Wavrin KM, Kurz SG, Colletti JA, Seiler NL, Warren CL et al (2013) Mapping of three genetic determinants of susceptibility to estrogen-induced mammary cancer within the Emca8 locus on rat chromosome 5. Cancer Prev Res (Phila) 6 (1):59–69. https://doi.org/10.1158/1940-6207.CAPR-12-0346-T
78. Gould KA, Tochacek M, Schaffer BS, Reindl TM, Murrin CR, Lachel CM et al (2004) Genetic determination of susceptibility to estrogen-induced mammary cancer in the ACI rat: mapping of Emca1 and Emca2 to chromosomes 5 and 18. Genetics 168 (4):2113–2125. https://doi.org/10.1534/genetics.104.033878
79. Schaffer BS, Lachel CM, Pennington KL, Murrin CR, Strecker TE, Tochacek M et al (2006) Genetic bases of estrogen-induced tumorigenesis in the rat: mapping of loci controlling susceptibility to mammary cancer in a Brown Norway x ACI intercross. Cancer

Res 66(15):7793–7800. https://doi.org/10.1158/0008-5472.CAN-06-0143

80. Colletti JA II, Leland-Wavrin KM, Kurz SG, Hickman MP, Seiler NL, Samanas NB et al (2014) Validation of six genetic determinants of susceptibility to estrogen-induced mammary cancer in the rat and assessment of their relevance to breast cancer risk in humans. G3 4(8):1385–1394. https://doi.org/10.1534/g3.114.011163
81. Wendell DL, Gorski J (1997) Quantitative trait loci for estrogen-dependent pituitary tumor growth in the rat. Mamm Genome 8 (11):823–829
82. Shull JD, Lachel CM, Murrin CR, Pennington KL, Schaffer BS, Strecker TE et al (2007) Genetic control of estrogen action in the rat: mapping of QTLs that impact pituitary lactotroph hyperplasia in a BN x ACI intercross. Mamm Genome 18(9):657–669. https://doi.org/10.1007/s00335-007-9052-2
83. Holtzman S, Stone JP, Shellabarger CJ (1979) Influence of diethylstilbestrol treatment on prolactin cells of female ACI and Sprague-Dawley rats. Cancer Res 39 (3):779–784
84. Kurz SG, Dennison KL, Samanas NB, Hickman MP, Eckert QA, Walker TL et al (2014) Ept7 influences estrogen action in the pituitary gland and body weight of rats. Mamm Genome 25(5–6):244–252. https://doi.org/10.1007/s00335-014-9504-4
85. Kurz SG, Hansen KK, McLaughlin MT, Shivaswamy V, Schaffer BS, Gould KA et al (2008) Tissue-specific actions of the Ept1, Ept2, Ept6, and Ept9 genetic determinants of responsiveness to estrogens in the female rat. Endocrinology 149(8):3850–3859. https://doi.org/10.1210/en.2008-0173
86. Spady TJ, Pennington KL, McComb RD, Shull JD (1999) Genetic bases of estrogen-induced pituitary growth in an intercross between the ACI and Copenhagen rat strains: dominant mendelian inheritance of the ACI phenotype. Endocrinology 140 (6):2828–2835. https://doi.org/10.1210/endo.140.6.6757
87. Strecker TE, Spady TJ, Schaffer BS, Gould KA, Kaufman AE, Shen F et al (2005) Genetic bases of estrogen-induced pituitary tumorigenesis: identification of genetic loci determining estrogen-induced pituitary growth in reciprocal crosses between the ACI and Copenhagen rat strains. Genetics 169 (4):2189–2197. https://doi.org/10.1534/genetics.104.039370
88. Spady TJ, McComb RD, Shull JD (1999) Estrogen action in the regulation of cell proliferation, cell survival, and tumorigenesis in the rat anterior pituitary gland. Endocrine 11(3):217–233. https://doi.org/10.1385/ENDO:11:3:217
89. Hennighausen L, Robinson GW (2005) Information networks in the mammary gland. Nat Rev Mol Cell Biol 6:715–725. https://doi.org/10.1038/nrm1714
90. Shea MP, O'Leary KA, Fakhraldeen SA, Goffin V, Friedl A, Wisinski KB et al (2018) Antiestrogen therapy increases plasticity and cancer stemness of prolactin-induced ERalpha + mammary carcinomas. Cancer Res 78 (7):1672–1684. https://doi.org/10.1158/0008-5472.CAN-17-0985
91. Tikk K, Sookthai D, Fortner RT, Johnson T, Rinaldi S, Romieu I et al (2015) Circulating prolactin and in situ breast cancer risk in the European EPIC cohort: a case-control study. Breast Cancer Res 17:49. https://doi.org/10.1186/s13058-015-0563-6
92. Tworoger SS, Eliassen AH, Zhang X, Qian J, Sluss PM, Rosner BA et al (2013) A 20-year prospective study of plasma prolactin as a risk marker of breast cancer development. Cancer Res 73(15):4810–4819. https://doi.org/10.1158/0008-5472.CAN-13-0665
93. Meites J (1972) Relation of prolactin and estrogen to mammary tumorigenesis in the rat. J Natl Cancer Inst 48(4):1217–1224
94. Dennison KL, Samanas NB, Harenda QE, Hickman MP, Seiler NL, Ding L et al (2015) Development and characterization of a novel rat model of estrogen-induced mammary cancer. Endocr Relat Cancer 22(2):239–248. https://doi.org/10.1530/ERC-14-0539
95. Dunning WF, Curtis MR, Segaloff A (1953) Strain differences in response to estrone and the induction of mammary gland, adrenal, and bladder cancer in rats. Cancer Res 13 (2):147–152
96. Spady TJ, Lemus-Wilson AM, Pennington KL, Blackwood DJ, Paschall TM, Birt DF et al (1998) Dietary energy restriction abolishes development of prolactin-producing pituitary tumors in Fischer 344 rats treated with 17beta-estradiol. Mol Carcinog 23(2):86–95
97. Liu W, Vivian CJ, Brinker AE, Hampton KR, Lianidou E, Welch DR (2014) Microenvironmental influences on metastasis suppressor expression and function during a metastatic cell's journey. Cancer Microenviron 7 (3):117–131. https://doi.org/10.1007/s12307-014-0148-4
98. Olson OC, Joyce JA (2013) Microenvironment-mediated resistance to

anticancer therapies. Cell Res 23 (2):179–181. https://doi.org/10.1038/cr.2012.123

99. Quail DF, Joyce JA (2013) Microenvironmental regulation of tumor progression and metastasis. Nat Med 19(11):1423–1437. https://doi.org/10.1038/nm.3394
100. Polyak K, Kalluri R (2010) The role of the microenvironment in mammary gland development and cancer. Cold Spring Harb Perspect Biol 2(11):a003244. https://doi.org/10.1101/cshperspect.a003244
101. McAllister SS, Weinberg RA (2014) The tumour-induced systemic environment as a critical regulator of cancer progression and metastasis. Nat Cell Biol 16(8):717–727. https://doi.org/10.1038/ncb3015
102. Liu H, Dowdle JA, Khurshid S, Sullivan NJ, Bertos N, Rambani K et al (2017) Discovery of stromal regulatory networks that suppress Ras-sensitized epithelial cell proliferation. Dev Cell 41(4):392–407. https://doi.org/10.1016/j.devcel.2017.04.024
103. Saleh SMI, Bertos N, Gruosso T, Gigoux M, Souleimanova M, Zhao H et al (2017) Identification of interacting stromal axes in triple-negative breast cancer. Cancer Res 77 (17):4673–4683. https://doi.org/10.1158/0008-5472.CAN-16-3427
104. Finak G, Bertos N, Pepin F, Sadekova S, Souleimanova M, Zhao H et al (2008) Stromal gene expression predicts clinical outcome in breast cancer. Nat Med 14(5):518–527. https://doi.org/10.1038/nm1764
105. Kalluri R (2016) The biology and function of fibroblasts in cancer. Nat Rev Cancer 16 (9):582–598. https://doi.org/10.1038/nrc.2016.73
106. Hanahan D, Folkman J (1996) Patterns and emerging mechanisms of the angiogenic switch during tumorigenesis. Cell 86 (3):353–364
107. Bergers G, Benjamin LE (2003) Tumorigenesis and the angiogenic switch. Nat Rev Cancer 3(6):401–410. https://doi.org/10.1038/nrc1093
108. Ran S, Volk L, Hall K, Flister MJ (2010) Lymphangiogenesis and lymphatic metastasis in breast cancer. Pathophysiology 17 (4):229–251. https://doi.org/10.1016/j.pathophys.2009.11.003
109. Fridman WH, Pages F, Sautes-Fridman C, Galon J (2012) The immune contexture in human tumours: impact on clinical outcome. Nat Rev Cancer 12(4):298–306. https://doi.org/10.1038/nrc3245
110. Nico B, Benagiano V, Mangieri D, Maruotti N, Vacca A, Ribatti D (2008) Evaluation of microvascular density in tumors: pro and contra. Histol Histopathol 23 (5):601–607
111. Xu X, Powell DW, Lambring CJ, Puckett AH, Deschenes L, Prough RA et al (2014) Human MCS5A1 candidate breast cancer susceptibility gene FBXO10 is induced by cellular stress and correlated with lens epithelium-derived growth factor (LEDGF). Mol Carcinog 53 (4):300–313. https://doi.org/10.1002/mc.21977
112. Flister MJ, Tsaih SW, Stoddard A, Plasterer C, Jagtap J, Parchur AK et al (2017) Host genetic modifiers of nonproductive angiogenesis inhibit breast cancer. Breast Cancer Res Treat 165(1):53–64. https://doi.org/10.1007/s10549-017-4311-8
113. Jagtap J, Sharma G, Parchur AK, Gogineni V, Bergom C, White S et al (2018) Methods for detecting host genetic modifiers of tumor vascular function using dynamic near-infrared fluorescence imaging. Biomed Opt Express 9 (2):543–556. https://doi.org/10.1364/boe.9.000543
114. Geurts AM, Cost GJ, Freyvert Y, Zeitler B, Miller JC, Choi VM et al (2009) Knockout rats via embryo microinjection of zinc-finger nucleases. Science 325(5939):433. https://doi.org/10.1126/science.1172447
115. Mashimo T, Takizawa A, Voigt B, Yoshimi K, Hiai H, Kuramoto T et al (2010) Generation of knockout rats with X-linked severe combined immunodeficiency (X-SCID) using zinc-finger nucleases. PLoS One 5(1):e8870. https://doi.org/10.1371/journal.pone.0008870
116. Nehls M, Pfeifer D, Schorpp M, Hedrich H, Boehm T (1994) New member of the winged-helix protein family disrupted in mouse and rat nude mutations. Nature 372 (6501):103–107. https://doi.org/10.1038/372103a0
117. Blunt T, Finnie NJ, Taccioli GE, Smith GC, Demengeot J, Gottlieb TM et al (1995) Defective DNA-dependent protein kinase activity is linked to V(D)J recombination and DNA repair defects associated with the murine scid mutation. Cell 80(5):813–823
118. Shultz LD, Lyons BL, Burzenski LM, Gott B, Chen X, Chaleff S et al (2005) Human lymphoid and myeloid cell development in NOD/LtSz-scid IL2R gamma null mice engrafted with mobilized human hemopoietic stem cells. J Immunol 174(10):6477–6489

119. Volk LD, Flister MJ, Bivens CM, Stutzman A, Desai N, Trieu V et al (2008) Nab-paclitaxel efficacy in the orthotopic model of human breast cancer is significantly enhanced by concurrent anti-vascular endothelial growth factor A therapy. Neoplasia 10(6):613–623

120. Volk LD, Flister MJ, Chihade D, Desai N, Trieu V, Ran S (2011) Synergy of nab-paclitaxel and bevacizumab in eradicating large orthotopic breast tumors and preexisting metastases. Neoplasia 13(4):327–338

121. Whitehurst B, Flister MJ, Bagaitkar J, Volk L, Bivens CM, Pickett B et al (2007) Anti-VEGF-A therapy reduces lymphatic vessel density and expression of VEGFR-3 in an orthotopic breast tumor model. Int J Cancer 121(10):2181–2191. https://doi.org/10.1002/ijc.22937

Chapter 13

Rat Models of Metabolic Syndrome

Anne E. Kwitek

Abstract

Metabolic syndrome is a complex disorder that comprises several other complex disorders, including obesity, hypertension, dyslipidemia, and diabetes. There are several rat models that encompass component features of MetS. Some models are inbred strains selected for one or more traits underlying MetS; others are population models with genetic risk for MetS traits, are induced by environmental stressors such as diet, are spontaneous monogenic mutant models, or are congenic strains derived from a combination of these models. Together they can be studied to identify the genetic and physiological underpinnings of MetS to identify candidate genes or mechanisms for study in human MetS subjects.

Key words Rat genetic model, Rat inbred model, Rat disease model, Metabolic syndrome, MetS, Obesity, Hypertension

1 Introduction

The metabolic syndrome, sometimes referred to as Syndrome X or MetS, is a coincident occurrence of disorders that substantially increase the risk for mortality from heart disease, stroke, and renal failure. Syndrome X was described by Reaven in 1988 as the co-occurrence of insulin resistance, hyperglycemia, hyperinsulinemia, hyperlipidemia and hypertension [1]. While the specific clinical criteria have been debated over the past three decades, the internationally recognized clinical diagnosis for metabolic syndrome is three or more of the following: elevated waist circumference (a measure of abdominal obesity); elevated triglycerides; reduced high-density lipoprotein (HDL) cholesterol; elevated blood pressure; elevated fasting glucose [2]. Globally, the prevalence of MetS is estimated to be 25% [3] and its prevalence in children and young adults is estimated at 6.5% [4]. While these numbers vary by population and by criteria used to define MetS, the dramatic increase over the past two to three decades is a major health concern.

G. Thomas Hayman et al. (eds.), *Rat Genomics*, Methods in Molecular Biology, vol. 2018,
https://doi.org/10.1007/978-1-4939-9581-3_13,

While MetS is a syndrome in name, it is actually a complex disorder that comprises several other complex disorders, including obesity, hypertension, and diabetes. MetS is caused by both heritable and environmental influences [5–7]. Each defining feature of MetS also has a genetic component, with strong influences by environmental stimuli [5, 6, 8]. To effectively treat hypertension in MetS, it is important to determine what genes and mechanisms underlie MetS and its individual components, a goal that is substantially benefitted by physiological and genetic studies in animal models of MetS and/or its defining traits.

There are several rat models that encompass component features of MetS. Some models are inbred strains selected for one or more traits underlying MetS; others are population models with genetic risk for MetS traits, are induced by environmental stressors such as diet, are spontaneous monogenic mutant models, or are congenic strains derived from a combination of these models. Together they can be studied to identify the genetic and physiological underpinnings of MetS to identify candidate genes or mechanisms for study in human MetS subjects.

Some rat strains have several MetS traits, allowing for identification of putative pleiotropic genes while some have only a small subset, avoiding the confounding effects of other MetS traits. A comprehensive list of rat strains was annotated as obesity/metabolic syndrome models in the Rat Genome Database Obesity Portal [9]. Table 1 lists examples of different strains with *two or more* MetS traits, along with their defining features of MetS. Because waist circumference is not readily translatable to rats, obesity was included as a defining feature. Because of species differences in lipid profiles between human and rats [10–12], increased total cholesterol was included with low HDL as a defining feature of MetS for this review. Below are some examples of the different rat models of MetS.

2 Inbred Selection Models of MetS

Selective breeding followed by inbreeding has been used to generate numerous polygenic models of human behaviors or conditions in rats ranging from learning and alcohol preference to hypertension and diabetes [13–16]. Selective breeding involves identifying and breeding animals with a specific trait over several generations, thereby "fixing" the genetic factors contributing to the trait. Subsequent inbreeding while maintaining the trait results in a genetically predisposed inbred strain that can be used in physiological and genetic studies. There are several inbred selection rat models that have all or a subset of characteristic features of MetS.

The **Lyon Hypertensive** (LH/Mav; RGDID:10021) rat model was selectively bred for spontaneous hypertension from a

Table 1
Selected rat models of metabolic syndrome

			Traits defining metabolic syndrome				
Strain	Disease type	Strain type	Hypertension	Obesity/↑ adiposity	Hyperglycemia/glucose intolerance/ insulin resistance	Elevated triglycerides	Reduced HDL/increased TC
LH	Spontaneous	Inbred selection	√	√	√	√	√
WOKW	Spontaneous	Inbred selection	√	√	√	√	√
OLETF	Spontaneous	Inbred selection	√	√	√	√	√
SHR	Spontaneous	Inbred selection	√		√	√	√
HTG	Spontaneous	Inbred selection	√		√	√	
PD/Cub	Spontaneous	Inbred selection		√	√	√	
GK	Spontaneous	Inbred selection	√		√		
SDT	Spontaneous	Inbred selection			√	√	
CRDH	Diet induced	Inbred	√		√		
Wistar	Diet induced	Outbred	√	√	√	√	√
SD	Diet induced	Outbred	√	√	√	√	√

(continued)

Table 1 (continued)

Strain	Disease type	Strain type	Traits defining metabolic syndrome				
			Hypertension	Obesity/↑ adiposity	Hyperglycemia/glucose intolerance/ insulin resistance	Elevated triglycerides	Reduced HDL/increased TC
Obese Koletsky	Monogenic	Inbred	√	√	√	√	√
Obese Zucker	Monogenic	Inbred		√	√	√	√
ZDF	Monogenic	Inbred		√	√		

TC Total cholesterol, *HDL* high-density lipoprotein cholesterol, *LDL* low-density lipoprotein cholesterol. Strain details are provided in the text

small founder population of six SD rats [17]. Selection for normotensive (LN/Mav; RGDID:10022) and hypotensive (LL/Mav; RGDID:1581645) rats were done in parallel from the same founders. After as few as three generations of selection, LH rats were indeed spontaneously hypertensive, LN rats were normotensive, and LL were hypotensive. However, upon further inbreeding, the LL rats lost the hypotensive phenotype but remained normotensive, providing a second control strain for the LH strain [18, 19]. LH rats are both spontaneously hypertensive and salt sensitive [20]. Further study determined that, in addition to hypertension, LH rats have increased body weight, serum lipids including triglycerides and cholesterol, and altered insulin/glucose ratios [21, 22] compared to the LN control strain. While initial studies indicate the LH rat is not diabetic or insulin resistant [23], we demonstrate male LH/MRrrcAek (RGDID:10755352) rats have altered glucose tolerance (*see* Fig. 1a) and are hyperinsulinemic (*see* Fig. 1b) compared to the control male LN (LN/MRrrcAek; RGDID:10755354) rats when fed a low-fat diet (D12450H, Research Diets, Inc.). Overall, the LH strain is a comprehensive MetS model for both physiological and genetic studies. Comprehensive genetic studies in the LH rat determined that MetS is polygenic, with both independent and pleiotropic genetic effects on MetS traits, and combined fine-mapping and systems genetic studies have identified multiple loci and candidate genes [22, 24–28].

The WOKW (**Wistar Ottawa Karlsburg W** (RT1u); RGDID:67935) and OLETF (**Otsuka Long-Evans Tokushima Fatty**; RGDID:61014) rat strains are also spontaneous genetic models of MetS with obesity, hypertension, hypertriglyceridemia, abnormal plasma cholesterol levels, glucose intolerance, and hyperinsulinemia [29–34]. Through genetic mapping and fine-mapping in congenic strains, several QTL and candidate genes for MetS traits have been identified in WOKW [35–37] and OLETF rats [33, 34, 38–40]. Both strains have spontaneous risk for all major features of MetS, although the progression can be associated with age and sex. Interestingly, the OLETF was found to have a spontaneous mutation in the CCK1 receptor that contributes to its hyperphagia and obesity [41].

The **Spontaneously Hypertensive Rat** (SHR; RGDID:61000) is arguably the most comprehensively studied model of genetic hypertension with over 19,000 publications in PubMed (query SHR AND rat). Like the LH rat, SHR rats are also genetically predisposed to several features of MetS including hypertension and dyslipidemia [42–47] although they are not generally considered obese. The breeding history of the SHR is complex, and over 40 substrains of SHR are annotated in the Rat Genome Database. The stroke-prone SHRSP strains are well-recognized SHR substrains derived from SHR progenitors with spontaneous stroke and then inbred. There

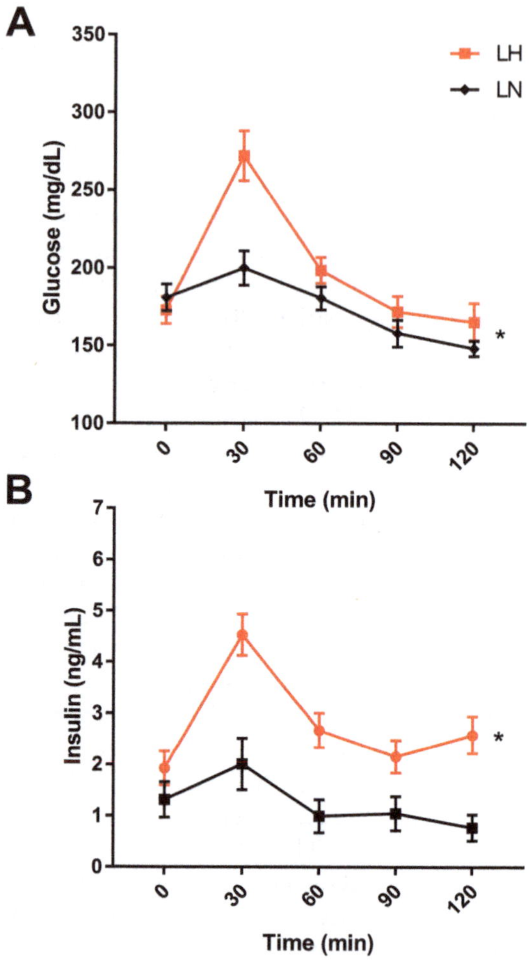

Fig. 1 Male LH/MavRrrcAek rats are glucose intolerant (**a**) and hyperinsulinemic (**b**) compared to LN/MavRrrcAek rats. Rats were fed a 10% fat diet (Research Diets D12450H) from wean. At 12 weeks of age Intraperitoneal Glucose Tolerance Tests (IPGTT) were performed on animals fasted 6 h ($\pm$0.5 h) then given 2 g glucose/kg body wt IP. Blood glucose and insulin levels were measured at 0, 30, 60, 90, and 120 min following glucose challenge. $*P < 0.05$, two-way ANOVA

are over 20 substrains of SHRSP. Gene identification using traditional positional cloning, comprehensive systems genetics, and gene targeting have been an active area of study in the SHR rat [45, 48, 49]. Dysregulation of several genes involved in MetS-related traits has been identified in the SHR rat through genetic and genomic approaches (e.g. *CD36, Ogn, Srebpf, Folr1*), and validated in transgenic and gene targeted models [49–52].

There are several additional inbred selection strains having a subset of MetS traits. For instance, the **Prague hereditary hypertriglyceridemic** (HTG; RGDID:1302795) strain is a nonobese model with high plasma triglycerides, hypertension and insulin resistance [53–55]. This model was selectively bred for high plasma triglycerides on a high-sucrose diet, suggesting possible pleiotropic genetic effects on regulation of lipids, blood pressure, and glucose, although genetic mapping studies failed to identify a single locus involved in all phenotypes [53, 56]. The **Polydactylous** rat strain (PD/Cub; RGDID:728161) has a spontaneous mutation causing polydactyly-luxate syndrome, but also shows features of MetS, including increased adiposity, insulin resistance, and hypertriglyceridemia [57]. A comprehensive genetic study that included dietary and pharmacologic stressors in an intercross between PD/Cub and BN/Cub identified several loci for MetS-related traits, some with independent effects and others showing epistatic interactions [58].

Two nonobese, spontaneously diabetic rat strains are the **Spontaneously Diabetic Torii** (SDT/Jcl; RGDID:631219) and **Goto-Kakizaki** (GK/Jcl; RGDID:13506737) strains. The SDT strain is a Type 2 diabetes model that also shows hypertriglyceridemia, severe diabetic retinopathy, nephropathy, and neuropathy [59, 60]. Several loci for hyperglycemia and glucose intolerance and diabetes have been identified in SDT rats [61, 62], and genetically induced islet inflammation is thought to play a role in developing diabetes [61]. Glucose intolerance and Type 2 diabetes in the GK rat have been extensively studied, and further studies revealed the GK strain also shows increased basal blood pressure compared to Wistar controls and salt sensitivity [63]. Several genetic loci have been identified that regulate not only the GK phenotypes but also its transcriptome and metabolome [64, 65]. Furthermore, the specific genetic contributions evolve over the lifetime of the animal and the disease progression [66, 67]. There are over 10 different substrains of GK rat annotated in the Rat Genome Database or reviewed in Portha and colleagues [66]; therefore, one must be cognizant of the substrain(s) being studied as there are substantive phenotype differences between them [66].

3 Diet-Induced Models of MetS

Most of the models described above show spontaneous features of MetS. However, induced models also exist whereby MetS traits are only evident after an environmental stressor such as diet. For example, when the inbred **Cohen Rosenthal Diabetic Hypertensive** rats (CDRH; RGDID:68019) are fed a high-sucrose, copper-poor diet, they become hypertensive and diabetic but not obese [68]. This strain was derived by crossing Cohen Diabetic Rats (CDR; RGDID:1357178) with SHR rats, then selectively breeding for both high blood glucose and high blood pressure for several generations while being fed the copper-poor, high-sucrose diet. This unique model has been used to study the effects of various anti-hypertensives on metabolic parameters in diabetic hypertension [69–71].

The **Dahl Salt-Sensitive** rat (SS; RGDID:69369) is likely the most studied salt-induced hypertension model, with nearly 3000 publication (query Dahl AND rat). While SS rats are a well-established inbred model of salt-sensitive hypertension, studies have also found SS rats display insulin resistance and dyslipidemia on normal and high-salt diets ([72, 73]; further reviewed in [74]). Furthermore, SS rats can develop hypertension, insulin resistance, and dyslipidemia when fed diets high in fat, fructose and/or sucrose [75–78]. Finally, dietary protein levels and even protein source influences the phenotypes of SS rats. A low-protein diet or replacing casein with wheat gluten as the dietary protein source lowered both blood pressure and body weight, potentially through modulation of the immune system [79, 80].

Some outbred strains also show diet-induced MetS. The two most prevalent models are **Sprague Dawley** (SD; RGDID:70508) and **Wistar** (WI; RGDID:13508588) rats fed high-energy (fat and/or sugar) diets, sometimes called DIO rats [81–83]. The SD DIO rats were developed after the observation that SD rat populations show a bimodal body weight distribution when fed a high-energy diet [84]. Levin and colleagues performed several generations of selective breeding in the obesity-prone SD rats to develop the DIO and in the obesity-resistant SD to develop the DR rats, with both lines being fed a diet high in fat and sucrose [81]. While they went through selection, there is not clear evidence that isogenic inbred strains were developed. Nevertheless, in addition to increased body weight and adiposity, most features of MetS have been observed in SD-derived DIO rats including glucose intolerance, dyslipidemia, and hypertension [81, 85]. Another outbred strain identified to have a bimodal weight distribution on high-fat diet is the Wistar strain. Studies of diet-induced obesity using this model continue to be performed in commercially available outbred Wistar rats, stratified into diet-induced or diet-resistant categories after exposure to the obesogenic moderately high-fat diet with or without high sugar [86–88]. Again, all features defining MetS have been observed in the Wistar DIO rats (reviewed in [89]). Of note, the Wistar DIO rats show evident sex differences [86].

Outbred rats such as SD and Wistar are often studied to recapitulate the heterogeneity of human populations. However, we and others have shown that phenotypes in outbred strains vary widely by source due to differing breeding histories of the strains; therefore, care must be taken when generalizing results from strains such as SD from differing vendors as they do not represent the same population sampling [90, 91].

4 Population Models of MetS

Another unique source of outbred rats was first developed by Hansen and colleagues at the National Institutes of Health (NIH), called **N/NIH heterogeneous stock** (HS) (RGDID:728185; [92]). HS rats are descendants of founders from eight inbred strains—ACI/N, BN/SsN, BUF/N, F344/N, M520/N, MR/N, WKY/N, and WN/N—that represent high genetic diversity and span the phylogenetic tree of the laboratory rat [93]. Crossbreeding between these eight founder strains was performed to minimize inbreeding and thus maximizing genetic diversity in the heterogeneous stock, making them a strong population model which is also amenable to genetic studies. While no intentional phenotypic selection was performed in the development of N/NIH, they have been shown to segregate many traits [94–96]. An advantage of the HS rats is that they segregate many traits without intentional selection. They are thus a strong model for human population studies of exposures, for example, to drugs and chemicals. The conditions most studied in HS rats include those related to psychiatric disorders and substance abuse, but also include symptoms of MetS. Solberg Woods and colleagues have performed extensive phenotyping and genotyping of HS rats, followed by genome-wide association studies to identify loci and candidate genes for adiposity and glucose homeostasis in HS rats [97, 98].

In the mid-1990s Drs. Koch and Britton selectively bred the N/NIH rats for low and high capacity exercise endurance, respectively [99]. Different from many selection models, the **Low Capacity Runners** (LCR; RGDID:2314396) and **High Capacity Runners** (HCR; RGDID:2314397) rats have been intentionally maintained as outbred stock. These rats have been extensively studied at the physiological level and studies determined the LCR rats are a robust MetS model [100, 101]. Genetic studies suggest high heritability of several metabolic traits and the potential of QTL mapping in crosses between LCR and HCR [102].

5 Genetic Models of MetS

Models of human disease often arise through spontaneous or targeted mutation of a gene that is sufficient to cause a phenotype. The most prominent monogenic rat models of MetS arose through two independent spontaneous mutations in the gene encoding the leptin receptor (*Lepr*). The **Obese Zucker** rat (ZUC-Lepr^{fa}; RGDID:629464) was described in the early 1960s [103] as a spontaneously obese rat in an outbred colony that was later found to be due to a homozygous missense mutation (Gln269Pro) in *Lepr* [104], called *fatty* or *fa* that resulted in constitutive receptor activity

[105]. Several other rat strains were derived from the Obese Zucker rat that included other MetS traits. The **Zucker Diabetic Fatty Rat** (ZDF-Lepr^{fa}/Drt; RGDID:12859287) originated from the original Zucker colony by selecting breeding rats that were both obese and diabetic [106]. Inbred MetS strains with the *fa* mutation include the obese diabetic **KZF** (KZ-Lepr^{fa}/Tky; RGDID:1302693) and models where the *fa* allele was bred onto hypertensive strains such as the SS (**SS.ZUC-Lepr$^{fa-/-}$/Slc**; RGDID:13432148) and SHRSP (**SHRSP.ZUC-(D5Rat4-D5Rat36)/IzmDmcr**; RGDID:2300018) that show all defining MetS features [107, 108].

The *cp* or *corpulent* allele of *Lepr* was identified in the **Koletsky** or **Obese SHR** rat strain (SHROB/KolGmiCrl-Lepr^{cp}; RGDID:2311049) which arose as a spontaneous *Lepr* mutation in a cross between a hypertensive SHR female and a SD male [109]. These rats were obese, hypertensive, and dyslipidemic but not diabetic [110]. The *cp* mutation was found to be a nonsense mutation, causing truncation at amino acid position 763 [111]. The *cp* mutation has also been bred into the **Diabetes Resistant BioBreeding** (BBDR/Rhw; RGDID:10003) rat to develop an obese, hypertensive, dyslipidemic, diabetic model (**BBDR.LA-(D5Rat98-D5Rat233)/Rhw**; RGDID:6893530), where the phenotype was found to be sex-dependent [112].

While the leptin receptor mutant rats were the only monogenic model of MetS for decades, transgenic rat models were developed to study MetS traits, for example, transgenic expression of *Cd36* and *Srebf1* in the SHR (RGDIDs:2302148–2302151; 2300216, 2313693) [113, 114] and SS rats overexpressing human *CETP* (RGDID:2290429) [115]. Furthermore, the successful culture of rat ES lines [116, 117] and other applications of targeted mutagenesis (e.g., using Zinc Finger Nucleases, TALENS, or CRISPR-Cas9) are opening up new opportunities to study pathobiological mechanisms in single gene rat MetS models [118–121], including melanocortin receptor (*Mc3r* and *Mc4r*) and ghrelin receptor (*Ghsr*) mutations in outbred Wistar rats and Angiopoietin-like protein 8 (*Angptl8*) in inbred F344 rats [122–124].

6 Notes

Herein we described numerous rat models that display key features of the human metabolic syndrome. The models range from single gene mutations, to inbred selection models with polygenic inheritance, to population models that perhaps best mimic the heterogeneity of human MetS. Each model has its own benefits and limitations. Some models have all defining traits of MetS and could be useful for identifying gene pleiotropy or common underlying pathophysiology linking seemingly independent traits.

However, identifying which mechanism(s) are causal may be confounded by the presence of other traits. The population models may better represent the general human clinical condition and can map traits to smaller genomic intervals than is possible with other approaches such as F2 intercrosses. A disadvantage of these models is that once a locus is identified, determining the causal variant (s) can be confounded by a mixed genomic background. Single gene models can facilitate thorough mechanistic understanding of the resulting phenotype. However, single gene MetS disorders are extremely rare in humans and do not represent the highly complex genetic, epigenetic, and environmental contributions to MetS. As such, one must carefully consider which type of model answers the research question at hand. Regardless, the rat is a rich resource for polygenic models such as the metabolic syndrome.

References

1. Reaven GM (1988) Banting lecture 1988. Role of insulin resistance in human disease. Diabetes 37(12):1595–1607
2. Alberti KG, Eckel RH, Grundy SM, Zimmet PZ, Cleeman JI, Donato KA et al (2009) Harmonizing the metabolic syndrome: a joint interim statement of the International Diabetes Federation Task Force on Epidemiology and Prevention; National Heart, Lung, and Blood Institute; American Heart Association; World Heart Federation; International Atherosclerosis Society; and International Association for the Study of Obesity. Circulation 120(16):1640–1645. https://doi.org/10.1161/CIRCULATIONAHA.109.192644
3. Saklayen MG (2018) The global epidemic of the metabolic syndrome. Curr Hypertens Rep 20(2):12. https://doi.org/10.1007/s11906-018-0812-z
4. Nolan PB, Carrick-Ranson G, Stinear JW, Reading SA, Dalleck LC (2017) Prevalence of metabolic syndrome and metabolic syndrome components in young adults: a pooled analysis. Prev Med Rep 7:211–215. https://doi.org/10.1016/j.pmedr.2017.07.004
5. Lin HF, Boden-Albala B, Juo SH, Park N, Rundek T, Sacco RL (2005) Heritabilities of the metabolic syndrome and its components in the Northern Manhattan Family Study. Diabetologia 48(10):2006–2012. https://doi.org/10.1007/s00125-005-1892-2
6. Henneman P, Aulchenko YS, Frants RR, van Dijk KW, Oostra BA, van Duijn CM (2008) Prevalence and heritability of the metabolic syndrome and its individual components in a Dutch isolate: the Erasmus Rucphen Family study. J Med Genet 45(9):572–577. https://doi.org/10.1136/jmg.2008.058388
7. Khan RJ, Gebreab SY, Sims M, Riestra P, Xu R, Davis SK (2015) Prevalence, associated factors and heritabilities of metabolic syndrome and its individual components in African Americans: the Jackson Heart Study. BMJ Open 5(10):e008675. https://doi.org/10.1136/bmjopen-2015-008675
8. van Dongen J, Willemsen G, Chen WM, de Geus EJ, Boomsma DI (2013) Heritability of metabolic syndrome traits in a large population-based sample. J Lipid Res 54(10):2914–2923. https://doi.org/10.1194/jlr.P041673
9. Hayman GT, Laulederkind SJ, Smith JR, Wang SJ, Petri V, Nigam R et al (2016) The disease portals, disease-gene annotation and the rgd disease ontology at the rat genome database. Database 2016:baw034. https://doi.org/10.1093/database/baw034
10. Oschry Y, Eisenberg S (1982) Rat plasma lipoproteins: re-evaluation of a lipoprotein system in an animal devoid of cholesteryl ester transfer activity. J Lipid Res 23(8):1099–1106
11. Bergen WG, Mersmann HJ (2005) Comparative aspects of lipid metabolism: impact on contemporary research and use of animal models. J Nutr 135(11):2499–2502. https://doi.org/10.1093/jn/135.11.2499
12. Yin W, Carballo-Jane E, McLaren DG, Mendoza VH, Gagen K, Geoghagen NS et al (2012) Plasma lipid profiling across species for the identification of optimal animal models of human dyslipidemia. J Lipid Res 53(1):51–65. https://doi.org/10.1194/jlr.M019927

13. Tryon RC (1940) Genetic differences in maze-learning ability in rats. In: Thirty-ninth yearbook of the National Society for the Study of Education. Intelligence: its nature and nurture, part 1. Comparative and critical exposition. Public School Publishing Co., Bloomington, IN, pp 111–119
14. Eriksson K (1968) Genetic selection for voluntary alcohol consumption in the albino rat. Science 159(3816):739–741. https://doi.org/10.1126/science.159.3816.739
15. Rapp JP (2000) Genetic analysis of inherited hypertension in the rat. Physiol Rev 80 (1):135–172
16. Goto Y, Kakizaki M, Masaki N (1976) Production of spontaneous diabetic rats by repetition of selective breeding. Tohoku J Exp Med 119(1):85–90
17. Dupont J, Dupont JC, Froment A, Milon H, Vincent M (1973) Selection of three strains of rats with spontaneously different levels of blood pressure. Biomedicine 19(1):36–41
18. Su DF, Cerutti C, Barres C, Vincent M, Sassard J (1986) Blood pressure and baroreflex sensitivity in conscious hypertensive rats of Lyon strain. Am J Phys 251(6 Pt 2): H1111–H1117
19. Sassard J, Vincent M, Orea V, Privat P, Bataillard A (1997) Genetics of blood pressure and associated phenotypes in the Lyon rat. Clin Exp Hypertens 19(5–6):567–575
20. Florin M, Lo M, Liu KL, Sassard J (2001) Salt sensitivity in genetically hypertensive rats of the Lyon strain. Kidney Int 59(5):1865–1872
21. Sassolas A, Vincent M, Benzoni D, Sassard J (1981) Plasma lipids in genetically hypertensive rats of the Lyon strain. J Cardiovasc Pharmacol 3(5):1008–1014
22. Vincent M, Boussairi EH, Cartier R, Lo M, Sassolas A, Cerutti C et al (1993) High blood pressure and metabolic disorders are associated in the Lyon hypertensive rat. J Hypertens 11(11):1179–1185
23. Boulanger M, Duhault J, Broux O, Bataillard A, Sassard J (1997) Lack of insulin resistance in the Lyon hypertensive rat. Fundam Clin Pharmacol 11(6):546–549
24. Vincent M, Cartier R, Privat P, Benzoni D, Samani NJ, Sassard J (1996) Major cardiovascular risk factors in Lyon hypertensive rats. A correlation analysis in a segregating population. J Hypertens 14(4):469-474
25. Dubay C, Vincent M, Samani NJ, Hilbert P, Kaiser MA, Beressi J-P et al (1993) Genetic determinants of diastolic and pulse pressure map to different loci in Lyon hypertensive rats. Nat Genet 3(4):354–357
26. Bilusic M, Bataillard A, Tschannen MR, Gao L, Barreto NE, Vincent M et al (2004) Mapping the genetic determinants of hypertension, metabolic diseases, and related phenotypes in the lyon hypertensive rat. Hypertension 44(5):695–701. https://doi.org/10.1161/01.HYP.0000144542.57306.5e
27. Ma MCJ, Pettus JM, Jakoubek JA, Traxler MG, Clark KC, Mennie AK et al (2017) Contribution of independent and pleiotropic genetic effects in the metabolic syndrome in a hypertensive rat. PLoS One 12(8): e0182650. https://doi.org/10.1371/journal.pone.0182650
28. Wang J, Ma MC, Mennie AK, Pettus JM, Xu Y, Lin L et al (2015) Systems biology with high-throughput sequencing reveals genetic mechanisms underlying the metabolic syndrome in the Lyon hypertensive rat. Circ Cardiovasc Genet 8(2):316–326. https://doi.org/10.1161/CIRCGENETICS.114.000520
29. Kloting N, Bluher M, Kloting I (2006) The polygenetically inherited metabolic syndrome of WOKW rats is associated with insulin resistance and altered gene expression in adipose tissue. Diabetes Metab Res Rev 22 (2):146–154. https://doi.org/10.1002/dmrr.582
30. van den Brandt J, Kovacs P, Kloting I (2002) Metabolic syndrome and aging in Wistar Ottawa Karlsburg W rats. Int J Obes Relat Metab Disord 26(4):573–576
31. van den Brandt J, Kovacs P, Kloting I (2000) Features of the metabolic syndrome in the spontaneously hypertriglyceridemic Wistar Ottawa Karlsburg W (RT1u Haplotype) rat. Metabolism 49(9):1140–1144. https://doi.org/10.1053/meta.2000.8610
32. Kovacs P, Voigt B, Berg S, Vogt L, Kloting I (1997) WOK.1W rats. A potential animal model of the insulin resistance syndrome. Ann N Y Acad Sci 827:94–99
33. Yagi K, Kim S, Wanibuchi H, Yamashita T, Yamamura Y, Iwao H (1997) Characteristics of diabetes, blood pressure, and cardiac and renal complications in Otsuka Long-Evans Tokushima Fatty rats. Hypertension 29 (3):728–735
34. Kawano K, Hirashima T, Mori S, Saitoh Y, Kurosumi M, Natori T (1992) Spontaneous long-term hyperglycemic rat with diabetic complications. Otsuka Long-Evans Tokushima Fatty (OLETF) strain. Diabetes 41 (11):1422–1428
35. Weingarten A, Turchetti L, Krohn K, Kloting I, Kern M, Kovacs P et al (2016)

Novel genes on rat chromosome 10 are linked to body fat mass, preadipocyte number and adipocyte size. Int J Obes 40 (12):1832–1840. https://doi.org/10.1038/ijo.2016.127

36. Baguhl R, Wilke B, Kloting N, Kloting I (2009) Genes on rat chromosomes 3, 5, 10, and 16 are linked with facets of metabolic syndrome. Obesity (Silver Spring) 17 (6):1215–1219. https://doi.org/10.1038/oby.2008.658
37. Kloting N, Wilke B, Kloting I (2007) Triplet repeat in the Repin1 3′-untranslated region on rat chromosome 4 correlates with facets of the metabolic syndrome. Diabetes Metab Res Rev 23(5):406–410. https://doi.org/10.1002/dmrr.713
38. Kose H, Moralejo DH, Ogino T, Mizuno A, Yamada T, Matsumoto K (2002) Examination of OLETF-derived non-insulin-dependent diabetes mellitus QTL by construction of a series of congenic rats. Mamm Genome 13 (10):558–562. https://doi.org/10.1007/s00335-002-2199-y
39. Muramatsu Y, Yamada T, Taniguchi Y, Ogino T, Kose H, Matsumoto K et al (2005) Pnlip encoding pancreatic lipase is possible candidate for obesity QTL in the OLETF rat. Biochem Biophys Res Commun 331(4):1270–1276. https://doi.org/10.1016/j.bbrc.2005.04.040
40. Watanabe TK, Suzuki M, Yamasaki Y, Okuno S, Hishigaki H, Ono T et al (2005) Mutated G-protein-coupled receptor GPR10 is responsible for the hyperphagia/dyslipidaemia/obesity locus of Dmo1 in the OLETF rat. Clin Exp Pharmacol Physiol 32 (5–6):355–366. https://doi.org/10.1111/j.1440-1681.2005.04196.x
41. Moran TH (2008) Unraveling the obesity of OLETF rats. Physiol Behav 94(1):71–78. https://doi.org/10.1016/j.physbeh.2007.11.035
42. Swales JD (1994) Textbook of hypertension. Blackwell Scientific Publications, Oxford
43. de Jong W (1984) Experimental and genetic models of hypertension. In: Birkenhager WH, Reid JL (eds) Handbook of hypertension, vol 4. Elsevier, Amsterdam
44. Coan PM, Hummel O, Garcia Diaz A, Barrier M, Alfazema N, Norsworthy PJ et al (2017) Genetic, physiological and comparative genomic studies of hypertension and insulin resistance in the spontaneously hypertensive rat. Dis Model Mech 10(3):297–306. https://doi.org/10.1242/dmm.026716
45. Pravenec M, Kren V, Landa V, Mlejnek P, Musilova A, Silhavy J et al (2014) Recent progress in the genetics of spontaneously hypertensive rats. Physiol Res 63(Suppl 1): S1–S8
46. Tofovic SP, Jackson EK (2003) Rat models of the metabolic syndrome. Methods Mol Med 86:29–46. https://doi.org/10.1385/1-59259-392-5:29
47. Aitman TJ, Gotoda T, Evans AL, Imrie H, Heath KE, Trembling PM et al (1997) Quantitative trait loci for cellular defects in glucose and fatty acid metabolism in hypertensive rats. Nat Genet 16(2):197–201. https://doi.org/10.1038/ng0697-197
48. Doris PA (2017) Genetics of hypertension: an assessment of progress in the spontaneously hypertensive rat. Physiol Genomics 49 (11):601–617. https://doi.org/10.1152/physiolgenomics.00065.2017
49. Pravenec M, Kurtz TW (2010) Recent advances in genetics of the spontaneously hypertensive rat. Curr Hypertens Rep 12 (1):5–9. https://doi.org/10.1007/s11906-009-0083-9
50. Aitman TJ, Glazier AM, Wallace CA, Cooper LD, Norsworthy PJ, Wahid FN et al (1999) Identification of Cd36 (Fat) as an insulin-resistance gene causing defective fatty acid and glucose metabolism in hypertensive rats. Nat Genet 21(1):76–83
51. Pravenec M, Kozich V, Krijt J, Sokolova J, Zidek V, Landa V et al (2016) Genetic variation in renal expression of folate receptor 1 (Folr1) gene predisposes spontaneously hypertensive rats to metabolic syndrome. Hypertension 67(2):335–341. https://doi.org/10.1161/HYPERTENSIONAHA.115.06158
52. Pravenec M, Zidek V, Landa V, Mlejnek P, Silhavy J, Simakova M et al (2017) Mutant Wars2 gene in spontaneously hypertensive rats impairs brown adipose tissue function and predisposes to visceral obesity. Physiol Res 66(6):917–924
53. Zicha J, Pechanova O, Cacanyiova S, Cebova M, Kristek F, Torok J (2006) Hereditary hypertriglyceridemic rat: a suitable model of cardiovascular disease and metabolic syndrome? Physiol Res 55(Suppl 1):S49–S63
54. Klimes I, Vrana A, Kunes J, Sebokova E, Dobesova Z, Stolba P et al (1995) Hereditary hypertriglyceridemic rat: a new animal model of metabolic alterations in hypertension. Blood Press 4(3):137–142
55. Stolba P, Dobesova Z, Husek P, Opltova H, Zicha J, Vrana A et al (1992) The hypertriglyceridemic rat as a genetic model of hypertension and diabetes. Life Sci 51(10):733–740
56. Ueno T, Tremblay J, Kunes J, Zicha J, Dobesova Z, Pausova Z et al (2004) Rat

model of familial combined hyperlipidemia as a result of comparative mapping. Physiol Genomics 17(1):38–47

57. Sedova L, Kazdova L, Seda O, Krenova D, Kren V (2000) Rat inbred PD/cub strain as a model of dyslipidemia and insulin resistance. Folia Biol (Praha) 46(3):99–106
58. Seda O, Liska F, Krenova D, Kazdova L, Sedova L, Zima T et al (2005) Dynamic genetic architecture of metabolic syndrome attributes in the rat. Physiol Genomics 21 (2):243–252. https://doi.org/10.1152/physiolgenomics.00230.2004
59. Sasase T, Ohta T, Masuyama T, Yokoi N, Kakehashi A, Shinohara M (2013) The spontaneously diabetic Torii rat: an animal model of nonobese type 2 diabetes with severe diabetic complications. J Diabetes Res 2013:976209. https://doi.org/10.1155/2013/976209
60. Shinohara M, Masuyama T, Shoda T, Takahashi T, Katsuda Y, Komeda K et al (2000) A new spontaneously diabetic non-obese Torii rat strain with severe ocular complications. Int J Exp Diabetes Res 1 (2):89–100
61. Fuse M, Yokoi N, Shinohara M, Masuyama T, Kitazawa R, Kitazawa S et al (2008) Identification of a major locus for islet inflammation and fibrosis in the spontaneously diabetic Torii rat. Physiol Genomics 35(1):96–105. https://doi.org/10.1152/physiolgenomics.90214.2008
62. Masuyama T, Fuse M, Yokoi N, Shinohara M, Tsujii H, Kanazawa M et al (2003) Genetic analysis for diabetes in a new rat model of nonobese type 2 diabetes, Spontaneously Diabetic Torii rat. Biochem Biophys Res Commun 304(1):196–206
63. Cheng ZJ, Vaskonen T, Tikkanen I, Nurminen K, Ruskoaho H, Vapaatalo H et al (2001) Endothelial dysfunction and salt-sensitive hypertension in spontaneously diabetic Goto-Kakizaki rats. Hypertension 37 (2 Pt 2):433–439
64. Kaisaki PJ, Otto GW, Argoud K, Collins SC, Wallis RH, Wilder SP et al (2016) Transcriptome profiling in rat inbred strains and experimental cross reveals discrepant genetic architecture of genome-wide gene expression. G3 (Bethesda) 6(11):3671–3683. https://doi.org/10.1534/g3.116.033274
65. Dumas ME, Wilder SP, Bihoreau MT, Barton RH, Fearnside JF, Argoud K et al (2007) Direct quantitative trait locus mapping of mammalian metabolic phenotypes in diabetic and normoglycemic rat models. Nat Genet 39 (5):666–672. https://doi.org/10.1038/ng2026
66. Portha B, Giroix MH, Tourrel-Cuzin C, Le-Stunff H, Movassat J (2012) The GK rat: a prototype for the study of non-overweight type 2 diabetes. Meth Mol Biol 933:125–159. https://doi.org/10.1007/978-1-62703-068-7_9
67. Nobrega MA, Woods LC, Fleming S, Jacob HJ (2009) Distinct genetic regulation of progression of diabetes and renal disease in the Goto-Kakizaki rat. Physiol Genomics 39 (1):38–46. https://doi.org/10.1152/physiolgenomics.90389.2008
68. Cohen AM, Rosenmann E, Rosenthal T (1993) The Cohen diabetic (non-insulin-dependent) hypertensive rat model. Description of the model and pathologic findings. Am J Hypertens 6(12):989–995
69. Younis F, Leor J, Abassi Z, Landa N, Rath L, Hollander K et al (2018) Beneficial effect of the SGLT2 inhibitor empagliflozin on glucose homeostasis and cardiovascular parameters in the Cohen Rosenthal Diabetic Hypertensive (CRDH) rat. J Cardiovasc Pharmacol Ther 23(4):358–371. https://doi.org/10.1177/1074248418763808
70. Younis F, Stern N, Limor R, Oron Y, Zangen S, Rosenthal T (2010) Telmisartan ameliorates hyperglycemia and metabolic profile in nonobese Cohen-Rosenthal diabetic hypertensive rats via peroxisome proliferator activator receptor-gamma activation. Metabolism 59(8):1200–1209. https://doi.org/10.1016/j.metabol.2009.11.013
71. Rosenthal T, Erlich Y, Rosenmann E, Cohen A (1997) Effects of enalapril, losartan, and verapamil on blood pressure and glucose metabolism in the Cohen-Rosenthal diabetic hypertensive rat. Hypertension 29 (6):1260–1264
72. Reaven GM, Twersky J, Chang H (1991) Abnormalities of carbohydrate and lipid metabolism in Dahl rats. Hypertension 18 (5):630–635
73. Somova L, Channa ML (1999) Glucose metabolism and insulin sensitivity in Dahl hypertensive rats. Methods Find Exp Clin Pharmacol 21(6):421–425
74. Zicha J, Dobesova Z, Vokurkova M, Rauchova H, Hojna S, Kadlecova M et al (2012) Age-dependent salt hypertension in Dahl rats: fifty years of research. Physiol Res 61(Suppl 1):S35–S87
75. Donnelly R, Ho H, Reaven GM (1995) Effects of low sodium diet and unilateral nephrectomy on the development of

carbohydrate-induced hypertension. Blood Press 4(3):164–169

76. Kotchen TA, Zhang HY, Covelli M, Blehschmidt N (1991) Insulin resistance and blood pressure in Dahl rats and in one-kidney, one--clip hypertensive rats. Am J Phys 261(6 Pt 1): E692–E697
77. Nagae A, Fujita M, Kawarazaki H, Matsui H, Ando K, Fujita T (2009) Effect of high fat loading in Dahl salt-sensitive rats. Clin Exp Hypertens 31(5):451–461
78. Zhang HY, Reddy S, Kotchen TA (1999) A high sucrose, high linoleic acid diet potentiates hypertension in the Dahl salt sensitive rat. Am J Hypertens 12(2 Pt 1):183–187
79. De Miguel C, Lund H, Mattson DL (2011) High dietary protein exacerbates hypertension and renal damage in Dahl SS rats by increasing infiltrating immune cells in the kidney. Hypertension 57(2):269–274. https://doi.org/10.1161/HYPERTENSIONAHA.110.154302
80. Mattson DL, Meister CJ, Marcelle ML (2005) Dietary protein source determines the degree of hypertension and renal disease in the Dahl salt-sensitive rat. Hypertension 45 (4):736–741. https://doi.org/10.1161/01.HYP.0000153318.74544.cc
81. Levin BE, Dunn-Meynell AA, Balkan B, Keesey RE (1997) Selective breeding for diet-induced obesity and resistance in Sprague-Dawley rats. Am J Phys 273(2 Pt 2): R725–R730. https://doi.org/10.1152/ajpregu.1997.273.2.R725
82. Chang S, Graham B, Yakubu F, Lin D, Peters JC, Hill JO (1990) Metabolic differences between obesity-prone and obesity-resistant rats. Am J Phys 259(6 Pt 2):R1103–R1110. https://doi.org/10.1152/ajpregu.1990.259.6.R1103
83. Levin BE, Triscari J, Hogan S, Sullivan AC (1987) Resistance to diet-induced obesity: food intake, pancreatic sympathetic tone, and insulin. Am J Phys 252(3 Pt 2): R471–R478. https://doi.org/10.1152/ajpregu.1987.252.3.R471
84. Schemmel R, Mickelsen O, Gill JL (1970) Dietary obesity in rats: body weight and body fat accretion in seven strains of rats. J Nutr 100(9):1041–1048. https://doi.org/10.1093/jn/100.9.1041
85. Dobrian AD, Davies MJ, Prewitt RL, Lauterio TJ (2000) Development of hypertension in a rat model of diet-induced obesity. Hypertension 35(4):1009–1015
86. Giles ED, Jackman MR, MacLean PS (2016) Modeling diet-induced obesity with obesity-prone rats: implications for studies in females. Front Nutr 3:50. https://doi.org/10.3389/fnut.2016.00050
87. Panchal SK, Poudyal H, Iyer A, Nazer R, Alam A, Diwan V et al (2011) High-carbohydrate high-fat diet-induced metabolic syndrome and cardiovascular remodeling in rats. J Cardiovasc Pharmacol 57(1):51–64. https://doi.org/10.1097/FJC.0b013e3181feb90a
88. Moreno-Fernandez S, Garces-Rimon M, Vera G, Astier J, Landrier JF, Miguel M (2018) High fat/high glucose diet induces metabolic syndrome in an experimental rat model. Nutrients 10(10):E1502. https://doi.org/10.3390/nu10101502
89. Wong SK, Chin KY, Suhaimi FH, Fairus A, Ima-Nirwana S (2016) Animal models of metabolic syndrome: a review. Nutr Metab (Lond) 13:65. https://doi.org/10.1186/s12986-016-0123-9
90. Kwitek AE, Jacob HJ, Baker JE, Dwinell MR, Forster HV, Greene AS et al (2006) BN phenome: detailed characterization of the cardiovascular, renal, and pulmonary systems of the sequenced rat. Physiol Genomics 25 (2):303–313. https://doi.org/10.1152/physiolgenomics.00288.2005
91. Brower M, Grace M, Kotz CM, Koya V (2015) Comparative analysis of growth characteristics of Sprague Dawley rats obtained from different sources. Lab Anim Res 31 (4):166–173. https://doi.org/10.5625/lar.2015.31.4.166
92. Hansen C, Spuhler K (1984) Development of the National Institutes of Health genetically heterogeneous rat stock. Alcohol Clin Exp Res 8(5):477–479
93. Hermsen R, de Ligt J, Spee W, Blokzijl F, Schafer S, Adami E et al (2015) Genomic landscape of rat strain and substrain variation. BMC Genomics 16:357. https://doi.org/10.1186/s12864-015-1594-1
94. Baud A, Guryev V, Hummel O, Johannesson M, Rat Genome Sequencing and Mapping Consortium, Flint J (2014) Genomes and phenomes of a population of outbred rats and its progenitors. Sci Data 1:140011. https://doi.org/10.1038/sdata.2014.11
95. Woods LC, Mott R (2017) Heterogeneous stock populations for analysis of complex traits. Meth Mol Biol 1488:31–44. https://doi.org/10.1007/978-1-4939-6427-7_2
96. Alam I, Koller DL, Canete T, Blazquez G, Mont-Cardona C, Lopez-Aumatell R et al (2015) Fine mapping of bone structure and

strength QTLs in heterogeneous stock rat. Bone 81:417–426. https://doi.org/10.1016/j.bone.2015.08.013

97. Holl K, He H, Wedemeyer M, Clopton L, Wert S, Meckes JK et al (2017) Heterogeneous stock rats: a model to study the genetics of despair-like behavior in adolescence. Genes Brain Behav 17(2):139–148. https://doi.org/10.1111/gbb.12410
98. Keele GR, Prokop JW, He H, Holl K, Littrell J, Deal A et al (2017) Genetic fine-mapping and identification of candidate genes and variants for adiposity traits in outbred rats. Obesity (Silver Spring) 26(1):213–222. https://doi.org/10.1002/oby.22075
99. Koch LG, Britton SL (2001) Artificial selection for intrinsic aerobic endurance running capacity in rats. Physiol Genomics 5 (1):45–52. https://doi.org/10.1152/physiolgenomics.2001.5.1.45
100. Koch LG, Britton SL, Wisloff U (2012) A rat model system to study complex disease risks, fitness, aging, and longevity. Trends Cardiovasc Med 22(2):29–34. https://doi.org/10.1016/j.tcm.2012.06.007
101. Koch LG, Britton SL (2008) Development of animal models to test the fundamental basis of gene-environment interactions. Obesity (Silver Spring) 16(Suppl 3):S28–S32. https://doi.org/10.1038/oby.2008.513
102. Ren YY, Overmyer KA, Qi NR, Treutelaar MK, Heckenkamp L, Kalahar M et al (2013) Genetic analysis of a rat model of aerobic capacity and metabolic fitness. PLoS One 8 (10):e77588. https://doi.org/10.1371/journal.pone.0077588
103. Zucker LM, Zucker TF (1961) FATTY, a new mutation in the rat. J Hered 52(6):275–278. https://doi.org/10.1093/oxfordjournals.jhered.a107093
104. Takaya K, Ogawa Y, Isse N, Okazaki T, Satoh N, Masuzaki H et al (1996) Molecular cloning of rat leptin receptor isoform complementary DNAs—identification of a missense mutation in Zucker fatty (fa/fa) rats. Biochem Biophys Res Commun 225(1):75–83. https://doi.org/10.1006/bbrc.1996.1133
105. White DW, Wang DW, Chua SC Jr, Morgenstern JP, Leibel RL, Baumann H et al (1997) Constitutive and impaired signaling of leptin receptors containing the Gln --> Pro extracellular domain fatty mutation. Proc Natl Acad Sci U S A 94(20):10657–10662
106. Peterson RG, Shaw WN, Neel M-A, Little LA, Eichberg J (1990) Zucker diabetic fatty rat as a model for non-insulin-dependent diabetes mellitus. ILAR J 32(3):16–19. https://doi.org/10.1093/ilar.32.3.16
107. Hattori T, Murase T, Ohtake M, Inoue T, Tsukamoto H, Takatsu M et al (2011) Characterization of a new animal model of metabolic syndrome: the DahlS.Z-Lepr(fa)/Lepr (fa) rat. Nutr Diabetes 1:e1. https://doi.org/10.1038/nutd.2010.1
108. Hiraoka-Yamamoto J, Nara Y, Yasui N, Onobayashi Y, Tsuchikura S, Ikeda K (2004) Establishment of a new animal model of metabolic syndrome: SHRSP fatty (fa/fa) rats. Clin Exp Pharmacol Physiol 31 (1–2):107–109
109. Koletsky S (1973) Obese spontaneously hypertensive rats—a model for study of atherosclerosis. Exp Mol Pathol 19(1):53–60
110. Koletsky S (1975) Pathologic findings and laboratory data in a new strain of obese hypertensive rats. Am J Pathol 80(1):129–142
111. Takaya K, Ogawa Y, Hiraoka J, Hosoda K, Yamori Y, Nakao K et al (1996) Nonsense mutation of leptin receptor in the obese spontaneously hypertensive Koletsky rat. Nat Genet 14(2):130–131. https://doi.org/10.1038/ng1096-130
112. Moralejo DH, Hansen CT, Treuting P, Hessner MJ, Fuller JM, Van Yserloo B et al (2010) Differential effects of leptin receptor mutation on male and female BBDR Gimap5−/Gimap5− spontaneously diabetic rats. Physiol Genomics 41(1):9–20. https://doi.org/10.1152/physiolgenomics.00186.2009
113. Qi NR, Wang J, Zidek V, Landa V, Mlejnek P, Kazdova L et al (2005) A new transgenic rat model of hepatic steatosis and the metabolic syndrome. Hypertension 45(5):1004–1011
114. Pravenec M, Landa V, Zidek V, Musilova A, Kazdova L, Qi N et al (2003) Transgenic expression of CD36 in the spontaneously hypertensive rat is associated with amelioration of metabolic disturbances but has no effect on hypertension. Physiol Res 52 (6):681–688
115. Herrera VL, Makrides SC, Xie HX, Adari H, Krauss RM, Ryan US et al (1999) Spontaneous combined hyperlipidemia, coronary heart disease and decreased survival in Dahl salt-sensitive hypertensive rats transgenic for human cholesteryl ester transfer protein. Nat Med 5(12):1383–1389. https://doi.org/10.1038/70956
116. Buehr M, Meek S, Blair K, Yang J, Ure J, Silva J et al (2008) Capture of authentic embryonic stem cells from rat blastocysts. Cell 135 (7):1287–1298

117. Li P, Tong C, Mehrian-Shai R, Jia L, Wu N, Yan Y et al (2008) Germline competent embryonic stem cells derived from rat blastocysts. Cell 135(7):1299–1310

118. Tesson L, Usal C, Menoret S, Leung E, Niles BJ, Remy S et al (2011) Knockout rats generated by embryo microinjection of TALENs. Nat Biotechnol 29(8):695–696. https://doi.org/10.1038/nbt.1940

119. Geurts AM, Cost GJ, Freyvert Y, Zeitler B, Miller JC, Choi VM et al (2009) Knockout rats via embryo microinjection of zinc-finger nucleases. Science 325(5939):433. pii: 325/5939/433. https://doi.org/10.1126/science.1172447

120. Li W, Teng F, Li T, Zhou Q (2013) Simultaneous generation and germline transmission of multiple gene mutations in rat using CRISPR-Cas systems. Nat Biotechnol 31 (8):684–686. https://doi.org/10.1038/nbt.2652

121. Meek S, Mashimo T, Burdon T (2017) From engineering to editing the rat genome. Mamm Genome 28(7):302–314. https://doi.org/10.1007/s00335-017-9705-8

122. Izumi R, Kusakabe T, Noguchi M, Iwakura H, Tanaka T, Miyazawa T et al (2018) CRISPR/Cas9-mediated Angptl8 knockout suppresses plasma triglyceride concentrations and adiposity in rats. J Lipid Res 59(9):1575–1585. https://doi.org/10.1194/jlr.M082099

123. Zallar LJ, Tunstall BJ, Richie CT, Zhang YJ, You ZB, Gardner EL et al (2019) Development and initial characterization of a novel ghrelin receptor CRISPR/Cas9 knockout wistar rat model. Int J Obes 43:344–354. https://doi.org/10.1038/s41366-018-0013-5

124. You P, Hu H, Chen Y, Zhao Y, Yang Y, Wang T et al (2016) Effects of melanocortin 3 and 4 receptor deficiency on energy homeostasis in rats. Sci Rep 6:34938. https://doi.org/10.1038/srep34938

Chapter 14

Genomic Research in Rat Models of Kidney Disease

Yoram Yagil, Ronen Levi-Varadi, and Chana Yagil

Abstract

Current understanding of the mechanisms underlying renal disease in humans is incomplete. Consequently, our ability to prevent the occurrence of renal disease or treat established kidney disease is limited. Investigating kidney disease directly in humans poses objective difficulties, which has led investigators to seek experimental animal models that simulate renal disease in humans. Animal models have thus become a tool of major importance in the study of renal physiology and have been crucial in shedding light on the complex mechanisms involved in kidney function and in our current understanding of the pathophysiology of renal disease. Among animal models, the rat has been the preferred and most commonly used species for the investigation of renal disease. This chapter reviews what has been achieved over the years, using the rat as a tool for the investigation of renal disease in humans, focusing on the contribution of rat genetics and genomics to the elucidation of the mechanisms underlying the pathophysiology of the major types of renal disease, including primary and secondary renal diseases.

Key words Kidneys, Disease, Primary kidney disease, Secondary kidney disease, Function, Experimental models, Rodents, Rat, Pathophysiology, Mechanisms, Genetic basis, Genomics, Genes, Expression, Transcriptomics, Proteomics

1 Introduction

The investigation of renal disease, whether primary or secondary, has been ongoing for several decades and a vast amount of knowledge has accrued over the years, some of which has been successfully translated into clinical practice. Current understanding of the mechanisms underlying renal disease remains, nonetheless, incomplete. Consequently, our ability to prevent the occurrence of renal disease or treat renal disease once it develops is limited, as is evident from the continuously increasing number of patients who develop chronic kidney disease that often leads to end stage renal failure (http://www.usrds.org). There is, therefore, dire need to continue in-depth investigation of the mechanisms underlying kidney disease, the predisposition (susceptibility and resistance) to develop the disease, and the cause of its frequent, relentless progression.

G. Thomas Hayman et al. (eds.), *Rat Genomics*, Methods in Molecular Biology, vol. 2018,
https://doi.org/10.1007/978-1-4939-9581-3_14,

There are objective difficulties in the continuing investigation of kidney disease in humans. One is inherent in the sheer definition of "kidney disease," a general term that encompasses a large variety of clinical and pathological entities that affect the kidneys, all of which ultimately culminate in renal failure. There are many types and subtypes of renal disease, and although there may be common pathophysiological pathways, each needs to be investigated in separate, a huge task by itself. A second difficulty is the fact that the kidneys are not readily accessible for studying, and much of the data reflecting renal disease are derived from surrogates of renal disease, such as protein excretion or indices of glomerular filtration rate, all of which are not truly satisfactory. Histological assessment of renal tissue is possible through kidney biopsy, but that is an invasive procedure that in most cases cannot be ethically justified, if performed solely for research purposes. The difficulty in investigating renal disease directly in humans has led investigators to resort to experimental models of renal disease, in the hope that such models simulate and represent as closely as possible renal disease in humans.

Animal models have indeed been a tool of major importance in the study of normal renal physiology and have been crucial in shedding light on the complex mechanisms involved in the normal daily function of the kidney, whether at the whole organ, cellular or molecular level. Animal models simulating renal disease have also contributed immensely to our understanding of renal disease and to our current ability to treat it. Among the animal models that have been used to study renal disease stand out rodents, dogs, and pigs, in that order of importance. Among rodents, traditional physiological and pathophysiological studies have been successfully carried out in the rat, and to a lesser extent in the mouse. The ease of use of the rat for phenotyping of the normal and diseased kidney has rendered it as the preferred species for investigating the kidney. Technological advances that have improved the ability to phenotype the mouse kidney and the long-standing availability of knockout technology in the mouse have rendered the mouse as an attractive alternative to the rat. Over the past decade, however, gene knockout and knock-in technologies have become readily available also in the rat and the rat continues to be the preferred and most commonly used experimental model for the investigation of renal disease.

Research of kidney disease in experimental models has been based in the past on traditional physiology and histopathology, which has had a major impact on our knowledge of how the kidney functions in both health and disease. Our current level of understanding of kidney disease, however, remains incomplete and largely unsatisfactory. With the advent of genomics and the sequencing of several mammalian genomes, there has been much hope that advanced technological tools that have come along with the genomic revolution would allow a major leap forwards in our understanding of disease in all its forms, including kidney disease. Over the years, there has been extensive investigation of renal disease using the traditional "deductive" candidate gene approach, based largely on what is already known about the pathophysiology of renal disease. There has subsequently been abundant "inductive" research, using the genomic and transcriptomic approaches based on linkage analysis in crosses between informative rat strains, targeted construction of consomic and congenic strains, differential gene expression profiling and more recently proteomics. Gain-of-function studies by over-expressing select candidate genes in the form of transgenic rats have been successfully achieved [1]. With the advent of knockout technology in the rat, loss of function studies have also become feasible, using targeted tissue-specific zinc-finger nucleases (ZFNs) and CRISPR-based technologies [2–5].

In this chapter, we aim to review what has been achieved using the rat as a tool for the investigation of renal disease in humans, focusing on the contribution of rat genetics and genomics to the elucidation of the mechanisms underlying the pathophysiology of the major types of renal disease, including primary and secondary renal diseases. We will not address in this review the important extensive genetic work that has been carried out deductively through the candidate gene approach, as such discussion is beyond the scope of this chapter. We will focus instead primarily on the inductive type of research that has incorporated genomics, transcriptomics, proteomics, and complex bioinformatic tools. Furthermore, it is clear that a large number of studies have been carried out in models of kidney disease not mentioned by us in this chapter, and it would be presumptuous for us to claim to have covered them all. We do hope, however, to provide sufficient coverage to provide the reader with an objective overview of the range of research that has been carried out in this field, and of the achievements attained so far.

2 Pitfalls and Limitations in the Investigation of Renal Disease

Prior to evaluating the contribution of genetic research to our understanding of renal disease, it is important to comprehend the limitations of such research in rodent models and its applicability to human disease.

The study of renal disease in experimental animal models has been based on the premise that it reproduces, at least to some degree, the disease in humans. To prove reproducibility of the human disease in the animal model, one has to show similarity to the human disease in the inciting etiological factors, in the pathophysiology of the disease, in the resulting histopathology of the lesion within the affected organ, in the level of renal dysfunction and in the course of the disease. One prerequisite in showing such reproducibility is the ability to accurately define the disease phenotype and translate it from the animal model to clinical terms in humans. A detailed discussion of these features is of essence.

In terms of inciting etiological factors, it is difficult to generalize as to the similarity between animal models and kidney disease in humans, as very often the etiology of the disease is unknown in either humans or animal models. It is also unclear as to whether the controlled environment and diet in experimental models of kidney disease, or the insults that generate the renal injury in animal models, truly reproduce etiological factors in humans. Thus, in terms of etiology, experimental animal models are of potential value, but findings in the animal must be validated back in humans.

In terms of pathophysiology of renal disease, even though the anatomy of the rat kidney is somewhat different from that of humans, the similarity in the structure and function of the nephron in humans and rats is remarkable. Therefore, the pathophysiology of renal disease in the rat may be highly relevant to the human disease, although validation in humans again remains a necessary step.

As a correlate to the similarity of the human and rat nephron, the histology of the kidney in the rat, in terms of differentiation between the medulla and the cortex and the appearance of the glomerulus and tubules, is also remarkably similar to that in humans. The histopathology of renal lesions that typify human disease may thus be sought in rat models that simulate glomerular, tubular and interstitial disease in humans. A significant difficulty arises, however, when one realizes that the histology of the renal lesion in the rat does not necessarily reproduce the exact or the full lesion in humans. This holds true for diabetic nephropathy, in which some of the models reproduce part but not all the features of the diabetic histopathological lesions. Although some experimental models of diabetes exhibit thickening of the basement membrane and mild diffuse focal glomerulosclerosis, which represent some of the changes observed in diabetic nephropathy in humans, rat models of diabetic nephropathy do not consistently develop glomerular hypertrophy, mesangial expansion or progress to severe global glomerulosclerosis with nodule formation leading to end-stage renal failure. A similar argument also applies to experimental models of focal and segmental glomerulosclerosis, in which the histopathological lesions are not uniform in all rat models, exhibiting at least three different types of glomerular lesions [6]. Thus, experimental models of glomerulosclerosis can reproduce parts of but often not the full spectrum of the lesions observed in the human diseased kidney.

Beyond the issue of whether histopathological changes in rat models accurately and fully reproduce those changes seen in the human kidney disease, genomic studies often require a quantitative assessment of such changes. A quantitative assessment of the extent of renal damage can be attained both in animal models and in humans by computerized morphometry which provides quantitative measures of normal and disease components of the kidney histology. The difficulty in applying this methodology to genomic studies, however, lies in its labor-intensiveness, high cost and low availability for both human and animal studies.

In terms of accurately defining the phenotype of the disease under investigation, assessing the level of renal function or dysfunction in the rat presents a certain problem. In humans, renal function is usually assessed by the glomerular filtration rate, which can either be measured and derived from plasma creatinine and creatinine excretion in 24 h urine collections or be estimated by using formulas such as the Cockcroft and Gault [7] or Modification of Diet in Renal Disease (MDRD) [8] or the Chronic Kidney Disease Epidemiology Collaboration (CKD-EPI) [9]. Acute studies of renal function in humans, which are rarely carried out nowadays, can be carried out by using inulin for glomerular filtration rate measurements and PAH for renal plasma flow. Cystatin C is an additional relatively new method for assessing renal function, essentially by measuring cystatin C plasma levels, without the need for urine measurements. In the rat, the assessment of glomerular filtration rate is similar to that in humans and can be derived from 24 h urine collections and measurement of plasma and urine creatinine levels. The inherent difficulties and inaccuracies of such measurements in rodents are, however, well recognized [10, 11]. Unfortunately, in the rat no formulas are available for estimating glomerular filtration rate based on plasma creatinine. Cystatin C levels may equally serve as a useful tool in the rat as in humans [12] and multiple ELISA kits have been developed to measure rat cystatin C, but the utility of this mode of measurement of glomerular filtration in experimental models remains to be validated. Assessment of renal dysfunction, a surrogate and consequence of renal disease, can be obtained in the rat by following sequentially serum creatinine levels, or by documenting a decline in creatinine clearance. As it is not possible to derive estimated glomerular filtration rate in the rat, creatinine clearance relies heavily on the accuracy of urine collection, which is often inaccurate in the rat. The use of blood urea nitrogen (BUN) as an alternative measure reflecting glomerular filtration is highly inaccurate, as urea levels are dependent on a number of variables that do not necessarily reflect renal function.

Another important surrogate of renal disease is proteinuria, reflecting either the leak of protein from the glomerular microcirculation into the tubular space, lack of adequate reabsorption of protein from Bowman's parietal epithelial cells or the tubular lumen into the tubular epithelial cells, or a combination of the above. Just as proteinuria is an important surrogate of renal disease in humans, it is used similarly in the rat. One problem in the study of proteinuria in the rat is the rather low absolute level of protein excretion, even in pathological states. Another problem is the length of time required for proteinuria to develop in experimental models simulating human disease. Although in humans, renal disease is usually progressive over many months to many years while in rat models, proteinuria usually develops within weeks or months, the cost of maintaining laboratory animals for this period of time may be prohibitive. Investigators have resorted to uninephrectomy, as uninephrectomy has been shown to both augment the level of proteinuria and to accelerate its development [13, 14]. A difficulty may arise in interpreting the results obtained in this model of renal disease, as one has to differentiate between the contribution of uninephrectomy per se to the renal disease from that of the underlying process initiating the renal disease, a significant confounder.

Despite all the reservations as to the use of rat models to investigate human kidney disease, the rat continues to serve as an important tool for genetic and genomic studies of the kidney. Among the methods that are most commonly used in defining the extent of the renal lesion, proteinuria is used the most often as the surrogate of renal injury, followed by glomerular filtration rate, and to an even lesser extent the histopathological changes within the kidney, which are scarcely used.

3 Genomic Investigation of Kidney Disease in the Rat

3.1 Primary Renal Disease

There are several primary renal diseases in humans that have been reproduced, to a greater or lesser extent, in the rat, most notable of which are glomerulosclerosis, IgA nephropathy, and membranous nephropathy (Heymann nephritis). Genomic tools have been extensively applied in several animal models in the investigation of the pathophysiology underlying glomerulosclerosis, and to a far lesser extent in one model of IgA nephropathy.

3.1.1 Glomerulosclerosis

In this clinical entity, histopathological features are common to many nephropathies which eventually lead to end stage renal failure. Glomerulosclerosis may be truly a primary renal disease or develop secondarily to or in conjunction with hypertension, depending on the model. In some rat models of glomerulosclerosis, the renal lesions develop spontaneously without hypertension, whereas in others, glomerulosclerosis develops along with the appearance of hypertension and it is then difficult to differentiate between the two entities. Irrespective of whether glomerulosclerosis is related or not to hypertension, the contribution of genomic studies in the rat to our understanding of glomerulosclerosis has been very substantial. The most commonly used experimental models of glomerulosclerosis have been the Fawn-hooded Hypertensive (FHH) rat, the Munich Wistar Frömter (MWF) rat, the Sabra Salt-sensitive Hypertension (SBH/y) prone rat, the Dahl Salt-sensitive (DS) rat, and the Buffalo Mna (Buffalo/Mna) rat.

Fawn-Hooded Rat

This model expresses a renal lesion that is manifested histologically by focal and segmental glomerulosclerosis [15]. The animals become hypertensive, develop proteinuria and eventually renal failure. Genetic research in this model has been directed at detecting the mechanisms underlying proteinuria, renal failure, and focal and segmental glomerulosclerosis. In a backcross involving the fawn-hooded hypertensive rats (FHH) and normotensive August-Copenhagen-Irish (ACI) rats, Brown et al. [16] reported genetic linkage between a quantitative trait locus (QTL) on chromosome 1 (RNO1) and markers of renal disease, including proteinuria, plasma creatinine, and a macroscopic renal index suggestive of renal damage. They termed this locus Rf-1, Rf standing for renal failure [16]. Provoost et al. [17] subsequently confirmed this locus by constructing a congenic strain in which part of the QTL interval from FHH was introgressed onto the ACI genetic background and reporting that a subsection of their QTL (Rf1B) was indeed associated with proteinuria and glomerulosclerosis. Brown et al. [16] reported in the same cross the presence of a second locus on RNO1, Rf-2, which was weakly linked with plasma creatinine and the macroscopic renal index. Rf-2 is concordant to human chromosome 19q13, which incorporates a yet unidentified gene underlying a monogenic form of focal segmental glomerulosclerosis [18]. The homologous region in mice is linked to glomerulonephritis susceptibility in a model for systemic lupus erythematosus [19]. Shiozawa et al. [20] repeated the cross and established that Rf-1 and Rf-2 were strongly linked to proteinuria and glomerulosclerosis. They detected this time three additional QTLs: Rf-3 on RNO3, Rf-4 on RNO1, and Rf-5 on RNO17. Rf-3 and Rf-4 showed suggestive linkage to proteinuria but only hints of linkage to glomerulosclerosis. The rat QTL on RNO3 is concordant with kidney disease QTL identified in Pima Indians [21]. Shiozawa et al. [20] went further and detected a significant and important interaction between these three loci, at least with respect to the level of proteinuria. Rangel-Filho et al. [22] identified within the Rf-2 region the *Rab38* gene as a highly likely candidate contributing to proteinuria in the FHH rat. They also found in FHH a spontaneous protein null mutation within *Rab38* which prevents translation of the protein. This was the first identification of a gene responsible for a QTL linked to proteinuria in an animal model of renal disease. Interestingly, in yet another study, Rangel-Filho et al. [23] showed that the mutation in the Rab38 protein affects urinary protein excretion through decreased proximal tubular protein reabsorption, rather than glomerular proteinuria.

Munich Wistar Frömter (MWF)

This model, which exhibits spontaneous glomerulosclerosis, early onset proteinuria, and moderate hypertension [24–26], all possibly related to an inherited inborn 30–50% reduction in the number of nephrons [27], has been extensively utilized in the genetic dissection of renal disease. Sexual dimorphism in this strain leads in males to more severe progression of proteinuria and kidney failure than in females [28]. Schulz et al. [26] investigated the genetic basis of proteinuria in this model by crossing MWF with Lewis (Lew) rats. They detected linkage for proteinuria at four loci: QTL1 on RNO1 which co-localizes with Rf-2, QTL2 on RNO6, QTL3 on RNO12 and QTL4 on RNO17 which co-localizes with Rf-5. Notable in this cross was the relatively low level of proteinuria (as assessed by urinary albumin excretion, UAE), which nonetheless allowed the detection of these four QTLs. In this cross, no putative linkage was detected for glomerulosclerosis. Yagil et al. [29] subsequently confirmed the presence of proteinuria QTLs on these chromosomes in consomic Sabra substrains [29]. In a separate study, Schulz et al. [30] used a different cross between MWF and SHR to dissect the genetic basis of proteinuria and renal interstitial fibrosis, the latter being a phenotypic feature consistent with evolving renal insufficiency. Eight QTLs were detected this time with suggestive or significant linkage on RNO1, 4, 6, 7, 8, 9, 15, and X. Once again, no linkage was detected to the glomerular sclerosis index, but suggestive linkage was detected for interstitial fibrosis on RNO6. Schulz et al. [31] pursued these findings by developing a congenic strain in which RNO6 from SHR was introgressed onto the genetic background of MWF, resulting in marked attenuation of proteinuria and renal interstitial fibrosis and demonstrating the importance of this genetic locus with respect to these two phenotypes. Schulz et al. [32] constructed an additional consomic strain in which RNO8 from SHR was introgressed into the genetic background of MWF, resulting in a marked reduction in proteinuria. Kreutz et al. [33] used a microarray platform to compare glomerular gene expression between MWF and Wistar and found that diminished expression of *C1q* and *CD24* in the glomerular endothelium was associated with glomerular proteinuria, identifying them as two candidate genes for glomerular dysfunction and proteinuria in this model. How and whether these two genes indeed induce impairment of renal function and proteinuria, or whether they merely serve as markers of proteinuria, remain to be clarified.

Sabra Salt-Sensitive Hypertension Prone (SBH/y)

This is a model of spontaneous glomerulosclerosis and proteinuria that is unrelated to salt loading or hypertension [29]. Both inbred strains are normotensive when fed normal diet, but SBH/y develops spontaneous proteinuria while SBN/y does not. Following salt loading, SBH/y animals become hypertensive while SBN/y remain normotensive. Yagil et al. [29] studied the level of proteinuria in consomic strains in which RNO1 and RNO17 had been introgressed from SBN/y (salt and hypertension resistant strain) onto the genetic background of SBH/y for another unrelated project, and found a marked reduction in the level of proteinuria without affecting the level of glomerulosclerosis in the RNO1 consomic. These findings suggest that RNO1 and RNO17 harbor genes that affect proteinuria, and that proteinuria is dissociated from glomerulosclerosis in this model. Yagil et al. [34] later dissected the genetic basis of proteinuria in the Sabra model using an F2 cross between SBH/y and SBN/y, detecting three QTLs that were associated with the development of proteinuria: on RNO2 (SUP2), RNO17 (SUP17) and RNO20 (SUP20). Interestingly, no proteinuria-related QTL was detected on RNO1. An additional QTL was detected on RNO3 (SUP3) which conferred protection from proteinuria. In a study aimed at detecting sex-specific proteinuria-related QTLs, Yagil et al. [35] demonstrated such QTLs on chromosomes RNO2 and RNO20 in male rats and on chromosomes RNO11, 13, and 20 in females. The only QTL overlap between the sexes was on RNO20. Consomic strains in which RNO2, RNO17, or RNO20 were introgressed from SBN/y onto the genetic background of SBH/y rats confirmed the presence of proteinuria-related genes on RNO2 and RNO20. Four genes that mapped to the QTL on RNO20 were found to be differentially expressed in SBN/y vs. SBH/y in both males and females: *Tubb5*, *C2*, *Ubd* and *Psmb8*.

Dahl Salt Sensitive (SS)

The Dahl rat, which is salt sensitive similar to the Sabra rat, develops proteinuria at a young age and prior to the development of hypertension, without any impairment in renal function [36]. The development of hypertension aggravates proteinuria and causes an impairment in renal function [36]. The glomerular damage in SS is characterized by mesangial expansion and fibrinoid arteriolar and glomerular necrosis and is thus not a true model of glomerulosclerosis. Poyan Mehr et al. [37] studied the genetic basis of proteinuria in the SS rat using an F2 cross between SS and the spontaneously hypertensive rat (SHR) which is resistant to proteinuria. They identified seven suggestive or significant proteinuria-related QTLs on RNO2, RNO6, RNO8, RNO9, RNO10, RNO11, and RNO19, demonstrating the polygenetic basis of proteinuria in the SS rat as in all other models of glomerulosclerosis. The QTL for albuminuria on RNO19 is concordant with the mouse region containing the *Os* mutation, which causes glomerular hypertrophy, severe glomerulosclerosis, and a 50% reduction in nephron number in mice [38]. Garrett et al. [39] investigated in parallel the genetic basis of proteinuria in SS but using a backcross with SHR and found nine proteinuria-related QTLs on RNO1, RNO2, RNO6, RNO8, RNO9, RNO10, RNO11, RNO13, and RNO19. Most of the proteinuria-related QTLs co-localized with QTLs for kidney lesions. Cowley et al. [40] confirmed the relevance of the QTL on RNO13 by constructing a consomic strain in which RNO13 was introgressed from the Brown Norway (BN) rat into the genomic SS background, resulting in a marked reduction in the level of proteinuria. Interestingly, the consomic strain was not protected from glomerular injury as evidenced by histology. The albuminuria-related QTL on RNO13 is concordant with a systemic lupus erythematosus glomerulonephritis susceptibility locus (*Sle1*) in mice [19] and a monogenic form of familial focal segmental glomerulosclerosis in humans [41]. Further studies in consomic/congenic strains confirmed the functional significance of RNO2, RNO6, RNO8, RNO9, RNO11, RNO13, and RNO19 proteinuria/albuminuria QTLs by introgressing those QTLs from SHR onto the SS genomic background [39, 42–46]. Moreno et al. [47] focused on the gender effect and identified in F2 females, resulting from a cross between SS and Brown Norway (BN) rats, proteinuria and glomerular injury-related QTLs on RNO2 and RNO11, which overlapped with some of the QTLs detected in males. It is noteworthy that despite abundant work using the Dahl model of glomerulosclerosis, none of these studies in this model have yet yielded any definitive candidate genes for proteinuria or glomerulosclerosis.

Buffalo/Mna Rat

This model develops spontaneous focal and segmental glomerulosclerosis, along with proteinuria [6, 48]. In a cross between BUF/Mna and WKY rats, a proteinuria-related QTL was detected by linkage analysis on RNO13 [49]. Akiyama et al. [50] subsequently fine-mapped the QTL with SNPs and identified 38 genes, 25 of which became potential candidate genes for proteinuria. In *Arp3* (actin-related protein 3), one of the candidate genes, a missense mutation was detected, causing actin assembly abnormalities within the podocytes. This mutation was related to the development of both proteinuria and FSGS in the BUF rat.

Lyon Hypertensive Rat

This model demonstrates mild spontaneous, salt-sensitive hypertension, proteinuria, and albuminuria and exhibits some features that resemble the human metabolic syndrome, including dyslipidemia, insulin resistance, and glucose intolerance [51]. Several QTLs related to blood pressure were identified on RNO2, RNO13, and RNO17 [52]. Using BN as a reference strain, the RNO13 carrying BP-QTLs was introgressed from BN/NHsdMcwi onto the LH background [53]. The resulting consomic strain, LH-13^{BN} exhibited a 50% reduction in proteinuria compared with LH. In contrast, the reciprocal consomic strain, BN-13^{LH} obtained by LH-RNO13 transfer onto the BN background, did not develop proteinuria.

3.1.2 Experimental Crescentic Glomerulonephritis

Two models of crescentic nephritis have been described—the nephrotoxic type (NTN) and the experimental autoimmune nephritis type (EAG).

The model of NTN s was described in the Wistar Kyoto (WKY) rat strain in which a single injection of nephrotoxic serum led to an autoimmune type of glomerulonephritis with proteinuria, glomerular macrophage infiltration and glomerular crescent formation with progression to severe scarring, similar to human focal and segmental glomerulosclerosis [54]. Interestingly, Lewis, Brown Norway and Wistar rats injected with the same serum did not develop crescents nor proteinuria. Aitman et al. [55] used this model in an F2 cross between WKY and Lewis rats and detected by linkage analysis two QTLs which were associated with crescent formation and proteinuria, one on RNO13 (Crgn1) and the other on RNO16 (Crgn2). The investigators identified within the Crgn1 QTL the activator Fc receptor for IgG (*Fcgr3*) as a major candidate gene affecting the development of glomerulosclerosis. Reynolds et al. [56] showed that macrophages derived from the WKY. L*Crgn1* congenic strain showed significant reduction in Fc receptor-mediated oxidative burst, phagocytosis of opsonized polystyrene beads, LPS-mediated MCP-1 secretion and *iNOS* mRNA expression compared with WKY rats. An important progress in the research of this well-established rat model has recently been made with the generation of two novel transgenic strains on the genetic background of Wistar Kyoto and Spontaneous Hypertensive Rat strains using Sleeping Beauty transposon germline transgenesis, ubiquitously expressing green fluorescent protein (GFP) under the rat elongation factor 1 alpha (EF1a) promoter [57].

The EAG model requires induction of autoimmunity by immunizing rats with recombinant non-collagenous domain of the alpha 3 chain of type IV collagen (the Goodpasture autoantigen) [56], resulting in circulating and deposited autoantibodies to the glomerular basement membrane and consequently Crescentic Glomerulonephritis (CrGN). The LEW rat serves as a contrasting resistant strain to this model. D'Souza et al. [58] generated a bicongenic strain, LEW.W*Crgn1,2*, in which WKY *Crgn1* and *Crgn2* were both introgressed onto the CrGN-resistant LEW background.

Both TNT and EAG models were studied in the LEW. W*Crgn1,2*, demonstrating that both Crgn1 and Crgn2 are necessary to promote glomerular crescent. In the TNT model, the introgression of both WKY *Crgn1* and *Crgn2* resulted in the onset of disease, accompanied by increased macrophage activity in the nephritic glomeruli. In addition, the gene expression profile representing inflammatory cytokines in the inflamed glomeruli was significantly altered. Interestingly the new bicongenic strain did not show sign of disease activity in the EAG model (as is the case for F1 generated by WKY × LEW cross), indicating that the presence of both loci is required to promote glomerular crescent formation in this model.

3.1.3 IgA Nephropathy

IgA nephropathy is one of the most common types of primary glomerulopathies. Its pathophysiology is incompletely understood. An experimental model of IgA nephropathy is the Lewis rat which is injected a single dose of anti-Thy1 antibody [59]. The animal develops the so-called *anti-Thy-1 glomerulonephritis*, which is a model of complement-mediated mesangial damage that is followed by mesangial hypercellularity and extracellular matrix deposition, simulating IgA nephropathy in humans [59, 60]. In the Lewis/Mollegard (Lew/Moll) substrain, the lesion heals spontaneously within 4 weeks, whereas in the Lewis/Maastricht (Lew/Maa) substrain, proteinuria develops and the animals go on to progressive glomerulosclerosis [59]. Ijpelaar et al. [60] most recently investigated the genetic basis of susceptibility to this type of renal injury in a cross between the two substrains and identified a QTL on RNO1 (GS1) that was linked to progressive glomerulosclerosis after acute glomerulonephritis. No candidate genes have been positively identified yet within this QTL.

3.2 Secondary Renal Disease

3.2.1 Diabetic Nephropathy

This is the secondary renal disease that has been studied the most in the rat. Diabetic nephropathy develops in genetic models of spontaneous diabetes, in models of obesity-related diabetes, in models of diet-induced diabetes, and in models of chemically induced diabetes. The histological lesions within the glomerulus in those animal models are quite similar to the early stages of diabetic nephropathy in humans, with thickening of the basement membrane and mild glomerulosclerosis; the characteristic changes of later stages of diabetic nephropathy, including glomerular hypertrophy, mesangial expansion, and progressive glomerulosclerosis, are often present only in part [61], and do not exhibit the complete features of diabetic nephropathy as it is expressed in the diseased human kidney. Nonetheless, rodent models of diabetic nephropathy appear to be very useful and informative in the ongoing attempts to elucidate the pathophysiology of diabetic nephropathy.

Spontaneous Diabetes

Diabetes can develop spontaneously in the rat, or it can be induced by dietary manipulations or injection of pancreo-toxic agents. Both types of diabetes develop end-organ damage in the kidney, expressing various features of diabetic nephropathy as is observed in humans. Among the genetic models that develop spontaneous diabetes with nephropathy and in which genomic studies have been carried out that stand out are the Goto–Kakizaki (GK), the Zucker rats, the T2DN, and the BioBreeding rats.

Goto-Kakizaki Rat

This is a model of nonobese normotensive type 2 diabetes that develops glomerular hypertrophy and thickening of basement membrane but does not develop proteinuria, glomerulosclerosis, interstitial fibrosis, or impairment of renal function [62]. With superimposed hypertension, however, these animals develop progressive nephropathy [63]. The non-hypertensive diabetic GK model thus represents a type of "sub-clinical" nephropathy due to prolonged hyperglycemia, without overt renal disease [62]. Page et al. [64] studied the non-hypertensive GK model. They used differential display, prior to the advent of high throughput DNA microarrays, to isolate genes that show transcriptional changes in the kidney during the development of diabetes. They successfully identified eight candidate cDNA fragments, *CDK1–8* that are allegedly involved in the development of the diabetic changes within the kidney. Page et al. [65] pursued these findings and established the identity of *CDK4* as the rat beta-defensin-1 gene (*rBD-1*). They further studied gene expression of biglycan and TGF-β1 in the kidneys of GK rats and found they were upregulated, as they have been found in other experimental models of diabetes and in humans [65]. Malik et al. [66] identified the *CDK7* as a thiol-related gene with a putative role in oxidative stress. The role of these two candidate genes in the pathophysiology of diabetes requires further validation and investigation. Nobrega et al. [67] identified in a genome-wide analysis of a salt fed (1% NaCl) $GK_{FL} \times$ BN/Mcwi F2 population four QTLs on RNO1, RNO4, RNO5 and RNO10 linked to early diabetes phenotypes. Two proteinuria-related QTLs were identified on RNO5 and RNO7 and both were linked to glomerulosclerosis and tubular sclerosis. Only the QTL on chromosome 5 co-localized with diabetes. Taken together, these data suggest that diabetes and proteinuria in the GK rat model are controlled by different mechanisms.

Zucker (Fatty) Rat

This is a genetic model which expresses obesity, hyperphagia, and hyperinsulinemia [68] which have been attributed to a spontaneous mutation in the leptin receptor gene (*fa*) [69]. The Zucker Diabetic Fatty (ZDF) rat is a model of type 2 diabetes that was derived by selective inbreeding of the hyperglycemic Zucker Fatty rat and which develops spontaneous overt diabetes [70]. Schafer et al. [71] investigated the renal lesion in the ZDF rat and found histological lesions resembling the typical lesions in human diabetic nephropathy. Erdely et al. [72] also studied the renal injury in ZDF rats and found a progressive decline in glomerular filtration rate (creatinine clearance), increasing proteinuria and abnormal histology [72] which, however, simulated those seen in focal and segmental glomerulosclerosis and not those seen in diabetic nephropathy in humans [72]. The investigators concluded that the nephropathy in ZDF can therefore not be considered as a model simulating the typical human diabetic nephropathy. Interestingly, Baylis et al. [73] found that the insulin sensitizing agent rosiglitazone decreased proteinuria and attenuated the structural changes within the kidneys of diabetic ZDF rats [73], suggesting that the nephropathy in ZDF is related to diabetes, but perhaps not with the typical histopathological changes seen in humans. Using the ZDF rat as a model of kidney disease that simulates at least in part the renal lesions seen in diabetes, Niehof and Borlak [74] identified, using expression assays and si-RNA, *HNF4a* and the calcium channel *Trpc1* as novel candidate genes for diabetic nephropathy. *HNF4a* is an orphan nuclear receptor which is an essential transcription factor and master regulatory protein in the control of gene expression of a wide range of enzymes involved in metabolism [74]. This gene is known to control glucose metabolism and participates in glucose dependent insulin secretory pathways [75]. The second gene *Trpc1* codes for a nonselective cation channel [74]. How both of these genes are involved in the pathophysiology of diabetic nephropathy remains to be resolved. Zhang et al. [76] pursued the findings with respect to *TRPC1* and attempted, unsuccessfully so far, to confirm in humans the association between polymorphisms within the *TRPC1* gene and diabetic nephropathy.

T2DN

This strain is a genetic model of diabetic nephropathy that was derived by Nobrega et al. [77] by crossing strains FHH and GK. This strain develops spontaneous type 2 diabetes, along with progressive proteinuria and a decline in renal function. The histopathological changes within the kidney are reminiscent of diabetic nephropathy. Genomic studies are ongoing in this strain to elucidate the pathophysiological mechanisms involved.

BioBreeding (BB) Rats

This model of spontaneous diabetes develops mild histological changes that simulate human diabetic nephropathy, including increased mesangial volume and glomerular basement membrane thickness. These changes can be detected, however, in older animals only [78]. Hyperfiltration and the mild glomerular morphological changes in the diabetic BioBreeding rat are thus similar to abnormalities seen in early stages of human diabetic nephropathy. Hsieh et al. [79] used DNA microarrays to study gene expression in the proximal tubules of diabetic BB rats and found that Osteopontin (*OPN* or *Spp1*) was significantly upregulated in the diabetic rats and that this increased expression was mediated via reactive oxygen species (ROS), activation of the renin-angiotensin system (RAS), protein kinase C β1 signaling and TGF-β1 expression. Their findings suggest that OPN is directly related to tubulointerstitial injury in diabetic nephropathy. In addition, they reported that a mutation in *Ian5*, a member of the Immune-Associated Nucleotide IAN-related gene family on RNO4, resulted in loss of function of GTPase, a possible explanation for disordered T cell development and susceptibility to autoimmune diabetes [80].

Komeda Diabetes Prone (KDP) Rats

The characteristic features of this model of type 1 diabetes are a sudden onset of diabetes with lymphocyte infiltration into the islets of Langerhans, followed by destruction of the β-cells and lymphocyte infiltration of the salivary glands and lacrimal glands

[81, 82]. At least two recessive genes appear to be involved in the pathogenesis of the insulitis, one that is closely linked to the rat MHC $RT1^{u}$ locus and the other, *iddm/kdp1*, that maps to RNO11 [83]. Within the MHC, a nonsense mutation in the *Cblb* gene (member of the cbl/Sli family of ubiquitin-protein ligases) causes enhanced T cell activation, with subsequent lymphocyte infiltration of pancreatic islets, suggesting autoimmunity [84].

LEW.1AR1/Ztm-Iddm Rats

In this strain, type 1 diabetes develops at age 2 months as result of immune damage caused by heavy infiltration of the islets of Langerhans by B and T lymphocytes, macrophages and NK cells and beta cell destruction by apoptosis [85–87]. The mutation in this strain resides in the *Dock8* gene, which encodes a member of the DOCK180 protein superfamily of guanine nucleotide exchange factors that act as activators of Rac/Rho family GTPases [88].

Induced Diabetes

There are other models of diabetes in which the disease is induced experimentally and in which diabetic nephropathy develops. Among these models stands out the Cohen Diabetic rat in which diabetes is induced by a custom-prepared diet and the Streptozotocin rat in which diabetes is induced by streptozotocin injection, resulting in injury to the endocrine pancreas.

Cohen Diabetic Rat

This is a genetic model of diet-induced diabetes that is composed of two strains: the Cohen Diabetic sensitive (CDs) strain, which when provided regular rat chow does not develop diabetes, but which invariably develops diabetes when provided a custom-prepared "diabetogenic" diet, and the Cohen Diabetic resistant (CDr) strain [89]. The CDs strain fed a diabetogenic diet develops over time diabetic nephropathy that is expressed histologically and functionally, but with no proteinuria. Yagil et al. [90] described in this model glomerular histopathological changes that are reminiscent of human diabetic nephropathy, including thickening of the glomerular basement membrane, mesangial expansion and glomerulosclerosis with deterioration of renal function over time. The outstanding feature in this model is that proteinuria does not develop [90]. This model thus reproduces the non-proteinuric variety of diabetic nephropathy described in humans [91]. Genomic studies on the genetic basis of the diabetic nephropathy in this model are currently ongoing [92].

Streptozotocin (STZ) Rat

Injection of Streptozotocin into the rat induces insulin-deficient type 1 diabetes. The diabetic animals develop a nephropathy which shows resemblance to diabetic nephropathy in humans, although the typical lesions seen in humans are not uniformly present, and their severity is usually milder in the rat [93]. STZ has been injected in Sprague Dawley (SD), Wistar Kyoto (WKY), and the spontaneously hypertensive rat (SHR), inducing in all diabetes and diabetic nephropathy which can usually begin to be studied, with all its limitations, already 3 weeks after the injection. Morrison et al. [94] used the STZ-SD rat to validate the identification of dysregulated genes detected by DNA microarrays in mesangial cells in culture that had been exposed in vitro to glucose. They identified upregulation of the thiol anti-oxidative pathway, which they interpreted as an adaptational response of mesangial cells to glucose, which may fulfill a critical role in the development of diabetic nephropathy. Hsieh et al. [79] injected STZ into Wistar rats, inducing diabetic nephropathy; they studied gene expression in the proximal tubules using DNA microarrays and found upregulation of OPN, as in the BB rat. Han et al. [95] used the STZ-SD rat to validate gene expression patterns in glucose-stimulated podocytes; they demonstrated upregulation of heme oxygenase-1 (*Hmox1*), vascular endothelial growth factor A (*Vegfa*) and thrombospondin (*Thbs1*) and down regulation of angiotensin converting enzyme 2 (*Ace2*) and peroxisomal proliferator activator receptor γ (*Pparg*). It is yet unclear how these candidate genes are involved in the pathophysiology of diabetic nephropathy.

3.3 Other Kidney Diseases

There are other kidney diseases that do not fall exactly into the category of primary or secondary renal disease in which investigators have made use of genomic tools to investigate the underlying pathophysiology. An example is renal stone disease. Nephrolithiasis is a major health problem in humans that is often due to excessive calcium excretion (idiopathic hypercalciuria) and that carries a family history and therefore a genetic component in a substantial number of patients. To investigate the pathophysiology of hypercalciuria, Hoopes et al. [96] used an F2 cross between the genetic hypercalciuric stone-forming (GSH) rats [97], and the normocalciuric WKY rat. They detected significant linkage for hypercalciuria on RNO1 (HC1), and suggestive linkage on RNO4, 7, 10 and 14 [96]. Hoopes et al. [98] subsequently confirmed the validity of the hypercalciuria-related HC1 QTL on RNO1 by constructing a congenic strain in which the QTL was introgressed from GSH onto the WKY background. Candidate genes remain to be identified within this QTL.

4 Conclusions and Perspectives

The promise of animal research has been that genetic studies in the rat (and mouse) would render identification of candidate genes easier and more successful because of a seemingly lesser genetic complexity and the availability of improved genetic tools and a controllable environment [99, 100]. Has this promise materialized? Genetic and genomic studies in the rat have identified multiple QTLs for a number of phenotypes that are related to several kidney diseases. In a number of studies, QTLs have led to the identification of high priority candidate genes and proteins. Several of these kidney disease-related QTLs in the rat are concordant with kidney disease loci in mice and humans, suggesting that conserved disease genes underlie these QTLs and implying that kidney disease QTLs in animal models can predict the locations of disease genes in humans [98]. This concordance further validates the use of animal models in searching for disease-related genes in humans. And yet, have genomic studies in animal models truly furthered our understanding of disease afflicting the kidneys, beyond the level of understanding achieved by traditional physiology or studies based on the traditional candidate gene approach? A lucid assessment of the gains achieved so far from studies in the rat (and mouse) reveals that we are still very far from unraveling the complexity of kidney structure and function in health and disease. It is now more clear than ever that the pathophysiology of underlying renal disease in the mammalian organism is highly complex and novel approaches, as well as application of advanced technologies, are required in the study of kidney disease before the expectations from genetic research in animal models can be more satisfactorily fulfilled.

References

1. Tesson L, Cozzi J, Ménoret S, Rémy S, Usal C, Fraichard A et al (2005) Transgenic modifications of the rat genome. Transgenic Res 14:531–546
2. Brown AJ, Fisher DA, Kouranova E, McCoy A, Forbes K, Wu Y et al (2013) Whole-rat conditional gene knockout via genome editing. Nat Methods 10:638–640
3. Geurts AM, Cost GJ, Rémy S, Cui X, Tesson L, Usal C et al (2010) Generation of gene-specific mutated rats using zinc-finger nucleases. Methods Mol Biol 597:211–225
4. Rémy S, Tesson L, Menoret S, Usal C, Scharenberg AM, Anegon I (2010) Zinc-finger nucleases: a powerful tool for genetic engineering of animals. Transgenic Res 19 (3):363–371
5. Lu J, Shao Y, Qin X, Liu D, Chen A, Li D et al (2017) CRISPR knockout rat cytochrome P450 3A1/2 model for advancing drug metabolism and pharmacokinetics research. Sci Rep 7:42922
6. Howie AJ, Kizaki T, Beaman M, Morland CM, Birtwistle RJ, Adu D et al (1989) Different types of segmental sclerosing glomerular lesions in six experimental models of proteinuria. J Pathol 157:141–151
7. Cockcroft DW, Gault MH (1976) Prediction of creatinine clearance from serum creatinine. Nephron 16:31–41
8. Levey AS, Bosch JP, Lewis JB, Greene T, Rogers N, Roth D (1999) A more accurate method to estimate glomerular filtration rate from serum creatinine: a new prediction equation. Modification of Diet in Renal Disease Study Group. Ann Intern Med 130:461–470
9. Levey AS, Stevens LA, Schmid CH, Zhang YL, Castro AF 3rd, Feldman HI (2009) A new equation to estimate glomerular filtration rate. Ann Intern Med 150:604–612
10. Pestel S, Krzykalla V, Weckesser G (2007) Measurement of glomerular filtration rate in the conscious rat. J Pharmacol Toxicol Meth 56:277–289
11. de Castro BB, Colugnati FA, Cenedeze MA, Suassuna PG, Pinheiro HS (2014) Standardization of renal function evaluation in Wistar rats (*Rattus norvegicus*) from the Federal University of Juiz de Fora's colony. J Bras Nefrol 36:139–149
12. Bokenkamp A, Ciarimboli G, Dieterich C (2001) Cystatin C in a rat model of end-stage renal failure. Ren Fail 23:431–438
13. Chamberlain RM, Shirley DG (2007) Time course of the renal functional response to partial nephrectomy: measurements in conscious rats. Exp Physiol 92:251–262
14. Griffin KA, Picken M, Bidani AK (1994) Method of renal mass reduction is a critical modulator of subsequent hypertension and

glomerular injury. J Am Soc Nephrol 4:2023–2031
15. Simons JL, Provoost AP, Anderson S, Troy JL, Rennke HG, Sandstrom DJ et al (1993) Pathogenesis of glomerular injury in the fawn-hooded rat: early glomerular capillary hypertension predicts glomerular sclerosis. J Am Soc Nephrol 3:1775–1782
16. Brown DM, Provoost AP, Daly MJ, Lander ES, Jacob HJ (1996) Renal disease susceptibility and hypertension are under independent genetic control in the fawn-hooded rat. Nat Genet 12:44–51
17. Provoost AP, Shiozawa M, Van Dokkum RPE, Jacob HJ (2002) Transfer of the Rf-1 region from FHH onto the ACI background increases susceptibility to renal impairment. Physiol Genomics 8:123–129
18. Mathis BJ, Kim SH, Calabrese K, Haas M, Seidman JG, Seidman CE et al (1998) A locus for inherited focal segmental glomerulosclerosis maps to chromosome 19q13. Kidney Int 53:282–286
19. Morel L, Rudofsky UH, Longmate JA, Schiffenbauer J, Wakeland EK (1994) Polygenic control of susceptibility to murine systemic lupus erythematosus. Immunity 1:219–229
20. Shiozawa M, Provoost AP, van Dokkum RP, Majewski RR, Jacob HJ (2000) Evidence of gene-gene Interactions in the genetic susceptibility to renal impairment after unilateral nephrectomy. J Am Soc Nephrol 11:2068–2078
21. Imperatore G, Hanson RL, Pettitt DJ, Schiffenbauer J, Wakeland EK (1998) Sib-pair linkage analysis for susceptibility genes for microvascular complications among Pima Indians with type 2 diabetes. Pima Diabetes Genes Group. Diabetes 47:821–830
22. Rangel-Filho A, Sharma M, Datta YH, Moreno C, Roman RJ, Iwamoto Y et al (2005) RF-2 gene modulates proteinuria and albuminuria independently of changes in glomerular permeability in the fawn-hooded hypertensive rat. J Am Soc Nephrol 16:852–856
23. Rangel-Filho A, Lazar J, Moreno C, Geurts A, Jacob HJ (2013) Rab38 modulates proteinuria in model of hypertension-associated renal disease. J Am Soc Nephrol 24:283–292
24. Remuzzi A, Puntorieri S, Mazzoleni A, Remuzzi G (1988) Sex related differences in glomerular ultrafiltration and proteinuria in Munich-Wistar rats. Kidney Int 34:481–486
25. Fassi A, Sangalli F, Maffi R, Colombi F, Mohamed EI, Brenner BM (1998) Progressive glomerular injury in the MWF rat is predicted by inborn nephron deficit. J Am Soc Nephrol 9:1399–1406
26. Schulz A, Litfin A, Kossmehl P, Kreutz R (2002) Genetic dissection of increased urinary albumin excretion in the Munich Wistar Frömter Rat. J Am Soc Nephrol 13:2706–2714
27. Hackbarth H, Gwinner W, Alt JM, Hagemann I, Thiemann A, Finke B (1991) The Munich Wistar Frömter rat: proteinuria and blood pressure in correlation to the number of superficial glomeruli. Ren Physiol Biochem 14:246–252
28. Schulz A, Schlesener M, Weiss J, Hänsch J, Wendt N, Kossmehl P (2008) Protective effect of female gender on the development of albuminuria in a polygenetic rat model is enhanced further by replacement of a major autosomal QTL. Clin Sci 114:305–311
29. Yagil C, Sapojnikov M, Katni G, Ilan Z, Zangen SW, Rosenmann E et al (2002) Proteinuria and glomerulosclerosis in the Sabra genetic rat model of salt susceptibility. Physiol Genomics 9:167–178
30. Schulz A, Standke D, Kovacevic L, Mostler M, Kossmehl P, Stoll M et al (2003) A major gene locus links early onset albuminuria with renal interstitial fibrosis in the MWF rat with polygenetic albuminuria. J Am Soc Nephrol 14:3081–3089
31. Schulz A, Weiss J, Schlesener M et al (2007) Development of overt proteinuria in the Munich Wistar Frömter rat is suppressed by replacement of chromosome 6 in a consomic rat strain. J Am Soc Nephrol 18:113–121
32. Schulz A, Hansch J, Kuhn K, Hänsch J, Wehland M, Wendt N et al (2008) Nephron deficit is not required for progressive proteinuria development in the Munich Wistar Frömter rat. Physiol Genomics 35:30–35
33. Kreutz R, Schulz A, Sietmann A, Stoll M, Daha MR, de Heer E et al (2007) Induction of C1q expression in glomerular endothelium in a rat model with arterial hypertension and albuminuria. J Hypertens 25:2308–2316
34. Yagil C, Sapojnikov M, Wechsler A, Korol A, Yagil Y (2006) Genetic dissection of proteinuria in the Sabra rat. Physiol Genomics 25:121–133
35. Yagil Y, Hessner M, Schulz H, Gosele C, Lebedev L, Barkalifa R et al (2010) Geno-transcriptomic dissection of proteinuria in the uninephrectomized rat uncovers a

molecular complexity with sexual dimorphism. Physiol Genomics 42A:301–316

36. Sterzel RB, Luft FC, Gao Y, Schnermann J, Briggs JP, Ganten D et al (1988) Renal disease and the development of hypertension in salt-sensitive Dahl rats. Kidney Int 33:1119–1129
37. Poyan Mehr A, Siegel AK, Kossmehl P, Schulz A, Plehm R, de Bruijn JA et al (2003) Early onset albuminuria in Dahl rats is a polygenetic trait that is independent from salt loading. Physiol Genomics 14:209–216
38. He C, Esposito C, Phillips C, Zalups RK, Henderson DA, Striker GE et al (1996) Dissociation of glomerular hypertrophy, cell proliferation, and glomerulosclerosis in mouse strains heterozygous for a mutation (Os) which induces a 50% reduction in nephron number. J Clin Invest 97:1242–1249
39. Garrett MR, Dene H, Rapp JP (2003) Time-course genetic analysis of albuminuria in Dahl Salt-Sensitive rats on low-salt diet. J Am Soc Nephrol 14:1175–1187
40. Cowley AW Jr, Roman RJ, Kaldunski ML, Dumas P, Dickhout JG, Greene AS et al (2001) Brown Norway chromosome 13 confers protection from high salt to consomic Dahl S rat. Hypertension 37:456–461
41. Winn MP, Conlon PJ, Lynn KL, Howell DN, Slotterbeck BD, Smith AH et al (1999) Linkage of a gene causing familial focal segmental glomerulosclerosis to chromosome 11 and further evidence of genetic heterogeneity. Genomics 58:113–120
42. Kumarasamy S, Gopalakrishnan K, Toland EJ, Yerga-Woolwine S, Farms P, Morgan EE et al (2011) Refined mapping of blood pressure quantitative trait loci using congenic strains developed from two genetically hypertensive rat models. Hypertens Res 34:1263–1270
43. Toland EJ, Yerga-Woolwine S, Farms P, Cicila GT, Saad Y, Joe B (2008) Blood pressure and proteinuria effects of multiple quantitative trait loci on rat chromosome 9 that differentiate the spontaneously hypertensive rat from the Dahl salt-sensitive rat. J Hypertens 26:2134–2141
44. Garrett MR, Joe B, Yerga-Woolwine S (2006) Genetic linkage of urinary albumin excretion in Dahl salt-sensitive rats: influence of dietary salt and confirmation using congenic strains. Physiol Genomics 25:39–49
45. Packard M, Saad Y, Gunning WT, Gupta S, Shapiro J, Garrett MR (2009) Investigating the effect of genetic background on proteinuria and renal injury using two hypertensive strains. Am J Physiol Renal Physiol 296: F839–F846
46. Wendt N, Schulz A, Siegel AK, Weiss J, Wehland M, Sietmann A et al (2007) Rat chromosome 19 transfer from SHR ameliorates hypertension, salt-sensitivity, cardiovascular and renal organ damage in salt-sensitive Dahl rats. J Hypertens 25:95–102
47. Moreno C, Dumas P, Kaldunski ML, Tonellato PJ, Greene AS, Roman RJ et al (2003) Genomic map of cardiovascular phenotypes of hypertension in female Dahl S rats. Physiol Genomics 15:243–257
48. Nakamura T, Oite T, Shimizu F, Matsuyama M, Kazama T, Koda Y et al (1986) Sclerotic lesions in the glomeruli of Buffalo/Mna rats. Nephron 43:50–55
49. Murayama S, Yagyu S, Higo K, Mizuno T, Oyabu A et al (1998) A genetic locus susceptible to the overt proteinuria in BUF/Mna rat. Mamm Genome 9:886–888
50. Akiyama K, Morita H, Suetsugu S, Kuraba S, Numata Y, Yamamoto Y et al (2008) Actin-related protein 3 (Arp3) is mutated in proteinuric BUF/Mna rats. Mamm Genome 19:41–50
51. Sassard J, Vincent M, Oréa V, Privat P, Bataillard A (1997) Genetics of blood pressure and associated phenotypes in the Lyon rat. Clin Exp Hypertens 19:567–575
52. Bilusic M, Bataillard A, Tschannen MR, Gao L, Barreto NE, Vincent M et al (2004) Mapping the genetic determinants of hypertension, metabolic diseases, and related phenotypes in the Lyon hypertensive rat. Hypertension 44:695–701
53. Gilibert S, Bataillard A, Nussberger J, Sassard J, Kwitek AE (2009) Implication of chromosome 13 on hypertension and associated disorders in Lyon hypertensive rats. J Hypertens 27:1186–1193
54. Muhammad S (2014) Nephrotoxic nephritis and glomerulonephritis: animal model versus human disease. Br J Biomed Sci 71:168–171
55. Aitman TJ, Dong R, Vyse TJ, Norsworthy PJ, Johnson MD, Smith J et al (2006) Copy number polymorphism in Fcgr3 predisposes to glomerulonephritis in rats and humans. Nature 439:851–855
56. Reynolds J, Cook PR, Behmoaras J, Smith J, Bhangal G, Tadros S et al (2012) Genetic susceptibility to experimental autoimmune glomerulonephritis in the Wistar Kyoto rat. Am J Pathol 180:1843–1851
57. Garcia Diaz AI, Moyon B, Coan PM, Alfazema N, Venda L, Woollard K et al (2016) New Wistar Kyoto and spontaneously hypertensive rat transgenic models with

ubiquitous expression of green fluorescent protein. Dis Model Mech 9:463–471

58. D'Souza Z, McAdoo SP, Smith J, Pusey CD, Cook HT, Behmoaras J et al (2013) Experimental crescentic glomerulonephritis: a new bicongenic rat model. Dis Model Mech 6:1477–1486
59. Ketteler M, Westenfeld R, Gawlik A, Bachmann S, Frey A, Schönfelder G et al (2002) Nitric oxide synthase isoform expression in acute versus chronic anti-Thy 1 nephritis. Kidney Int 61:826–833
60. Ijpelaar DHT, Schulz A, Aben J, van der Wal A, Bruijn JA, Kreutz R et al (2008) Genetic predisposition for glomerulonephritis-induced glomerulosclerosis in rats is linked to chromosome 1. Physiol Genomics 35:173–181
61. Susztak K, Sharma K, Schiffer M, McCue P, Ciccone E, Böttinger EP (2003) Genomic strategies for diabetic nephropathy. J Am Soc Nephrol 14:S271–S278
62. Phillips AO, Baboolal K, Riley S, Gröne H, Janssen U, Steadman R et al (2001) Association of prolonged hyperglycemia with glomerular hypertrophy and renal basement membrane thickening in the Goto Kakizaki model of non-insulin-dependent diabetes mellitus. Am J Kidney Dis 37:400–410
63. Janssen U, Riley SG, Vassiliadou A, Floege J, Phillips AO (2003) Hypertension superimposed on type II diabetes in Goto Kakizaki rats induces progressive nephropathy. Kidney Int 63:2162–2170
64. Page R, Morris C, Williams J, von Ruhland C, Malik AN (1997) Isolation of diabetes-associated kidney genes using differential display. Biochem Biophys Res Commun 232:49–53
65. Page RA, Malik AN (2003) Elevated levels of beta defensin-1 mRNA in diabetic kidneys of GK rats. Biochem Biophys Res Commun 310:513–521
66. Malik AN, Rossios C, Al Kafaji G, Shah A, Page RA (2007) Glucose regulation of CDK7, a putative thiol related gene, in experimental diabetic nephropathy. Biochem Biophys Res Commun 357:237–244
67. Nobrega MA, Solberg Woods LC, Fleming S, Jacob HJ (2009) Distinct genetic regulation of progression of diabetes and renal disease in the Goto-Kakizaki rat. Physiol Genomics 39:38–46
68. Bray GA (1977) The Zucker-fatty rat: a review. Fed Proc 36:148–153
69. Phillips MS, Liu Q, Hammond HA, Dugan V, Hey PJ, Caskey CJ et al (1996) Leptin receptor missense mutation in the fatty Zucker rat. Nat Genet 13:18–19
70. Etgen GJ, Oldham BA (2000) Profiling of Zucker diabetic fatty rats in their progression to the overt diabetic state. Metabolism 49:684–688
71. Schafer S, Linz W, Bube A, Gerl M, Huber J, Kürzel GU et al (2003) Vasopeptidase inhibition prevents nephropathy in Zucker diabetic fatty rats. Cardiovasc Res 60:447–454
72. Erdely A, Freshour G, Maddox DA, Olson JL, Samsell L, Baylis C (2004) Renal disease in rats with type 2 diabetes is associated with decreased renal nitric oxide production. Diabetologia 47:1672–1676
73. Baylis C, Atzpodien EA, Freshour G, Engels K (2003) Peroxisome proliferator-activated receptor [gamma] agonist provides superior renal protection versus angiotensin-converting enzyme inhibition in a rat model of type 2 diabetes with obesity. J Pharmacol Exp Ther 307:854–860
74. Niehof M, Borlak J (2008) HNF4{alpha} and the Ca-channel TRPC1 are novel disease candidate genes in diabetic nephropathy. Diabetes 57:1069–1077
75. Wang H, Maechler P, Antinozzi PA, Hagenfeldt KA, Wollheim CB (2000) Hepatocyte nuclear factor 4alpha regulates the expression of pancreatic beta-cell genes implicated in glucose metabolism and nutrient-induced insulin secretion. J Biol Chem 275:35953–35959
76. Zhang D, Freedman BI, Flekac M, Santos E, Hicks PJ, Bowden DW et al (2008) Evaluation of genetic association and expression reduction of TRPC1 in the development of diabetic nephropathy. Am J Nephrol 29:244–251
77. Nobrega MA, Fleming S, Roman RJ, Shiozawa M, Schlick N, Lazar J et al (2004) Initial characterization of a rat model of diabetic nephropathy. Diabetes 53:735–742
78. Feld LG, Cachero S, Ellis EN, Dedeoglu O, Ueda Y, Leonard C et al (1995) Resistance to glomerular injury in the diabetic biobreeding rat. Exp Physiol 80:991–1000
79. Hsieh TJ, Chen R, Zhang SL, Liu F, Brezniceanu ML, Whiteside CI et al (2006) Upregulation of osteopontin gene expression in diabetic rat proximal tubular cells revealed by microarray profiling. Kidney Int 69:1005–1015
80. MacMurray AJ, Moralejo DH, Kwitek AE, Rutledge EA, Van Yserloo B, Gohlke P et al (2002) Lymphopenia in the BB rat model of type 1 diabetes is due to a mutation in a novel

immune-associated nucleotide (Ian)-related gene. Genome Res 12:1029–1039
81. Kawano K, Hirashima T, Mori S, Saitoh Y, Kurosumi M, Natori T (1991) New inbred strain of Long-Evans Tokushima lean rats with IDDM without lymphopenia. Diabetes 40:1375–1381
82. Komeda K, Noda M, Terao K, Kuzuya N, Kanazawa M, Kanazawa Y (1998) Establishment of two substrains, diabetes-prone and non-diabetic, from Long-Evans Tokushima Lean (LETL) rats. Endocr J 45:737–744
83. Yokoi N, Kanazawa M, Kitada K, Tanaka A, Kanazawa Y, Suda S (1997) A non-MHC locus essential for autoimmune type I diabetes in the Komeda Diabetes-Prone rat. J Clin Invest 100:2015–2021
84. Yokoi N, Komeda K, Wang HY, Yano H, Kitada K, Saitoh Y (2002) Cblb is a major susceptibility gene for rat type 1 diabetes mellitus. Nat Genet 31:391–394
85. Lenzen S, Tiedge M, Elsner M, Lortz S, Weiss H, Jörns A et al (2001) The LEW.1AR1/Ztm-iddm rat: a new model of spontaneous insulin-dependent diabetes mellitus. Diabetologia 44:1189–1196
86. Jorns A, Günther A, Hedrich HJ, Wedekind D, Tiedge M, Lenzen S (2005) Immune cell infiltration, cytokine expression, and beta-cell apoptosis during the development of type 1 diabetes in the spontaneously diabetic LEW.1AR1/Ztm-iddm rat. Diabetes 54:2041–2052
87. Arndt T, Jörns A, Weiss H, Tiedge M, Hedrich HJ, Lenzen S et al (2013) A variable CD3(+) T-cell frequency in peripheral blood lymphocytes associated with type 1 diabetes mellitus development in the LEW.1AR1-iddm rat. PLoS One 8:e64305
88. Arndt T, Wedekind D, Jörns A, Tsiavaliaris G, Cuppen E, Hedrich HJ et al (2015) A novel Dock8 gene mutation confers diabetogenic susceptibility in the LEW.1AR1/Ztm-iddm rat, an animal model of human type 1 diabetes. Diabetologia 58:2800–2809
89. Weksler-Zangen S, Yagil C, Zangen DH, Ornoy A, Jacob HJ, Yagil Y (2001) The newly inbred Cohen diabetic rat: a nonobese normolipidemic genetic model of diet-induced type 2 diabetes expressing sex differences. Diabetes 50:2521–2529
90. Yagil C, Barak A, Ben Dor D, Rosenmann E, Bernheim J, Rosner M et al (2005) Nonproteinuric diabetes-associated nephropathy in the Cohen rat model of type 2 diabetes. Diabetes 54:1487–1496
91. MacIsaac RJ, Tsalamandris C, Panagiotopoulos S, Smith TJ, McNeil KJ, Jerums G (2004) Nonalbuminuric renal insufficiency in type 2 diabetes. Diabetes Care 27:195–200
92. Yagil Y, Roif D, Sapojnikov M, Ben-Dor D, Tobar A, Rosenmann E (2014) Unmasking of proteinuria in the course of genetic dissection of nonproteinuric diabetic nephropathy. Physiol Genomics 46:29–38
93. Tesch GH, Allen TJ (2007) Rodent models of streptozotocin-induced diabetic nephropathy. Nephrology (Carlton) 12:261–266
94. Morrison J, Knoll K, Hessner MJ, Liang M (2004) Effect of high glucose on gene expression in mesangial cells: upregulation of the thiol pathway is an adaptational response. Physiol Genomics 17:271–282
95. Han SH, Yang S, Jung DS, Li JJ, Kim JJ, Kwak SJ et al (2008) Gene expression patterns in glucose-stimulated podocytes. Biochem Biophys Res Commun 370:514–518
96. Hoopes RR Jr, Reid R, Sen S, Szpirer C, Dixon P, Pannett AA et al (2003) Quantitative trait loci for hypercalciuria in a rat model of kidney stone disease. J Am Soc Nephrol 14:1844–1850
97. Bushinsky DA, Favus MJ (1988) Mechanism of hypercalciuria in genetic hypercalciuric rats. Inherited defect in intestinal calcium transport. J Clin Invest 82:1585–1591
98. Hoopes RR Jr, Middleton FA, Sen S, Hueber PA, Reid R, Bushinsky DA et al (2006) Isolation and confirmation of a calcium excretion quantitative trait locus on chromosome 1 in genetic hypercalciuric stone-forming congenic rats. J Am Soc Nephrol 17:1292–1304
99. Korstanje R, DiPetrillo K (2004) Unraveling the genetics of chronic kidney disease using animal models. Am J Physiol Renal Physiol 287:F347–F352
100. Liang M, Cowley AW Jr, Hessner MJ, Lazar J, Basile DP, Pietrusz JL (2005) Transcriptome analysis and kidney research: toward systems biology. Kidney Int 67:2114–2122

Chapter 15

Rat Models of Exercise for the Study of Complex Disease

Lauren Gerard Koch and Steven L. Britton

Abstract

Variation in exercise capacity is a translationally powerful indicator for overall health and disease. Here we review the basic methods used for development of theoretically based and hypothesis-driven rat models that divide for both exercise capacity and numerous complex disease risks This rat model system was made by selectively breeding genetically heterogeneous rat populations for low and high performance on a speed ramped treadmill running test.

Key words Low Capacity Runners, LCR, High Capacity Runners, HCR, Rat models of exercise, Treadmill running test, Exercise capacity, Complex disease

1 Introduction

In the early 1990s, we initiated the development of genomic-based rat models for the study of complex diseases. Our motivation was that physical and chemical-based animal models are overly artificial to mimic underlying disease causation and that the increasingly popular single gene targeted rodent models on inbred genetic backgrounds were too simplistic to emulate the polygenic nature and inheritance of common chronic diseases [1]. Further was evidence that diseases emerge not as discrete events, but as complexes, such as the cascade represented by metabolic syndrome for which hypertension, diabetes, heart failure, and obesity tend to cluster. Most critically, was our view that selection models based on "trait modeling" or a single measure thought to characterize a complex disease (e.g., high blood pressure) would not necessarily reproduce the full array of underlying mechanisms to be useful for preclinical animal model exploration. In the 1980s a clinical literature began to grow in support of a strong association between exercise capacity and all-cause morbidity and mortality [2]. Evidence was building that low exercise capacity out-ranked other risk factors including history of hypertension, smoking, type 2 diabetes, and obesity, but mechanistically this connection between low exercise capacity and

G. Thomas Hayman et al. (eds.), *Rat Genomics*, Methods in Molecular Biology, vol. 2018,
https://doi.org/10.1007/978-1-4939-9581-3_15,

premature death was unresolved. What was also missing was a theoretical and predictive explanation for these links. Using a first principles approach, we put forth the hypothesis that: *Variation in metabolic capacity for energy transfer is the central mechanistic determinant between disease and health* (*Energy Transfer Hypothesis*: ETH).

By integrating ideas from Ilya Prigogine, Hans Krebs, and Peter Mitchell, we formulated a theoretical *explanation* for the ETH: (1) evolution was driven by nonequilibrium thermodynamic energy dissipation mechanisms (order from disorder and maximum entropy production) and (2) emergence of complexity was coupled to the high energetic nature afforded by the high electronegativity of oxygen [3]. As a *predictive* test of the ETH, we show that two-way artificial selective breeding of rats for low and high energy transfer, measured through maximal distance running capacity, also produces rats that differ for a multitude of disease risks, including metabolic syndrome [4], neurodegeneration [5, 6], cognitive decline [7], fatty liver disease [8, 9], susceptibility to cancer [10], age-related obesity [11], and reduced longevity [12]. The rat model system we developed is termed Low Capacity Runners (LCR) and High Capacity Runners (HCR) [1].

Our large-scale selection for intrinsic exercise capacity (i.e., not trained) was started using a founder population of 96 males and 96 females from the genetically heterogeneous N:NIH rats. Unlike outbred stocks such as Wistar, Sprague-Dawley, and Long-Evans rats made from a relatively small set of breeders, the N:NIH were purposely bred from eight phylogenetically distant inbred strains to produce 60 outcrossed families and maintained as a research resource from the National Institutes of Health (NIH) until 2003 [13, 14]. We continue to use a rotational with-in breeding pattern to minimize inbreeding. Here we detail the methodology for the selection criteria for exercise capacity that we have tested in over 40 generations of bi-directional selection and more than 20,000 offspring.

2 Materials

2.1 Motorized Rodent Treadmills

We use commercially available motorized animal treadmills (Exer-4R, Exer 3/6, and Metabolic Modular Treadmill 1020-R1) by Columbus Instruments (Columbus, OH). The earliest treadmill design (Exer-4R) includes up to four individual running lanes equipped with an electric stimulus grid at the base of each lane and a programmable controller that automatically adjusts the speed of the moving treadmill belt. The maximal speed of the belt is 99.9 m/min. The inclination of the running surface is also manually adjustable up to 25°. The treadmill is equipped with a stimulus controller and toggle switches that allow each electric grid to be enabled or disabled individually. An electrical stimulus at the back of each exercise lane can be set to any desired intensity from 0 to 160 V, a maximum shocker current from 0.34 to 1.6 mA, and Stimulus Repetition Range between 1.4 to 3 Hz. When enabled, a red LED blinks every time the corresponding stimulus grid is energized, at a frequency based on the REP. RATE setting. The overall specifications for this unit is Rat System Overall Dimensions: 32″*L* × 28″*W* × 24″*H*, Running Surface Dimensions: 18″*L* × 5″*W* × 6.5″*H*, Shocker Grid Dimensions: 5″*L* × 5″*W*. Calibration of treadmills is done according to manufacturer's instructions. Daily calibration checks are done using a high precision digital contact tachometer (model 461891, ExTech Instruments by Flir Systems, Nashua, NH).

3 Methods

The protocol for estimation of exercise capacity requires 2 weeks to complete.

3.1 Introduction to Treadmill Running (Week 1)

The goal of week 1 is to educate each rat to treadmill running by gradual introduction to increasing speed and duration over a 5-day period. The treadmill running surface is set and maintained at a 15° incline. The shock grid is set to deliver 1.2 mA current at 3 Hz.

Speed and duration is gradually increased.

1. Day 1 and Day 2 consist of simply placing the rat on the belt that is moving at a low speed (15° slope) and picking the rat up and moving it forward if it sits and starts to slide off the back of the belt. Day 1: Speed is set @ 10 m/min for 1-min duration. Day 2: Speed is set @ 10 m/min for 2-min duration.
2. On Day 3 and Day 4, the shock grid is toggled to on position for the first time to encourage the rat to keep moving and stay walking on the treadmill belt for the duration of the set time. When rats slide off the belt onto the shock grid, limit the exposure to shock to no more than 2 s. Toggle the switch to disable the shock, and then gently prod or pick the rat up and move it forward onto the moving belt. Repeat this process until the rat learns to stay up on the moving belt in order to avoid the mild shock. Day 3: Speed is set @ 10 m/min for 3 min duration. Day 4: Speed is set @ 10 m/min for 4 min duration.
3. On Day 5, the rats learn to adjust to a speed acceleration of 1 m/min. The shock grid is on. Day 5: Speed is started @ 10 m/min and then manually accelerate the speed 1 m/min across a 5-min duration. At minute 1, increase to speed 11. At minute 2, increase to speed 12. At minute 3, increase to speed 13. At minute 4, increase to speed 14. Shut down upon reaching minute 5.

4. The ability to achieve a 5 min minimal level of running constitutes the threshold performance necessary for inclusion in evaluation with a speed-ramped treadmill running test the following week. If the rat is uncooperative after 5 min of gentle coaxing, pull the rat from the study and denote it as a non-runner (*see* **Notes 1** and **2**).

3.2 Speed Ramped Treadmill Running Test (Week 2)

Each rat runs to exhaustion on three alternate days (Monday--Wednesday–Friday or Tuesday–Thursday–Saturday) to allow 48 h of recovery between trials. The treadmill incline is set and maintained at a 15° angle. The exercise test protocol starts at the speed of 10 m/min and is increased 1 m every 2 min until "exhausted," as operationally defined below (*see* Fig. 1, *adapted from* [1]).

1. Load one rat per treadmill lane prepared with a corresponding timer.
2. Set treadmill to start at a pace of 10 m/min.
3. If the treadmill controller is equipped with an automatic acceleration feature, set to increase 1 m every 2 min.
4. At the beginning (first 30–60 s) of the test, make sure each rat is cooperating with the task of treadmill running, and then turn on the shock grid (set to deliver a shock with a 1.5 mA current at 3 Hz).
5. If the rat lands onto the stimulus grid and stays, gently pick the rat up after 2 s and encourage him to continue running. Continue to physically place the rat at the top of the belt and encourage running (*see* **Note 3**).
6. The end of test or "exhaustion" is determined if after three consecutive attempts to encourage, the rat continues to sit on the shock rather than run (i.e., refuses), stop the timer, and pull the rat out from the lane (*see* **Note 4**).

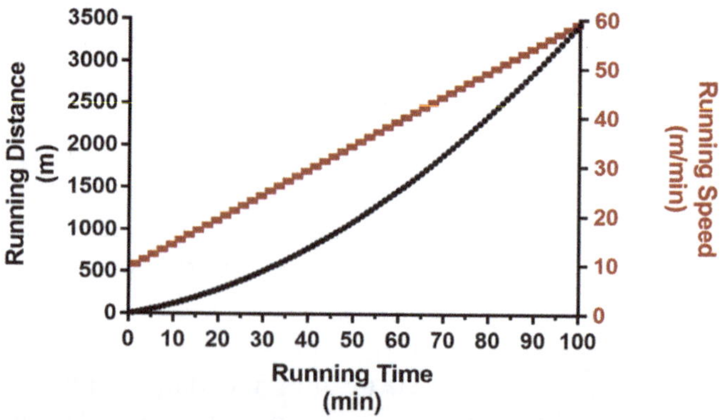

Fig. 1 Nomogram showing measured speed-ramped treadmill running time (*x*-axis) and corresponding running distance (*y*-left axis—black circles) and running speed (*y*-right axis—red squares). Adapted from [1]

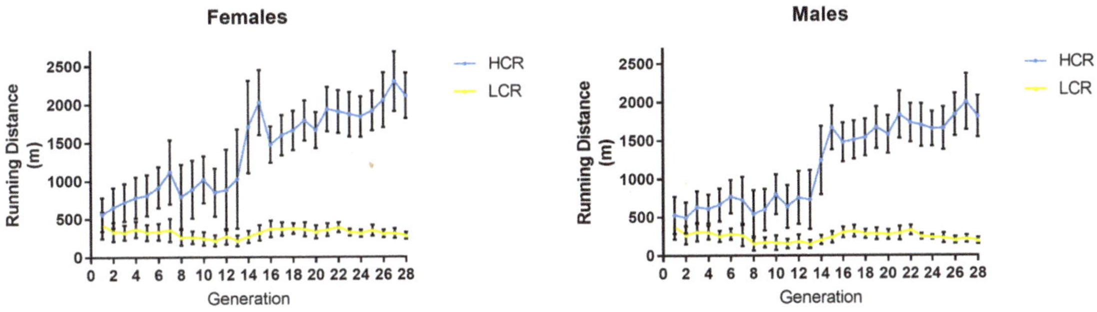

Fig. 2 Running distance on speed-ramped treadmill test was used as criteria for selection to develop rat models of low capacity runners (LCR—yellow triangles) and high capacity runners (HCR—blue circles). Data (mean, STD) show average running distance across 28 generations for females (left) and males (right) adapted from Table 1 in [15]. $n = 11{,}422$ total rats

7. Measure each rat's body weight, record, and return rat to home cage.
8. Document all data (time in min and s; body weight in g), as well as record individual notes for each rat. For example: "excellent" "very poor" "needed much encouragement."
9. Use the single best running trial out of three to estimate exercise capacity for each rat (*see* **Note 5**).
10. At each generation, use the "best" represented male and female from each of 13 families of low and high selected lines as breeders. Continued selection pressure will produce lines of low and high exercise capacity rats (*see* Fig. 2, *adapted from* [15]).

3.3 Calculations of Exercise Capacity

1. Total distance run. Total distance run in meters (m) calculates from the summation of an advancing belt speed (m/min) and duration distance (total min) of the running trial.

Velocity initial (m/min) =	10.00
Velocity step (m/min) =	1.00
Time step (min) =	2.00

The following formula written in Visual Basic programming language achieves this.

```
        Function distance(vel_init, vel_step, time_step,
duration)
        distance = 0
        n = CInt(duration / time_step)
        For i = 0 To n - 1
            distance = distance + time_step * (vel_init + i *
vel_step)
```

```
    Next i
      distance = distance + (duration - n * time_step) *
(vel_init + n * vel_step)
    End Function
```

2. Vertical work performed. Estimate the vertical work (joule) performed for each rat during each trial using the equation: Vertical work ($J = kg\ m^2/s^2$)=Total distance run (m) $\times$ body weight (kg) $\times$ sin(15°) $\times$ acceleration of gravity (9.8 m/s^2).
3. Maximal running speed. Maximal running speed is equivalent to the highest unit step reached based on the speed-ramped treadmill protocol of 1 m increase every 2 min.
4. Figure 3 demonstrates the utility of speed-ramped exercise test to estimate VO_2 max and calculate fuel use. Volume of oxygen consumed (VO_2) and carbon dioxide produced (VCO_2) are recorded using a Comprehensive Laboratory Animal Monitoring System (CLAMS, Columbus Instruments) equipped with Metabolic Modular Treadmill 1020-R1. Data show VO_2 differences between LCR males versus HCR males (Fig. 3 *adapted from* [16]).

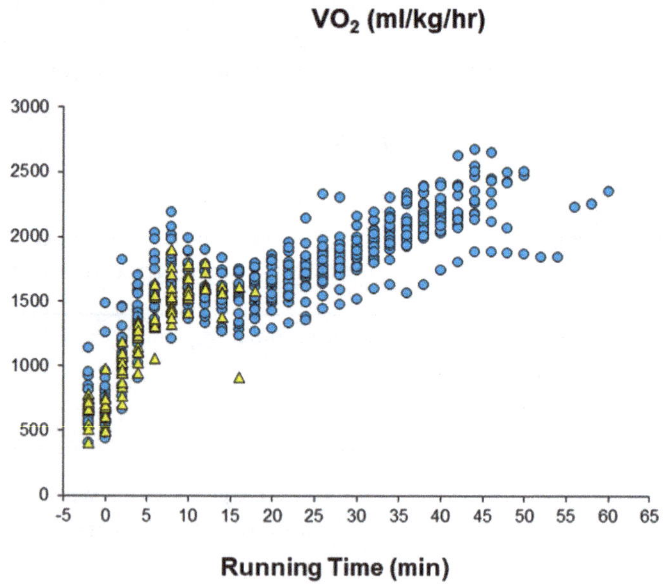

Fig. 3 Speed-ramped treadmill running test is used to assess energy expenditure using a Comprehensive Laboratory Animal Monitoring System (CLAMS, Columbus Instruments) equipped with a Metabolic Modular Treadmill. Data show significant differences in VO_2 between LCR males (yellow triangles, $n = 16$ rats) versus HCR males (blue circles, $n = 23$ rats) as a result of differential in fat oxidation. This experiment was performed in the University of Michigan Animal Phenotyping Core and data adapted from [16]

4 Notes

1. Rat running behavior. Each rat running behavior is unique and adds some subjectivity in the measure of exercise capacity. Use a scoring system from 1–5 to score each rat based on the following definitions (*see* Table 1).
2. Most rats take to treadmill running quite easily. From our experience, rats cooperate when the testing room is quiet and they are not bothered while running. Occasionally, rats do NOT take to treadmill running and/or freeze when placed on the treadmill. The following tips are for coaxing rats reluctant to treadmill running: lightly tapping the tail, briefly elevating tail, scratching along behind them with your hand, snapping fingers, or tapping on the back of the lane behind the rat. Always approach the rats from behind to encourage them to move forward. Never use loud banging noises or hard pinches to the tail or hindquarters.
3. Do not over-shock a rat! If a rat does not seem to be able to get off the electric grid, simply turn the shocker off. Usually the rat will then jump off the grid and keep running. Other tips to teach the rat to remove itself from the electric grid area include gently blowing on the rat, dangle a paper towel over the shock grid area, tap the electric grid from underneath, or touching the rats hindquarters and feet area with a wet sponge.
4. Signs that a rat is getting tired are: (a) starting to defecate and urinate, (b) begins running more toward the bottom half of the treadmill, no longer staying at the top of the running chute, (c) begins to look "heavy," sort of slow-motion, movements

Table 1
Rat running behavior scoring system

Score	Definition
1	*Excellent.* Made it to threshold time WITHOUT encouragement; did not step off onto the shock grid and stayed at the front of the treadmill
2	*Good.* Made it to threshold time WITHOUT encouragement; Stepped onto shock grid 1–3 times, but could easily remove self to continue running
3	*Poor.* Made it to threshold time WITH little encouragement; Stepped onto shock grid 3–5 times, operator manually assisted with keeping off the shock grid
4	*Very Poor.* Made it to threshold time WITH much encouragement; Stepped onto shock grid >5 times and operator required to manually assist with running and shock grid was turned to off position
5	*Non-Runner.* Did NOT make it to threshold time WITH encouragement; Rat refused to run, froze on treadmill and on shock grid; removed from study

are no longer springy, (d) gait falters, changes between trotting/galloping, and finally begins to scramble, (e) begins to vocalize, and (f) starts to step on and off (bouncing) on the electric grid area. A tired rat will most often simply slump onto the grid as a signal of exercise fatigue.

5. We consider the single best daily run of three trials for each rat as the trial most closely associated with the genetic component of intrinsic running capacity. We base this on the reasoning that environment factors such as subtle differences in feeding, housing, or daily handling can have a negative influence upon exercise capacity, reducing the distance run to zero. In other words, environment could cause a genetically superior rat to perform below its normal level on a given day. While selecting on the "best" day may reduce environmental variance, it creates the possibility that a component of selection may be related to a short-term adaptation to exercise. Indeed, at Generation 39, the average best day based on the algorithm for days of the week (1 = Mon, 3 = Wed, and 5 = Fri) shows low capacity runners' average best day is 2.99 ± 1.71 and the high capacity runners' best day averages to be on day 4.25 ± 1.19 ($P < 0.0001$). This fits with an operational definition that the trait of aerobic capacity is the result of two genetic expressions. First, is a complement of genes that determine intrinsic exercise capacity in the untrained state. On top of intrinsic capacity is another set of genes that dictate the adaptation response to exercise [17]. We have currently developed a second contrasting model system by selection on low and high response to 8 weeks of exercise training to test this definition more closely [18].

Acknowledgments

Development of the LCR-HCR rat model system was supported by NIH grant P40 OD012098A from the Office of Research Infrastructure Programs to L.G.K. and S.L.B. Contact L.G.K. (Lauren.Koch2@UToledo.Edu) or S.L.B. (brittons@umich.edu) for information on the LCR and HCR rats: these rat models are maintained as an international collaborative animal resource at The University of Toledo, Toledo, OH.

References

1. Koch LG, Britton SL (2001) Artificial selection for intrinsic aerobic endurance running capacity in rats. Physiol Genomics 5(1):45–52
2. Myers J, Prakash M, Froelicher V, Do D, Partington S, Atwood JE (2002) Exercise capacity and mortality among men referred for exercise testing. N Engl J Med 346 (11):793–801
3. Koch LG, Britton SL (2018) Theoretical and biological evaluation of the link between low

exercise capacity and disease risk. Cold Spring Harb Perspect Med 8(1):a029868. https://doi.org/10.1101/cshperspect.a029868

4. Wisloff U, Najjar SM, Ellingsen O, Haram PM, Swoap S, Al-Share Q et al (2005) Cardiovascular risk factors emerge after artificial selection for low aerobic capacity. Science 307 (5708):418–420
5. Chandrasekaran K, Muragundla A, Demarest TG, Choi J, Sagi AR, Najimi N et al (2017) mGluR2/3 activation of the SIRT1 axis preserves mitochondrial function in diabetic neuropathy. Ann Clin Transl Neurol 4 (12):844–858
6. Choi J, Chandrasekaran K, Demarest TG, Kristian T, Xu S, Vijaykumar K et al (2014) Brain diabetic neurodegeneration segregates with low intrinsic aerobic capacity. Ann Clin Transl Neurol 1(8):589–604
7. Feng X, Uchida Y, Koch L, Britton S, Hu J, Lutrin D et al (2017) Exercise prevents enhanced postoperative neuroinflammation and cognitive decline and rectifies the gut icrobiome in a rat model of metabolic syndrome. Front Immunol 8:1768
8. Thyfault JP, Rector RS, Uptergrove GM, Borengasser SJ, Morris EM, Wei Y et al (2009) Rats selectively bred for low aerobic capacity have reduced hepatic mitochondrial oxidative capacity and susceptibility to hepatic steatosis and injury. J Physiol 587(Pt 8):1805–1816
9. Morris EM, McCoin CS, Allen JA, Gastecki ML, Koch LG, Britton SL et al (2017) Aerobic capacity mediates susceptibility for the transition from steatosis to steatohepatitis. J Physiol 595(14):4909–4926
10. Thompson HJ, Jones LW, Koch LG, Britton SL, Neil ES, McGinley JN (2017) Inherent aerobic capacity-dependent differences in breast carcinogenesis. Carcinogenesis 38 (9):920–928
11. Park SJ, Gavrilova O, Brown AL, Soto JE, Bremner S, Kim J et al (2017) DNA-PK promotes the mitochondrial, metabolic, and physical decline that occurs during aging. Cell Metab 26(2):447
12. Koch LG, Kemi OJ, Qi N, Leng SX, Bijma P, Gilligan LJ et al (2011) Intrinsic aerobic capacity sets a divide for aging and longevity. Circ Res 109(10):1162–1172
13. Solberg Woods LC, Stelloh C, Regner KR, Schwabe T, Eisenhauer J, Garrett MR (2010) Heterogeneous stock rats: a new model to study the genetics of renal phenotypes. Am J Physiol Renal Physiol 298(6):F1484–F1491
14. Hansen C, Spuhler K (1984) Development of the National Institutes of Health genetically heterogeneous rat stock. Alcohol Clin Exp Res 8(5):477–479
15. Ren YY, Overmyer KA, Qi NR, Treutelaar MK, Heckenkamp L, Kalahar M et al (2013) Genetic analysis of a rat model of aerobic capacity and metabolic fitness. PLoS One 8(10): e77588
16. Overmyer KA, Evans CR, Qi NR, Minogue CE, Carson JJ, Chermside-Scabbo CJ et al (2015) Maximal oxidative capacity during exercise is associated with skeletal muscle fuel selection and dynamic changes in mitochondrial protein acetylation. Cell Metab 21(3):468–478
17. Bouchard C, Rankinen T, Chagnon YC, Rice T, Pérusse L, Gagnon J et al (2000) Genomic scan for maximal oxygen uptake and its response to training in the HERITAGE family study. J Appl Physiol 88(2):551–559
18. Koch LG, Pollott GE, Britton SL (2013) Selectively bred rat model system for low and high response to exercise training. Physiol Genomics 45(14):606–614

Chapter 16

Behavioral Genetic Studies in Rats

Yangsu Ren and Abraham A. Palmer

Abstract

In this chapter, we briefly review the use of rats as a genetic model for the study of behavior. Rats were the first mammalian species used for genetic and biological research. Since the development of the first inbred rat strain in 1909, more than 700 unique inbred and outbred rat lines have been generated. Although rats have been somewhat eclipsed by mice in the last few decades, a renewed appreciation of the advantages of rats for behavioral and other types of research is upon us. We briefly review the pertinent characteristics of the rat and highlight the key advantages of using the rat to examine behavioral phenotypes.

Key words Rats, Animal models, Inbred, Behavior model

1 Use of Rodent Model Organisms in Research

Over the past 100 years the Brown rat, sometimes called the Norway rat (*Rattus norvegicus*), has been one of the most widely used mammalian model organisms. Domesticated rats were first used for laboratory research in 1828; the first genetic studies using rats, which focused on coat color, were performed by Hugo Crampe from 1877 to 1885 [1]. The first inbred rat strain (known as King Albino or PA strain) was established by Helen Dean King in 1909 at the Wistar Institute in Philadelphia [2]. Since then, over 200 inbred rat strains have been created [3]. Behavioral studies of "cognition" and "emotionality" have used rats since at least the 1930s (e.g., [4, 5]).

In the last few decades, mice (*Mus musculus*) have become more widely used than rats for a number of reasons, including the ability to create transgenic mice [6], and the development of embryonic stem (ES) cells, which have allowed for targeted mutagenesis using homologous recombination ("knockout mice"; [7]). Indeed, after the sequencing of the human genome, the next genome sequenced was the mouse [8]. Resources such as the Knockout Mouse Project and the International Mouse Phenotyping Consortium have used mice rather than rats [9, 10]. Mice

G. Thomas Hayman et al. (eds.), *Rat Genomics*, Methods in Molecular Biology, vol. 2018,
https://doi.org/10.1007/978-1-4939-9581-3_16,

continue to be more heavily used than rats for genetic and other studies, in part due to the support of the Jackson Laboratory, but also because they are smaller and thus somewhat less expensive to maintain. Nevertheless, as we will describe in this chapter, there are compelling reasons why certain phenotypes, including many behavioral phenotypes, can benefit from the use of rats rather than mice [11].

2 The Rat's Growing Genetic Toolbox

Whereas the mouse genome was sequenced in 2002, the rat genome was sequenced shortly thereafter, in 2004, making it the third mammalian specie to be sequenced [12]. The inbred Brown Norway (BN) rat was selected as the reference strain. This allowed for a three-way mammalian evolutionary comparison, which showed that most human genes have orthologues in the rat and mouse [13]. Currently, many of the genetic tools available for mice are also available in rat, including *N*-ethyl-*N*-nitrosourea (ENU) mutagenesis [14], Zinc-finger nucleases (ZFNs) and transcription activator-like effector domains for efficient TALE nucleases (TALENs) [15, 16], homologous recombination [17], and clustered regularly interspaced short palindromic repeats (CRISPR)/Cas-mediated genome editing [18]; these techniques are reviewed in detail in another chapter of this book. Another chapter of this book describes the similarities between outbred heterogeneous stock rats (the N/NIH HS rats) and how they are very similar to the Diversity Outbred (DO) mice. Thus, many of the genetic resources that have been developed in mice are also available for use in rats.

3 Key Advantages to Using the Rat for Behavioral Studies

3.1 Rat Physiology

Rats are roughly ten times heavier than mice [19]. Their larger size makes rats easier to use for a variety of physiological phenotypes, including surgical interventions [19, 20]. Certain invasive procedures are much easier to perform in rats, such as implantation of venous catheters for drug self-administration or serial blood draws [21]. The larger size of the rat translates to a larger brain, which simplifies implantation of intracerebral cannula for in vivo microdialysis, voltammetry, electrophysiological recording and stimulation, optogenetics, and drug and viral microinjections [22]. The larger size of the rat brain also allows for higher spatial resolution for brain imaging, such as functional magnetic resonance imaging (fMRI) [23], and positron emission tomography (PET) [24], and near-infrared brain imaging [25]. Perhaps more importantly, rats can be trained to sit still during such imaging procedures, unlike mice, which removes the need for anesthesia that interferes with normal brain activity [26].

3.2 Brain Function and Cognition

Several studies have argued that certain features of the rat brain are more similar to humans when compared to mice, making them a better model for studying cognitive functions. For example, rats demonstrate very high levels of 5-HT6 mRNA and receptor-binding in the dorsal and ventral striatum, whereas 5-HT6 receptor levels in the same regions are very low in mice [27]. 5-HT6 plays a key role in reward perception, habit formation, cognitive flexibility, and motor functioning, which are involved in most psychiatric disorders, including schizophrenia, drug addiction, and attention deficit hyperactivity disorders (ADHD). Furthermore, the rates of neurogenesis in the adult hippocampus have been found to be significantly different between rats and mice. The hippocampus is one of the few brain regions where, even in adulthood, new neurons continue to be formed in humans, a process that may be involved in learning and memory [28]. When compared to mice, the rate of hippocampal neurogenesis was found to be much faster in rats, and these new cells matured about 2 weeks earlier, were twice as likely to escape cell death, and were ten times more likely to be activated during learning, which may make them a better model of human neurogenesis [29].

4 Behavioral Studies Using Rats

In the remainder of the chapter we will briefly review the use of rats in several categories of behavioral phenotypes, including addictive behavior, social behavior, impulsivity, and cognitive behavior.

4.1 Addictive Behavior

Substance use disorders (SUD) are defined by compulsive drug use despite adverse consequences for the user [30]. Rats have been shown to display several behaviors that are considered the hallmarks of SUD; (1) difficulty stopping drug use or limiting drug intake; (2) extremely high motivation to take drugs; and (3) substance use despite harmful consequences [31]. Studies on drug self-administration, conditioned place preference, and intracranial self-stimulation have all shed light on the etiology of drug reward and reinforcement; here we focus on self-administration studies to illustrate some of the advantages of rats [32–34]. Drug self-administration is an operant paradigm in which subjects perform operant tasks to obtain drugs via iv injection, drinking or access to vaporized drugs [35]. Rats given extended access to cocaine show steady increases in their intake across days, which is called escalation [36]. Rats with a history of escalated cocaine self-administration have also been shown have increased motivation for taking cocaine [37], increased reinstatement of cocaine seeking [31, 38], and reduced sensitivity to the negative effects of cocaine seeking [39]. These hallmarks of addiction have also been shown in rat self-administration studies using ethanol [40, 41], amphetamines [42], and opiates [43].

4.2 Social Behavior

Rats have been heavily used to study social behavior. Rats demonstrate a greater range of social behavior and are considered to be relatively resistant to the confounding effects of human experimenters [20]. In one study that explicitly compared mice and rats using dyadic interactions (two animals are placed in a large enclosure that is new to them), the average distance between the rat pairs was dramatically smaller than the distance between the mouse pairs, the latency to initiate social interactions was shorter, and the total frequency of interactions was higher in rats [44]. Conditioned place preference (CPP) is a form of Pavlovian conditioning used to measure the motivational effects of interoceptive and exteroceptive stimuli. While CPP is often used to assess the reinforcing effects of drugs, it can also be used to examine the rewarding value of social interactions when a drug is the alternative. Rats tested for preference for social interaction versus cocaine showed that both stimuli possess the same reward strength [45]. In contrast a similar study using mice found that cocaine was more rewarding than social interaction [46]. In addition, rats given access to social interaction stopped self-administration of methamphetamine and heroin, further showing the rewarding magnitude of social interaction in rats [47].

4.3 Impulsive Behavior

Impulsivity is defined as a predisposition toward rapid, unplanned reactions to internal or external stimuli without regard to the negative consequences of these reactions to themselves or others [48], and has been shown to influence behaviors including gambling [49], smoking [50], drug abuse [51], and Attention Deficit Hyperactivity Disorder (ADHD) [52]. Several operant-based paradigms have been used to test impulsivity, including delay-discounting, five-choice serial-reaction, go/no-go and stop-signal tasks [53]. Several studies have identified rat strains that have high levels of impulsivity, including Roman high avoidance (RHA) rats [54] and spontaneously hypertensive rats (SHRs) [55]. In a study comparing the impulsivity between the inbred SHR and Wistar Kyoto (WKY) rat strains, SHR showed greater impulsively compared to WKY, though there was also more within-strain variance among SHR rats, consistent with previous findings [56]. Delay-discounting is the most commonly used test, and involves individuals making a series of choices between small, immediate rewards versus larger rewards received after a longer delay, with impulsive individuals choosing the smaller reward even if it means they miss out on the large reward by waiting, which is called the impulsive choice [57]. Interestingly, whereas rats can learn the delay-discounting task fairly well, mice demonstrate substantially more difficulty; it takes significantly longer to train mice, and their response is much more variable than that in rats [11].

4.4 Cognitive Behavior

Cognitive function refers to the information-processing capacity of an animal, which includes learning, memory, attention, and performance [58]. Cognitive deficits are a core of many psychiatric and neurological disorders (e.g., schizophrenia and dementia). The Morris water maze (MWM) is one of the most common tests of spatial learning and memory for rodents that rely on distal cues to navigate from a start position around the perimeter of an open swimming area to locate a platform that is submerged underwater [59–61]. By placing the animals at different starting points in the water, the rodent's ability to find the platform based on external cues can be assessed. Compared to rats, mice have much greater difficulty in learning to find the platform [62]. In addition, rats typically require less training to perform this and other cognitive tasks compared to mice, resulting in less stress and anxiety in the process [63–65].

Aging research often compares the performance of young and old rats using tests including the eight-arm radial arm maze, MWM, and Barnes circular platform maze [66]. All three mazes require the rats to use spatial cues to navigate; in the eight-arm radial maze the rat must remember which parts of the maze it has already explored in order to gather food efficiently, and in the Barnes circular platform maze the rat must remember the position of an escape tunnel to avoid bright light. Aged rats showed impaired performances in all three maze tests compared to their younger counterparts, event after compensating for age-related physiological differences (e.g., distance traveled, errors or ambulatory time). Furthermore, aged rats showed reduced neuron densities in the CA1 and CA3 regions of the hippocampal complex, which has been shown to be important for spatial learning tasks [67, 68].

5 Conclusion

The rat is a valuable animal model for studying behavior. Many of the genetic and genomic technologies that have encouraged the use of mice are now available for rats. Because a wide variety of behavioral tests are easier and more informative in rats compared to mice, there is a growing appreciation of the continuing importance of rat genetic studies for understanding behavior. In this chapter we have highlighted examples of behavioral phenotypes for which rats have clear advantages over mice. While this list is not intended to be exhaustive, it is intended to illustrate some of the many situations in which the rat is an indispensable tool for behavioral genetic studies.

References

1. Lindsey JR (1979) Historical foundations in the laboratory rat. In: The laboratory rat, vol 1. Academic Press, New York, pp 1–36
2. Yamada J, Nikaido H, Matsumoto S (1979) Genetic variability within and between outbred Wistar strains of rats. Jikken Dobutsu 28 (2):259–265
3. Delves PJ, Roitt IM (1998) Encyclopedia of immunology, 2nd edn. Academic Press, San Diego, p 4
4. Shepard JF (1933) Higher processes in the behavior of rats. Proc Natl Acad Sci U S A 19:149–152
5. Hall C, Ballachey EL (1932) A study of the rat's behavior in a field. A contribution to method in comparative psychology, University of California publications in psychology, vol 6. University of California Press, Berkeley, CA, pp 1–12
6. Gordon JW, Ruddle FH (1981) Integration and stable germ line transmission of genes injected into mouse pronuclei. Science 214 (4526):1244–1246
7. Thomas KR, Capecchi MR (1987) Site-directed mutagenesis by gene targeting in mouse embryo-derived stem cells. Cell 51 (3):503–512
8. Waterston RH, Lindblad-Toh K, Birney E, Rogers J, Abril JF, Agarwal P et al (2002) Initial sequencing and comparative analysis of the mouse genome. Nature 420 (6915):520–562
9. Austin CP, Battey JF, Bradley A, Bucan M, Capecchi M, Collins FS et al (2004) The knockout mouse project. Nat Genet 36 (9):921–924
10. Munoz-Fuentes V, Cacheiro P, Meehan TF, Aguilar-Pimentel J, Brown SDM, Flenniken AM et al (2018) The International Mouse Phenotyping Consortium (IMPC): a functional catalogue of the mammalian genome that informs conservation. Conserv Genet 19 (4):995–1005
11. Parker CC, Chen H, Flagel SB, Geurts AM, Richards JB, Robinson TE et al (2014) Rats are the smart choice: rationale for a renewed focus on rats in behavioral genetics. Neuropharmacology 76(Pt B):250–258
12. Gibbs RA, Weinstock GM, Metzker ML, Muzny DM, Sodergren EJ, Scherer S et al (2004) Genome sequence of the Brown Norway rat yields insights into mammalian evolution. Nature 428(6982):493–521
13. Mullins LJ, Mullins JJ (2004) Insights from the rat genome sequence. Genome Biol 5(5):221
14. Zan Y, Haag JD, Chen KS, Shepel LA, Wigington D, Wang YR et al (2003) Production of knockout rats using ENU mutagenesis and a yeast-based screening assay. Nat Biotechnol 21(6):645–651
15. Geurts AM, Cost GJ, Freyvert Y, Zeitler B, Miller JC, Choi VM et al (2009) Knockout rats via embryo microinjection of zinc-finger nucleases. Science 325(5939):433
16. Cui X, Ji D, Fisher DA, Wu Y, Briner DM, Weinstein EJ (2011) Targeted integration in rat and mouse embryos with zinc-finger nucleases. Nat Biotechnol 29(1):64–67
17. Tong C, Huang G, Ashton C, Li P, Ying QL (2011) Generating gene knockout rats by homologous recombination in embryonic stem cells. Nat Protoc 6(6):827–844
18. Shao Y, Guan Y, Wang L, Qiu Z, Liu M, Chen Y et al (2014) CRISPR/Cas-mediated genome editing in the rat via direct injection of one-cell embryos. Nat Protoc 9(10):2493–2512
19. Kjell J, Olson L (2016) Rat models of spinal cord injury: from pathology to potential therapies. Dis Model Mech 9(10):1125–1137
20. Meijer MK, Sommer R, Spruijt BM, van Zutphen LF, Baumans V (2007) Influence of environmental enrichment and handling on the acute stress response in individually housed mice. Lab Anim 41(2):161–173
21. Feduccia AA, Duvauchelle CL (2010) Novel apparatus and method for drug reinforcement. J Vis Exp. https://doi.org/10.3791/1998
22. Kokare DM, Shelkar GP, Borkar CD, Nakhate KT, Subhedar NK (2011) A simple and inexpensive method to fabricate a cannula system for intracranial injections in rats and mice. J Pharmacol Toxicol Methods 64(3):246–250
23. Barandov A, Bartelle BB, Gonzalez BA, White WL, Lippard SJ, Jasanoff A (2016) Membrane-permeable Mn(III) complexes for molecular magnetic resonance imaging of intracellular targets. J Am Chem Soc 138(17):5483–5486
24. Zimmer ER, Leuzy A, Bhat V, Gauthier S, Rosa-Neto P (2014) In vivo tracking of tau pathology using positron emission tomography (PET) molecular imaging in small animals. Transl Neurodegener 3(1):6
25. Kim HY, Seo K, Jeon HJ, Lee U, Lee H (2017) Application of functional near-infrared spectroscopy to the study of brain function in humans and animal models. Mol Cells 40 (8):523–532
26. Febo M (2011) Technical and conceptual considerations for performing and interpreting

functional MRI studies in awake rats. Front Psychiatry 2:43

27. Hirst WD, Abrahamsen B, Blaney FE, Calver AR, Aloj L, Price GW et al (2003) Differences in the central nervous system distribution and pharmacology of the mouse 5-hydroxytryptamine-6 receptor compared with rat and human receptors investigated by radioligand binding, site-directed mutagenesis, and molecular modeling. Mol Pharmacol 64(6):1295–1308
28. Lazarov O, Hollands C (2016) Hippocampal neurogenesis: learning to remember. Prog Neurobiol 138–140:1–18
29. Snyder JS, Radik R, Wojtowicz JM, Cameron HA (2009) Anatomical gradients of adult neurogenesis and activity: young neurons in the ventral dentate gyrus are activated by water maze training. Hippocampus 19(4):360–370
30. Stieglitz RD (2000) Diagnostic and statistical manual of mental disorders. Z Klin Psychol-Forsc 29(1):63–64
31. Deroche-Gamonet V, Belin D, Piazza PV (2004) Evidence for addiction-like behavior in the rat. Science 305(5686):1014–1017
32. Wise RA (1996) Addictive drugs and brain stimulation reward. Annu Rev Neurosci 19:319–340
33. Tzschentke TM (2007) Measuring reward with the conditioned place preference (CPP) paradigm: update of the last decade. Addict Biol 12 (3–4):227–462
34. O'Connor EC, Chapman K, Butler P, Mead AN (2011) The predictive validity of the rat self-administration model for abuse liability. Neurosci Biobehav Rev 35(3):912–938
35. Panlilio LV, Goldberg SR (2007) Self-administration of drugs in animals and humans as a model and an investigative tool. Addiction 102(12):1863–1870
36. Ahmed SH, Koob GF (1999) Long-lasting increase in the set point for cocaine self-administration after escalation in rats. Psychopharmacology (Berl) 146(3):303–312
37. Paterson NE, Markou A (2003) Increased motivation for self-administered cocaine after escalated cocaine intake. Neuroreport 14 (17):2229–2232
38. Kippin TE, Fuchs RA, See RE (2006) Contributions of prolonged contingent and noncontingent cocaine exposure to enhanced reinstatement of cocaine seeking in rats. Psychopharmacology (Berl) 187(1):60–67
39. Vanderschuren LJ, Everitt BJ (2004) Drug seeking becomes compulsive after prolonged cocaine self-administration. Science 305 (5686):1017–1019
40. Wise RA (1973) Voluntary ethanol intake in rats following exposure to ethanol on various schedules. Psychopharmacologia 29 (3):203–210
41. Wolffgramm J, Heyne A (1995) From controlled drug intake to loss of control: the irreversible development of drug addiction in the rat. Behav Brain Res 70(1):77–94
42. Heyne A, Wolffgramm J (1998) The development of addiction to d-amphetamine in an animal model: same principles as for alcohol and opiate. Psychopharmacology 140(4):510–518
43. Heyne A (1996) The development of opiate addiction in the rat. Pharmacol Biochem Behav 53(1):11–25
44. Vestal BM (1977) Sociability and individual distance in four species of rodents. Proc Oklahoma Acad Sci 57:98–102
45. Fritz M, El Rawas R, Klement S, Kummer K, Mayr MJ, Eggart V et al (2011) Differential effects of accumbens core vs. shell lesions in a rat concurrent conditioned place preference paradigm for cocaine vs. social interaction. PLoS One 6(10):e26761
46. Kummer KK, Hofhansel L, Barwitz CM, Schardl A, Prast JM, Salti A et al (2014) Differences in social interaction- vs. cocaine reward in mouse vs. rat. Front Behav Neurosci 8:363
47. Venniro M, Zhang M, Caprioli D, Hoots JK, Golden SA, Heins C et al (2018) Volitional social interaction prevents drug addiction in rat models. Nat Neurosci 21(11):1520–1529
48. Moeller FG, Barratt ES, Dougherty DM, Schmitz JM, Swann AC (2001) Psychiatric aspects of impulsivity. Am J Psychiatry 158 (11):1783–1793
49. Alessi SM, Petry NM (2003) Pathological gambling severity is associated with impulsivity in a delay discounting procedure. Behav Process 64(3):345–354
50. Mitchell SH (1999) Measures of impulsivity in cigarette smokers and non-smokers. Psychopharmacology (Berl) 146(4):455–464
51. Bickel WK, Marsch LA (2001) Toward a behavioral economic understanding of drug dependence: delay discounting processes. Addiction 96(1):73–86
52. Neef NA, Marckel J, Ferreri SJ, Bicard DF, Endo S, Aman MG et al (2005) Behavioral assessment of impulsivity: a comparison of children with and without attention deficit hyperactivity disorder. J Appl Behav Anal 38 (1):23–37
53. Winstanley CA (2011) The utility of rat models of impulsivity in developing pharmacotherapies for impulse control disorders. Br J Pharmacol 164(4):1301–1321

54. Moreno M, Cardona D, Gómez MJ, Sánchez-Santed F, Tobeña A, Fernández-Teruel A et al (2010) Impulsivity characterization in the Roman high- and low-avoidance rat strains: behavioral and neurochemical differences. Neuropsychopharmacology 35(5):1198–1208
55. Adriani W, Caprioli A, Granstrem O, Carli M, Laviola G (2003) The spontaneously hypertensive-rat as an animal model of ADHD: evidence for impulsive and non-impulsive subpopulations. Neurosci Biobehav Rev 27(7):639–651
56. Aparicio CF, Hennigan PJ, Mulligan LJ, Alonso-Alvarez B (2019) Spontaneously hypertensive (SHR) rats choose more impulsively than Wistar-Kyoto (WKY) rats on a delay discounting task. Behav Brain Res 364:480–493. https://doi.org/10.1016/j.bbr.2017.09.040
57. Richards JB, Zhang L, Mitchell SH, de Wit H (1999) Delay or probability discounting in a model of impulsive behavior: effect of alcohol. J Exp Anal Behav 71(2):121–143
58. Bushnell PJ (2001) Advanced behavioral testing in rodents: assessment of cognitive function in animals. Curr Protoc Toxicol Chapter 11: Unit 11.4
59. Morris R (1984) Developments of a water-maze procedure for studying spatial learning in the rat. J Neurosci Methods 11(1):47–60
60. Vorhees CV, Williams MT (2006) Morris water maze: procedures for assessing spatial and related forms of learning and memory. Nat Protoc 1(2):848–858
61. D'Hooge R, De Deyn PP (2001) Applications of the Morris water maze in the study of learning and memory. Brain Res Brain Res Rev 36(1):60–90
62. Whishaw IQ, Tomie J (1996) Of mice and mazes: similarities between mice and rats on dry land but not water mazes. Physiol Behav 60(5):1191–1197
63. Colacicco G, Welzl H, Lipp HP, Wurbel H (2002) Attentional set-shifting in mice: modification of a rat paradigm, and evidence for strain-dependent variation. Behav Brain Res 132(1):95–102
64. Jaramillo S, Zador AM (2014) Mice and rats achieve similar levels of performance in an adaptive decision-making task. Front Syst Neurosci 8:173
65. Prusky GT, West PW, Douglas RM (2000) Behavioral assessment of visual acuity in mice and rats. Vis Res 40(16):2201–2209
66. McLay RN, Freeman SM, Harlan RE, Kastin AJ, Zadina JE (1999) Tests used to assess the cognitive abilities of aged rats: their relation to each other and to hippocampal morphology and neurotrophin expression. Gerontology 45 (3):143–155
67. Stubley-Weatherly L, Harding JW, Wright JW (1996) Effects of discrete kainic acid-induced hippocampal lesions on spatial and contextual learning and memory in rats. Brain Res 716 (1–2):29–38
68. Jarrard LE (1993) On the role of the hippocampus in learning and memory in the rat. Behav Neural Biol 60(1):9–26

Index

G. Thomas Hayman et al. (eds.), *Rat Genomics*, Methods in Molecular Biology, vol. 2018,
https://doi.org/10.1007/978-1-4939-9581-3,

H

I

K

L

M

N

O

P

Q

R

S

T

U

V

W

Z

Kolibri 360 GmbH
Ferdinand-Jühlke-Straße 7
99095 Erfurt, Deutschland
produktsicherheit@kolibri360.de